T0231727

Access for Dialysis:
Surgical and Radiologic Procedures,
2nd Edition

Ingemar J.A. Davidson, MD, PhD, FACS
Medical City Dallas Hospital
Dallas, Texas, U.S.A.

CRC Press
Taylor & Francis Group
Boca Raton London New York

CRC Press is an imprint of the
Taylor & Francis Group, an informa business

Dedication

The first edition of the vascular access book published in 1996 was dedicated to the patients at Parkland Memorial Hospital, Dallas, Texas, who, between 1982 and 1995 taught me more about medicine than all medical textbooks combined.

This edition is dedicated to Dorothy Griffin, RN, vascular access coordinator at Parkland between 1991-1998, retiring at the age of seventy. Prior to becoming the coordinator, Mrs. Griffin was the head nurse at Parkland outpatient dialysis unit, which made her uniquely fit to be the vascular access coordinator. I deeply respected her knowledge, principles, robust plain approach to life problems, founded in her southern Baptist principles. She was so good that some doctors couldn't quite stand up to her. She knew more about dialysis than the chief of nephrology. She knew more about vascular access than I did.

As the vascular access coordinator, Dorothy performed miracles. When she took the job, the "wait list" for vascular access overnight dropped from 50 patients to 3. The OR cancellation and rescheduling rate dropped from 110% to less than 15%. The list goes on. What was happening? It can be summarized in a few words: communication, system problem solving, authority, doing the right thing, etc. She did it all. I was told by a high ranking Parkland administrator at some time that Dorothy has saved Parkland several hundred thousand dollars.

Am I making my point? We need more people like Dorothy Griffin in vascular access. Fixing the vascular access system problems requires a dedicated vascular access coordinator. There are hundreds of capable and dedicated individuals, just like Dorothy, willing to meet the challenge. We, the medical community, must make the vascular access coordinator job description a reality.

You are what you repeatedly do;
then excellence is not an art
but a habit — Aristotle

Contents

Editor

Ingemar J.A. Davidson, MD, PhD, FACS
Medical City Dallas Hospital
Dallas, Texas, U.S.A.
Fax: (972) 566-3858
email: drd@medicalcity.com
Chapters 1-6, 8, 11, Appendices I, III-V

Contributors

Diana J. Adams, RHIA
Reimbursement Consultant
Dallas, Texas, U.S.A.
Chapter 11

W. Perry Arnold, MD
Interventional Radiologist
RMS Lifeline
Baltimore, Maryland, U.S.A.
Chapters 5, 7

Amer Ar'Rajab, MD, PhD
Division of Transplantation
The Ohio State University
Columbus, Ohio, U.S.A.
Chapter 10

Stephen T. Brown, MA
Medical Illustrator
Dallas, Texas, U.S.A.
Illustrations Chapters 4, 6

Kristin Davidson
Dallas, Texas, U.S.A.
Cover image design

Jacek Dmochowski, PhD
Department of Biomedical Statistics
State University of New York at Buffalo
Buffalo, New York, U.S.A.
Appendix V

Mark Durso, PharmD
Clinical Pharmacist
Medical City Dallas Hospital
Dallas, Texas, U.S.A.
Appendix II

Wilson V. Garrett, MD
Division of Vascular Surgery
Baylor University Medical Center
Dallas, Texas, U.S.A.
Appendix III

Mitchell Henry, MD
Division of Transplantation
The Ohio State University
Columbus, Ohio, U.S.A.
Chapter 10

Sandra Hinton, RN
Peritoneal Dialysis Supervisor
Dallas, Texas, U.S.A.
Chapter 9

Angela Kuhnel, RN, BSN
Transplant Clinical Coordinator
Medical City Dallas Hospital
Dallas, Texas, U.S.A.
Chapter 9, Appendix III, IV

Lisa M. McAdams, MD, MPH
Division of Clinical Standards and
    Quality
Centers for Medicare & Medicaid
    Services
Dallas, Texas, U.S.A.
Preface

Robert N. McClelland, MD, FACS
Division of GI/Endocrine Surgery
Department of Surgery
The University of Texas
Southwestern Medical Center at Dallas
Dallas, Texas, U.S.A.
Foreword

Carolyn E. Munschauer, BA
Vascular Access Coordinator
Dallas, Texas, U.S.A.
Chapters 1, 9, 11, Appendices I, III-V

Gregory J. Pearl, MD
Chief, Division of Vascular Surgery
Baylor University Medical Center
Dallas, Texas, U.S.A.
Appendices III, IV

Frank Rivera, MD
Division of Radiology
U. Texas Southwestern Med Center
Dallas, Texas, U.S.A.
Chapter 5

Dwight Shewchuk, MD
Staff Anesthesiologist
Medical City Dallas Hospital
Dallas, Texas, U.S.A.
Photography Appendix I

Bertram L. Smith, III, MD
Division of Vascular Surgery
Baylor University Medical Center
Dallas, Texas, U.S.A.
Appendix IV

Jonathan P. Young, MA
Medical Illustrator
Dallas, Texas, U.S.A.
Illustrations Chapter 3

A Letter From the Editor

This publication represents a comprehensive guide to current common diagnostic, operative and percutaneous techniques used in creating and maintaining access for dialysis. When writing the text, the authors have focused on surgeons in training, interventional radiologists and clinically active nephrologists and fellows, dialysis nurses and technicians, health professionals involved in the care of end stage renal disease. Many dialysis patients may also benefit from this handbook. The second edition also contains expanded sections on ESRD, access surveillance and surgical and diagnostic devices, as well as new sections on peritoneal dialysis and dual lumen catheter placement, hemo and peritoneal dialysis techniques and finally CPT and ICD coding for statistical and billing purposes. The new additions reflect the highly technical nature of clinical management in this rapidly evolving field. The radiologic procedures chapter is new and surgical management of access complications is dealt with in a separate chapter.

With more than 350,000 patients on dialysis in the US and about 75,000 new patients added every year, surgical vascular access procedures have become the most common intervention in the operating room in hospitals, not to mention radiologic diagnostic and therapeutic interventions, and the catheter placements and the dialysis needle punctures for the dialysis treatment itself. Therefore, this new edition also reaches out to dialysis RN's and technicians, with the intent to expand knowledge and understanding of each team member's expertise and roles. The style and general outline of this second edition has intentionally remained unchanged. Pictures often speak more than words. Only selected references are included after each chapter. Bear in mind, that many statements regarding technique and choice of equipment are the opinions of the authors. We certainly recognize that many similarly effective alternatives can achieve similar excellent results. However, the scope of this communication will not allow us to cover all variations. In a way, this publication represents our very personal and perhaps slightly biased experience on vascular access for dialysis. Should this second edition be met with the same enthusiasm as the first edition, we plan to make future updates. We again invite your comments and suggestions to solve specific or difficult problems. We welcome your comments, suggestions by fax or e-mail.

We are looking forward to hearing from you.

The Editor

Foreword

This second edition of Dr. Davidson's text *"Access For Dialysis: Surgical and Radiologic Procedures"* details both technical aspects and enduring tenets in this emerging subspecialty.

Since the publication of the first edition in 1996, there have been several significant changes in the population of end stage renal disease (ESRD) patients and the products available for dialysis access. Patients entering ESRD treatment programs are now ten years older than in the early 1990's. In addition, patients often suffer from comorbidities, most notably diabetes, hypertension, obesity and cardiac and peripheral vascular disease. The ESRD population poses a significant challenge to the vascular access surgeon, not just in the creation of an access, but in the maintenance of its function and patency. In the two prevalent dialysis modalities, hemo and peritoneal, technical advances in materials and design provide the treatment team with choices for access type among brands and styles. The downside to these developments is an increasing number of products, often without stratifiable data on their usefulness. Also, interventional radiologists have joined surgeons in the vascular access arena with a wide arsenal of techniques and devices to maintain patency, survey for impending access failure and correct access complications.

Within this framework of an increasing patient population, numerous products, additional practitioners and available treatment options, access for dialysis continues to attract attention from surgical institutions and training programs, individual surgeons, nephrologists, radiologists, patient advocacy groups and payors. As dialysis therapy continues to be refined, providing increased life span and quality of life, patients require more maintenance and repair procedures than creation of new access. The skillful management of access over time may have the greatest impact on long term patient outcome on dialysis.

It is in this forum that Dr. Davidson's extensive surgical experience and thoughtful attention to detail are most prominent. The teaching of surgical procedures and techniques is presented in a clear, accessible format that can be utilized by experienced staff surgeons and emerging fellows. At the core of this educational material is one fundamental principle that guides the methodology; that being to do the right thing for each individual patient in every unique situation.

This second edition has been expanded to include a great deal of new information on general surgical strategies, the use of protocols including and exceeding those based on the NKF-DOQI Guidelines, the efficacious use of dialysis catheters, a new, broad section on peritoneal dialysis catheters and an outstanding appendix presenting some complex vascular access challenges with stepwise treatment options and comments by Dr. Davidson.

This is a very ambitious book dealing with a vast and complex subject. It is a valuable resource for all health care professionals involved in the care of the ESRD patient, for although it is primarily a surgical text, it is a surgical text with a soul; a technical manual with a philosophical message not simply "First, do no harm", but to do the right thing.

Robert N. McClelland, M.D., F.A.C.S.
Professor of Surgery
Division of GI/Endocrine Surgery
Department of Surgery
The University of Texas
Southwestern Medical Center at Dallas
Dallas, Texas, U.S.A.

Preface

Imagine for a moment that you have just been diagnosed with renal failure. You are facing the prospect of dialysis and you feel that hemodialysis is the option that best suits you. This usually means for 3-4 hours three times a week you will be connected to a dialysis machine. Before you can begin dialysis, however, you have to see more doctors (a vascular surgeon) and undergo more tests (to assess your current vascular status and suitability for the different types of vascular access). This is followed by surgery to establish the vascular access. It may take several surgeries and some time with a catheter before you have a functioning access. Over the years, your access will need repairs, revisions and replacement - more visits to the surgeon, more testing, more surgery. Throughout this you hope that everything is being done to minimize the number of times that you have to undergo this cycle in the future.

Hemodialysis is truly a miracle of modern medicine. Without it, the nearly 300,000 individuals receiving it at the end of 2000 would have had little hope of survival. Yet, that does not relieve us, as physicians and other health care providers, from our duty to ensure that those patients are receiving the best care possible. For hemodialysis patients this includes attention to their vascular access.

Recommendations for vascular access were published by the National Kidney Foundation in 1997 with an update in 2000. These are, for the most part, evidence-based guidelines. In some instances where evidence is not available, the opinions of the expert panels serve as the basis for the recommendations. These experts have recommended the arteriovenous fistula (AVF) as the type of vascular access most closely meeting the features of the ideal access—one that delivers an adequate blood flow, has a long life-span, and a low complication rate (e.g., stenosis, thrombosis, and infection). Studies have shown that the AVF is superior to both grafts and catheters in 4-5 year patency rates, the number and cost of subsequent surgical interventions, infection rates, and hospitalizations for other complications.

So why have we seen in the United States a decline in the proportion of AVFs placed compared to grafts and catheters during the early 1990s? The reasons are numerous and include factors at every stage of caring for the dialysis patient (care of the impending end-stage renal disease patient by the primary care physician and others, care of the patient at the initiation of dialysis, and maintenance care of the dialysis patient). These factors also span the range of individuals involved in the patient's care—the patients themselves, primary care physicians, nephrologists, vascular surgeons, interventional radiologists, dialysis facility staff, hospital staff, quality improvement professionals, and others.

You might think that with this many factors that any endeavor to increase the number of AVFs is hopeless. Not so! Efforts are currently underway to better address these issues.

Under its "Core Indicators" project, the Centers for Medicare and Medicaid Services (CMS, formerly the Health Care Financing Administration) has been tracking quality measures in ESRD since 1994. In 1999, this project became the Clinical Performance Measurement Project and incorporated vascular access measures, including the prevalence of AVFs. These measures are an important component of CMS's ESRD Networks' (CMS contractors responsible for quality oversight for dialysis facilities) program. Beginning in 2002, all ESRD Networks will be conducting Vascular Access Quality Improvement Projects, some of them designed to increase the rate of AVFs.

For your part, as a vascular surgeon, nephrologist, interventional radiologist, dialysis facility staff, quality improvement professional, or other individual interested in vascular access you can learn all that you can about the types of access, the techniques involved in their placement, and their care and maintenance. A text such as this is a first step towards developing a solid foundation on which you can depend to make sound clinical decisions and provide valuable advice to hemodialysis patients, and is a valuable resource for anybody involved in the care of end stage renal disease patients. You can also share your knowledge with other surgeons and nephrologists and become actively involved in the team that supports these patients at both a local and national level.

Through our combined efforts we can improve the AVF rates in the United States, reaching levels that have been achieved in Europe and Canada. By doing so, we can rest assured that the dialysis patient's hope is not unfulfilled.

Lisa M. McAdams, MD, MPH
Medical Officer, Region VI
Division of Clinical Standards and Quality
Centers for Medicare & Medicaid Services
Dallas, Texas, U.S.A.

This preface was written by Dr. McAdams in her private capacity. No official support or endorsement by the Centers for Medicare and Medicaid Services is intended or should be inferred.

CHAPTER 1

The End Stage Renal Disease Patient as Related to Dialysis

Ingemar J.A. Davidson, Carolyn E. Munschauer

The End Stage Renal Disease (ESRD) Timeline

Many patients with impending renal failure are regularly seeing their primary physician or a nephrologist (Fig. 1.1, Phase 1). Interventions, such as maintaining optimal blood pressure control, may delay or halt the development of progressive renal disease. As the creatinine rises to the 6-7 mg/dl range (GFR 10-15 ml/min) in nondiabetics, or 4-5 mg/dl in diabetic patients (GFR 15-25 ml/min) a vascular access should be placed, since the need for dialysis may be close. (Phase 2). Every effort should be made to allow two or more weeks before the use of a PTFE graft, and 4-6 weeks for a primary AV fistula to allow for maturation. However, about 40% of patients have no permanent vascular access by the time dialysis treatment becomes necessary (Phase 3). Often, patients are referred too late and thus, require temporary central vein dual lumen catheters, with their inherent and associated morbidity.

Depending on disease process and concomitant medical problems, most patients will remain on some form of dialysis treatment for the rest of their lives. Others are evaluated for a possible renal transplant and placed on kidney transplant waiting lists, where they may remain for varying periods of time depending on their blood type, preformed antibody levels, HLA matching, medical urgency and the local organ bank efficiency (Phase 3). For example, the median waiting time to transplant for a blood type O kidney recipient is currently about 1213 days (United Network for Organ Sharing (UNOS) 2000). According to annual trends, of the more than 46,769 patients currently (8/01) on the waiting list only about 13,332 will receive a kidney transplant per year (UNOS, 2000). The transplant procedure includes organ procurement, in which the kidney is part of a multi-organ procurement effort (Phase 4). Typically, the organs are excised within 48 hours after a donor has been identified. Ideally, the kidney will be implanted and reperfused within 20 h of excision (the cold ischemia time). The transplant surgical procedure itself takes only 3-4 h (Phase 5), followed by 5-10 days in a hospital setting (Phase 6). The posttransplant close surgical surveillance period lasts for 90 days (Phase 7). During this time, immunosuppression is designed to prevent acute rejection episodes. After 3-6 months, the stable patients revert to their referring physician for follow-up, which lasts for the duration of the graft or the patient's life (Phase 8, 9). Graft failure, for whatever reason, brings the patient back to phase 2, either as a hemodialysis patient or back on the waiting list for a repeat transplant. Patients may be placed on the waiting list when the graft is failing and perhaps be transplanted before the need for dialysis.

Access for Dialysis: Surgical and Radiologic Procedures, 2nd ed., edited by Ingemar J.A. Davidson. ©2002 Landes Bioscience.

1

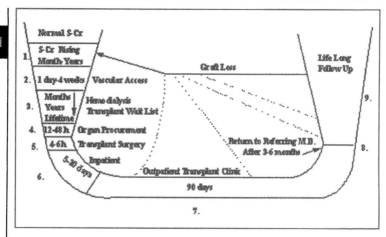

Fig. 1.1. The End stage renal disease (ESRD) timeline as it pertains to renal transplantation. Patients with renal disease develop decreased renal function (1) and are followed
by their primary physicians and later usually by a nephrologist. As serum creatinine rises
to 4-5 mg/dl in diabetics (GFR 15-20 ml/min) or 7-8 mg/dl in patients with no comorbidity
(GFR 10-15 ml/min) permanent vascular access or peritoneal catheter is placed in an-
ticipation of hemodialysis or continuous ambulatory peritoneal dialysis (CAPD) (2). Patients may remain on hemodialysis for life or be evaluated for transplantation and placed
on UNOS wait list (3). The transplant procedure involves donor maintenance and organ
excision (4), recipient surgery (5), a short 5-10 day hospital stay (6), followed by close
outpatient follow-up for 3-6 months (7), before reverting to their referring physician (8-
9). At any time patients may lose the organ to rejection or technical problems and revert
to dialysis treatment.

Practical Guidelines for Vascular Access Placement

At the beginning of 1999; 249,983 patients were on chronic hemodialysis. By
the end of the year, 66,964 patients died and had been replaced with 79,482 new
patients, resulting in an annual increase of about 12,000 dialysis patients. Of all patients starting dialysis 19.8% (15,737) will die within their first dialysis year. Of all
patients who die, it is estimated that 25% (19,870) will die with inadequate vascular
access or complications related to vascular access as a major contributing factor.

For patients on hemodialysis, the vascular access is literally their "lifeline." A
well functioning access, whether a primary fistula, synthetic graft or peritoneal catheter, will enable the patient to receive adequate dialysis treatment, minimizing metabolic complications associated with increased mortality.

Timing of Access

Vascular access and Peritoneal catheter insertion should be placed prior to the
need for initiation of dialysis to allow maturation. This timing is guided by the
individual patient's predicted progression or renal failure decline from repeated function tests. As a rule of thumb, diabetic patients will benefit from access placement
when the creatinine is 4-5 mg/dl. This corresponds to a creatinine clearance of about
20 ml/min. Patients with no additional contributing medical comorbidities should

have access placed when the creatinine approaches 6-7 mg/dl (creatinine clearance of 15 ml/min). Patients should be evaluated early, and measures taken to preserve native veins in an attempt to create primary AV fistulae. In fact, it is preferable to place primary (native) AV fistulae early (even 6 months to a year) before anticipated dialysis to allow time for optimal maturing. Even if a native vein does not enlarge enough for cannulation, the antecubital veins have now become more suitable for PTFE graft placement with perhaps greater than 90% one year graft patency (Appendix V, Fig. 5A).

History and Physical Exam

Previous central venous catheters or pacemaker placement is associated with radiologically significant central venous stenosis formation in 30-50% of cases. Five percent will have clinical symptoms, i.e., arm edema. In these cases, examination with duplex Doppler or venogram prior to access placement is recommended. An access placement in an arm with subclavian or innominate vein occlusion or severe stenosis is destined for a disastrous outcome, often requiring emergent ligation or removal.

The non-dominant arm is preferred for vascular access placement. A history of arterial lines or arterial needle punctures may have caused vascular injury. For instance, multiple radial artery sticks during previous ICU stays may preclude a native radial artery to cephalic vein fistula at the wrist.

Individuals with diabetes mellitus or advanced age with significant peripheral atherosclerosis are at increased risk of hand ischemia after access placement, often referred to as "arterial steal"(Appendix 1, Case #28 for diagnosis and treatment). Patients with major comorbidity and decreased life expectancy may benefit from temporary access placement, such as cuffed, tunneled dual lumen catheters (Appendix I, Case #48).

Previous vascular access surgeries will limit sites and options. This patient category may represent 50% of all vascular access cases. Preoperative duplex Doppler as a first line tool will help outline available vessels. If duplex Doppler is not conclusive, venogram of central veins is the next recommended step. Likewise, previous neck surgery or open chest surgery may have resulted in central vein anatomy distortion, excluding involved arm. A venogram will clarify such cases.

In patients with renal transplantation anticipated in the near future, i.e., living donor, temporary dialysis catheter access may suffice.

The Effective Dialysis Access Team

The end stage renal disease (ESRD) patient population is a captive audience and therefore an excellent study population. By survival instinct most patients come regularly three times weekly for follow-up and treatment. Most patients willingly admit themselves as research subjects to find new and better treatment regimens. We as the treatment team owe the patient the best medical management. The total care of the ESRD patient requires a team, where each member has a specific role or duty at certain times and under specific circumstances. Each member of the team often belongs to or represents a specific group or medical specialty or academic department with its own agenda, rules and regulations. This multifaceted structure inevitably lends itself to a number of system problems related to timing of service, overlapping duties or responsibilities. The current system does not work well in a hierarchy setting. A flat or web administrative structure may be more feasible where each member performs a specific task and is dismissed when no longer needed. Whatever system is used, the rules and regulations must be structured around the patient's needs.

Strategies to improve the outcome and longevity of access is deeply entrenched in the access team's level of knowledge, skills and attitude. The access team in this context includes the referring nephrologist or internist, the dialysis unit members, the surgeon and the interventional radiologist. The most important team member is the patient.

It is easy to understand the need for skills and knowledge in the context of choosing and creating optimal access. What is less obvious is that attitude may represent up to 80% of the outcome effectiveness. Attitude in this context includes factors such as willingness to go the "extra mile" in the "gray" area between team members' "territory" (duties). It involves spending time to educate the patient about treatment options, making the patient (and family) part of the decision making. Attitude is to make the right choice/recommendation for the patient based on patient, not on provider (team member), convenience or profit. How do we make this happen? It is a matter of attitude change, self-confidence, doing the right thing at the right time for the right reason, at the right moment, modeled by your skills and knowledge filtered through the laws and societal constraints in which you live.

Physical Exam

Assessment

Examination of the Arterial System

The arterial vasculature is examined by palpating available pulses and by using a hand held ultrasound device (i.e., Site Rite®) (Table 1.1) and a duplex Doppler as indicated. For instance, an abnormal Allen test may warrant a more extensive duplex Doppler examination and finger pressure determination. Bilateral blood pressures of upper arms are part of the arterial inflow examinations. The level of detailed knowledge of the extremity artery size and flow rate is determined based on patients' needs as part of preoperative evaluation for vascular access placement. Definitive determination whether an access will induce distal ischemia or arterial steal after placement is hard to assess or prevent with current practice in the author's experience. Should severe ischemic signs occur, the banding procedure is performed (Chapter 4, Fig. 4.29; and Appendix I, Case #28).

Examination of the Venous System

The lack of an adequate outflow vein is more often the limiting factor as opposed to an unacceptable artery. The presence of acute or chronic swelling or edema suggests outflow problems and warrants a venogram or duplex Doppler of central veins. The small, portable ultrasound device, Site-Rite® (Table 1.1) has a number of usable applications, in the office, operating room and radiology suite (Chapter 5, Figs. 5.4 and 5.8, Table 1.1). The authors strongly recommend the use of this device in all cases of dialysis dual lumen catheter placement; also to further increase safety, a so-called micropuncture set for placement of guidewires (Chapter 5, Fig. 5.2A-B). The presence of collateral veins in the upper arm and on the chest indicates a high likelihood for central vein stenosis or occlusion. Unless permanently corrected, central vein stenosis precludes ipsilateral peripheral access placement.

Table 1.1. Examples of useful applications of the portable ultrasound device SiteRite®

---

1. Establish patency of veins in neck (i.e., internal jugular vein)
2. "Vein mapping" for optimal fistula outflow site
3. Accessing veins for a variety of applications, including the placement of dual lumen catheters under direct vision
4. Anesthesia needle guidance for nerve blocks (i.e., axillary arm block)

Patients with previous prolonged or multiple hospital stays often lack superficial veins for native fistula placement.

Patients should be examined in a warm room under quiet circumstances to allow veins to dilate. Also, application of an upper arm tourniquet or blood pressure cuff (30-40 mm Hg) will induce vein dilatation for palpation and mapping, helping in selecting adequate veins for access.

Diagnostic Tests

With clinical history and examination, the majority of vascular access patients can be evaluated, needing no further testing. When in question, a duplex Doppler is used for arm vein mapping and to confirm patent subclavian veins and the availability of deep concomitant antecubital and upper arm adequate vasculature furthermore, a venogram is performed to assess central veins. Seldom is an MRI examination warranted.

Venogram is indicated in cases of extremity edema, collateral vein development, differential extremity size, current or previous subclavian catheter placement, including pacemakers, previous arm, neck or chest surgery and multiple previous accesses in the effected extremity. The radiologist must be prepared to treat a lesion when indicated in the same setting, avoiding a second procedure.

Selection Versus Type of Dialysis Access

Preferred AV Access Sites

The author's order of preference for placement of AV access for chronic hemodialysis is outlined in Table 1.2. The non-dominant arm should be considered first. As one goes down the list in Table 1.2, the suggested order becomes less firm. For instance, a forearm AV graft loop before an upper arm AV fistula may be chosen. Likewise, a groin AV graft may take precedence before a basilic vein transposition.

The first choice is a primary native AV fistula between the radial artery and cephalic vein in either arm. The next best option is a primary AV fistula in either upper arm between the brachial artery and cephalic vein. The third option is a loop PTFE in either forearm between the brachial artery and any antecubital vein. The fourth option is an upper arm PTFE AV graft where the venous outflow is placed as distal as possible to allow future revisions. Other access options include basilic vein transposition and an anastomosis to the brachial artery, either extremity. Still an option is a groin PTFE AV graft between the superficial femoral artery and the saphenous vein. In many dialysis units 5-20% of patients on chronic hemodialysis are maintained on various types of dual lumen catheters. In this context, the reader is reminded of the peritoneal dialysis option (Chapter 6).

1

Table 1.2. Type, site and order of permanent vascular access of preference

1. Native radial artery to cephalic vein in non-dominant arm.
2. Same as (1) in the dominant arm.
3. Native cephalic vein to brachial artery above antecubital fossa in non-dominant arm.
4. Same as (3) in dominant arm.
5. PTFE AV graft in non-dominant forearm.
6. PTFE in dominant forearm.
7. Upper arm PTFE AV graft in non-dominant arm; brachial artery to any vein.
8. Upper arm PTFE AV graft in dominant arm: brachial artery to any vein.
9. Transposition of basilic vein and anastomosis to brachial artery either arm.
10. Groin PTFE AV graft between femoral artery and saphenous vein.
11. Dual lumen or two line cuffed, tunneled catheters.

Type, Location and Material of AV Grafts

If a primary AV fistula cannot be placed at the wrist or upper arm, a polytetrafluoroethylene (PTFE) graft is preferred over other materials (Chapter 4). The first choice is a forearm graft in a loop configuration. A loop graft is usually associated with higher flows (> 500 ml/min) than a straight graft (< 500 ml/min) between the distal radial artery and the antecubital vein. When both forearm sites have been exhausted, an upper arm graft is indicated.

Site of Insertion and Type of Dual Lumen Catheters

The location, type and indications for the use of catheters for hemodialysis vary greatly between centers, and are dictated by experience, intended use of duration, morbidity and cost. Temporary vascular access is accomplished with percutaneous or tunneled and cuffed catheters. Temporary access with dual lumen cuffed catheters is placed in patients who have had a primary AV fistula or a PTFE graft, requiring some time to mature, or have experienced a non-salvageable failure (usually thrombosis) of a current access. Rarely, catheters are used for patients who have exhausted all other access options (Table 1.3).

Femoral Vein Dual Lumen

When immediate dialysis is necessary, a femoral vein non-cuffed dual lumen catheter is placed under local anesthesia. Such a catheter placed at the bedside or in the hemodialysis unit should not be left for more than 48-72 hours, since it leaves the patient bedridden. It is the author's opinion that the subclavian vein must never be used for temporary dialysis catheters, because of the high incidence of stenosis and occlusion (30-50%), ruining the upper extremities from future permanent access (details in Chapter 5).

The Right Internal Jugular Vein

The preferred insertion site for tunneled cuffed venous catheters is the right internal jugular (IJ) vein. Left sided neck catheters need to be longer and tend to involve more problems for anatomical reasons (See Chapter 5, Figs. 5.3 and 5.4). Other options include the external jugular veins (usually through cutdowns) either side and, rarely, translumbar access to the inferior vena cava. Again, subclavian veins should not be used when other options are available. Also, catheters of any kind

---

Table 1.3. Do's and dont's with respect to dialysis catheters

---

1. Don't ever use the subclavian veins for any kind of dialysis catheters
2. When both right and left IJ are occluded, use femoral vein rather than subclavian vein
3. Don't use the right IJ for temporary percutaneous catheters
4. Don't leave femoral percutaneous dialysis catheters in place for > 48-72 hours

---

should be avoided on the same side as a maturing AV access. Temporary percutaneous catheters to the right IJ should be avoided at any price because it jeopardizes the site for cuffed tunneled catheters if needed.

Fluoroscopy is needed for insertion of all neck cuffed tunneled dialysis catheters, with the tip of the catheter adjusted to the junction between the cava and the right atrium (Chapter 5, Fig. 5.3)

Ultrasound guided, i.e., Site-Rite® insertion of dual lumen catheters and use of the micropuncture technique is strongly recommended in order to reduce complications (See Chapter 5, Figs. 5.2 and 5.8, Table 5.4).

Type of Catheters

There is a large number of different catheters available (Chapter 5, Table 5.5). The choice of catheter is based on experience, intended use and cost. Currently, the authors prefer the dual lumen split ash catheter or the Tesio two line catheter because of the longevity and high efficiency. Other brands may stand up to the same standards, since no blinded comparative studies are available.

The Percutaneous Non-Cuffed Catheter

A non-cuffed, percutaneously placed catheter is indicated for very short term dialysis. Such catheters are used immediately after placement and should not be inserted before they are needed. The site chosen is based on several factors, including expected duration, patient clinical status and the operator's experience.

Non-cuffed catheters can be inserted at the bedside in the femoral or internal jugular position. The subclavian veins MUST NEVER be used in any patients who are expected to have future vascular access upper extremity placement (Table 1.3).

Cuffed Tunneled Catheters

Cuffed tunneled catheters are intended for temporary use but are often functional for many weeks or months without complication. The optimal insertion site is the right internal jugular vein; while the left side renders more problems for anatomical reasons.

Post catheter placement chest X-ray is done to assess catheter position and to exclude complications prior to starting dialysis. With a "perfect" ultrasound (Site-Rite®) guided fluoroscopy catheter position confirmation, the authors feel confident without post placement X-ray. Again, ultrasound (Site-Rite®) should be used to direct insertion.

Femoral catheters should be at least 19 cm in length and not left in place longer than 48-72 hours, since it requires patient be bedridden.

A non-functioning non-cuffed catheter can be exchanged over a guidewire (Chapter 5, Figs. 5.26-5.27) or treated with tissue plasminogen activator (t-PA) (Chapter 5, Table 5.6, Appendix IV).

Exit site or tunnel tract infections or positive blood cultures are strong indications for removal of catheters. When cuffed catheters are exchanged over guidewire technique, a new skin exit site is recommended.

Preservation of Veins for Future AV Access

The most common sin in the emergency rooms and on hospital floors is the (ab)use of the forearm antecubital and cephalic veins for blood draws or IV infusion. In the author's opinion this practice has contributed to the high percentage (80%) of PTFE graft placements in the US. Hospital workers (i.e., phlebotomists and nurses) should be instructed to instead use the dorsum of the hand.

Likewise, the use of the subclavian vein by physicians for IV access results in a 30-50% stenosis precluding that side extremity for future access for hemodialysis. In 5% the arm shows swelling suggesting total central vein occlusion. The most effective means of preventing unnecessary destruction of the cephalic and antecubital veins is to educate the conscious patient to decline the use of these areas for IV access.

Maturation of Access

A primary AV fistula at the wrist is ready to use when the diameter of the vein allows cannulation. This may take only a few days, but more often 3-4 weeks, or sometimes up to 3-4 months after its placement. The repeated, successful dialysis needle cannulation is highly "operator" dependent. This required skill level is evident more so with primary AV fistulae than with AV grafts. Continued access puncture skill improvement by using "expert stickers" to supervise and train young and less experienced dialysis staff is a highly desirable activity that fits in the attitude category of professional outcome effectiveness (Chapter 9).

The PTFE graft usually requires at least 14 days "healing" before cannulation. Many centers, including the authors' have experience using grafts early, even within 24 hours, with little morbidity. Longer time before cannulation may be desirable when swelling persists, especially if other means of dialysis are available, such as an unaffected dual lumen, tunneled, cuffed catheter. If swelling persists, a venogram to assess central veins is warranted. Exercising, making fists and arm elevation early after surgery, both for native and graft fistulae, will help decrease edema formation, but will not directly affect patency or maturation. However, starting 10-14 days after surgery, repeatedly applying slight pressure, i.e., a blood pressure cuff ~ 30-40 mm Hg at the upper arm in cases of primary AV fistulae at the wrist may help enlarge the cephalic vein.

Surveillance Programs of Hemodialysis Access

Physical examination of every access is performed weekly and includes inspection and palpation. Depending on each center's surveillance program, additional studies are performed, including fistula flow measurements, venous pressure measurements and recirculation studies. When indicated by changes in these parameters a duplex Doppler study, a venogram or a fistulagram is performed (Chapters 7 and 10).

Recirculation Studies

Any access is abnormal when urea-based recirculation exceeds 10% or when non-urea-based dilutional method is used exceeding 5%. This should initiate investigation as to its cause, using duplex Doppler or fistulagram according to the center's agreed upon algorithm (Chapter 10).

The Vascular Access Coordinator (VAC)

The role of a vascular access coordinator is to optimize communication between nephrology, surgery, radiology, dialysis personnel and the patient when the surveillance protocol requires intervention, or when acute access problems arise, i.e., clotting or infection.

While all forms of dialysis access are created in a surgical setting, they are utilized in a self-contained outpatient unit without continuous surgical evaluation. Once patients have had sutures and temporary catheters removed, they are instructed to return to the surgeon only if problems develop. In effect, then, the dialysis unit staff become the triage personnel for vascular access maintenance and complications. In a large dialysis unit, as many as 50 patients may dialyze per shift, with three shifts per day on an every other day rotation, or up to 300 total patients. The needs of these patients almost assures some degree of staff crossover during a dialysis session, which may reduce the opportunity for surveillance of access function.

Often, the first time a surgeon hears about a poorly functioning vascular access is when the patient presents to dialysis clotted and unable to dialyze. In the case of a surgical group practice, in excess of 1,000 vascular access procedures may be performed in a year, with a different surgeon seeing access complications each time they occur. The same variable continuity may occur in the dialysis unit and on the nephrology service.

The emergence of vascular interventional radiology as a source of diagnosis and treatment of the nonfunctioning vascular access has further confounded the continuity of care in the dialysis patient. A patient may be re-referred to the surgeon only after interventional radiology has failed to correct the access problem. Without effective communication between surgeon, radiologist, nephrologist and dialysis staff, the extent of the access problem may not be realized and addressed in a timely, cost effective fashion (Chapter 8). In the ideal world, the "access center" places the patient's needs in the center (Fig. 8.5B), where radiology and surgery work in conjunction based on individual skills, knowledge and resources. Market forces currently seem to be defining the roles and boundaries of the main parties. With the new interventional nephrology subspecialty evolving, the gray areas are becoming even larger; so does the space for attitude improvement.

While morbidity and mortality statistics are sobering, they point out the need for continuity of care in the end stage renal disease population. By virtue of their disease process, dialysis patients have frequent contact with many health care institutions.

The access coordinator functions in many roles, depending on the local structure and needs.

The access coordinator can be an RN, a dialysis staff member or other ancillary medical personnel. He or she must be familiar with all aspects of access creation and maintenance and educate patients and the physicians involved in their care. The coordinator should be highly motivated, task oriented, willing to see a problem through to completion, and familiar with the local hospital and referral systems in place. Since the most efficient method of tracking a large patient population is through the use of a database, computer knowledge is needed (Table 1.4).

1

Table 1.4. Examples of vascular access coordinator job duties

---

Scheduling elective access placement
Arranging emergency transport for patients
Arranging surgical declotting
Screening access questions from patients and dialysis units
Data collection
Interacting with outpatient and inpatient hospital units
Patient tracking/outcome reporting
Facilitate efficient patient movement through the hospital system
Improve preemptive access surveillance programs
Assist in education of patient and dialysis staff
Develop protocols

The most common role a vascular access coordinator occupies is to coordinate timely surgical or radiologic thrombectomy of a clotted access, enabling the patient to return to dialysis, avoid temporization with central venous dialysis catheters and associated morbidity.

Through the vascular access coordinator, hospital admissions are decreased for all types of vascular access procedures. Patients presenting for outpatient thrombectomy without being screened or given proper preoperative instructions often results in surgery cancellation and admission for emergent hemodialysis and central venous or femoral catheterization. The coordinator, gaining pertinent information on the patient's fluid and metabolic status, communicates to the nephrologist and surgeon, facilitating the timely and correct interventions.

The vascular access coordinator is not a panacea for all problems of the end stage renal disease population. The presence of a coordinator only improves continuity and quality of care in these complicated patients, which will improve vascular access longevity.

Selected References

1. NKF-DOQI Clinical Practice Guidelines for Vascular Access. New York: National Kidney Foundation, 1999.
2. Covey Stephen R. The Seven Habits of Highly Effective People. New York: Simon and Schuster, 1989.
3. Goleman Daniel. Working with Emotional Intelligence. New York: Bantam Books, 1998.
4. Davidson IJA, Ar'Rajab A, et al. Early Use of the Gore Tex Stretch Graft. In: Henry ML, Ferguson RM, eds. Vascular Access for Hemodialysis-IV. Chicago: W.L. Gore & Associates, Inc. and Precept Press, 1995; 4:109-117.
5. Munschauer CE, Gable DR. Vascular access coordination at a large tertiary care hospital. Presented to Vascular Access for Hemodialysis VIII, Rancho Mirage, CA, May, 2002.

CHAPTER 2

Vascular Access Surgery
—General Considerations

Ingemar J.A. Davidson

Background

An individualized treatment strategy should be carefully tailored for every patient needing permanent renal replacement therapy. Therefore, the surgeon must have a thorough knowledge of various treatment options and the patient's expected date of dialysis initiation. The patient must also understand short- and long-term goals and agree with the surgical treatment plan. There are currently three modes of therapy for patients with end-stage renal disease (ESRD): 1) renal transplantation, 2) peritoneal dialysis (PD) and 3) hemodialysis. Pre-emptive transplantation instead of dialysis, and the consequent avoidance of vascular access surgery, is especially suitable for patients with living related donors, since the transplant procedure can be planned in advance. PD may provide higher quality dialysis with a greater degree of freedom, and it should be offered to suitable patients, particularly individuals who want to be mobile and are reluctant to devote the many hours required for hemodialysis. Although an estimated 30-50% of ESRD patients may benefit from PD currently only 17% of patients undergoing dialysis in the US receive PD.

Hemodialysis is the most commonly employed modality. This book, which does not favor one form of treatment over another, presents a stepwise approach to operative procedures for the establishment of vascular access for hemodialysis. Also, the most common procedures for addressing complications are described, including radiologic diagnostic and therapeutic measures (Chapter 7).

All patients needing vascular access should initially be considered for placement of a primary arterio-venous fistula (PAVF) because of the relative freedom from short- and long-term complications associated with this procedure. Depending on patient population characteristics (i.e., age, diabetes, body mass index), at least 50% of patients may be candidates for a PAVF in the US. The most common limiting factor is the lack of a suitable native vein, often because of previous needle punctures by health professionals at medical institutions or by patients participating in IV drug abuse. In diabetic and elderly patients, arterial inflow may be a limiting factor. In rare cases, the artery may be so calcified that surgery becomes technically difficult or impossible to perform. The entire length of the intended vein (usually the cephalic vein) must be present, since the needle placements for hemodialysis are made along the vein in the forearm. A long segment of the cephalic (or basilic) vein is required to provide adequate length for rotation of needle puncture sites and also to avoid recirculation during the dialysis procedure (Chapter 9). A complete preoperative examination must therefore include evaluation of the entire vascular system of both upper extremities to determine the most beneficial access strategy for each patient. The nondominant extremity is selected if other factors are equal. Preoperative

Access for Dialysis: Surgical and Radiologic Procedures, 2nd ed., edited by Ingemar J.A. Davidson. ©2002 Landes Bioscience.

evaluation includes the Allen test to insure adequate blood flow to the hand, while the ulnar and radial arteries are sequentially manually occluded. When in question (older patients and diabetics) it is preferable to utilize duplex Doppler ultrasonography with finger pressures while testing each artery's capacity to supply blood flow to the hand. Duplex Doppler sonography is valuable in diagnosing access complications, i.e., abscess formation, hematoma infiltrates, stenosis and arterial steal phenomena.

An upper arm PAVF between the cephalic vein and the brachial artery should be considered in the setting of a thrombosed, nonsalvageable forearm access. This is especially useful when there is a dilated, and therefore ready to use upper arm cephalic vein following a failed forearm access.

Anesthesia

Axillary block is the preferred anesthetic modality. Premedication is avoided, and sedatives such as midazalom (Versed) or fentanyl should be administered with great caution since unwise use of these agents may result in intra- and/or postoperative cardiorespiratory compromise and even cardiopulmonary arrest. A rule of thumb is to give one-tenth the sedative dose given to a non-renal failure patient and repeat this dose until the desired effect is obtained. In cases of incomplete axillary anesthetic block (when both the patient and the surgeon are unhappy), the anesthesiologist may feel pressured into giving too much sedation. The surgeon must instead work with the anesthetist and use local anesthesia injections (1% lidocaine or 0.5% bupivacaine without epinephrine) as needed. Seldom is general anesthesia indicated, except for patients with positive HIV or hepatitis B status, in whom complete intraoperative immobilization is desired for protection of OR personnel.

Surgical Instruments and Tools

Use of surgical magnifying lenses of at least 2.5 times magnification is strongly recommended. Technical errors are less likely to occur using proper magnification (Fig. 2.1).

Several microsurgical instruments are also highly recommended, samples of which are shown in Figure 2.2. The authors prefer a nonlocking needle driver with a round handle that can be held as a pen and rolled in the surgeon's hand while suturing (Figs. 2.2A-B). The nonlocking needle driver allows continuous suturing without loss of eye contact with the small operating field. Figure 2.2B shows a close-up view of the nonlocking needle driver used by the authors. Needle holders with a locking mechanism often lead to loss of control while resetting the needle.

The authors recommend two types of microforceps. A small indented "eye" on the first type improves the grip for vascular adventitial tissues (Fig. 2.2C). The sharp tipped "Blue Darter" forceps are especially useful for handling very small structures and for dilating vessels during corner suture placement (Fig. 2.2D).

The author currently uses small metallic vascular clamps (Heifet's clips) of various sizes and configurations for both arteries and veins (Fig. 2.3). Even though a "softer" clamp might be preferable, the authors do not believe that the Heifet's clamps cause permanent injury to the vessels. The use of large vascular metallic clamps and double snaring of the vessels with a suture is strongly discouraged. Often the Heifets clips are not strong enough to occlude a larger brachial artery. The disposable felt covered (yellow) clamps are an excellent alternative (Medcomp, 1499 Delp Dr. Harleysville, PA 19438 Phone: (215) 256-4201 Fax: (215) 256-1787. www.medcompnet.com) (Fig 2.4).

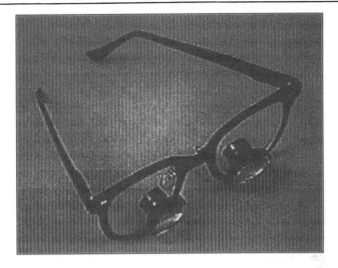

Fig. 2.1. The use of magnifying glasses greatly improves surgical accuracy.

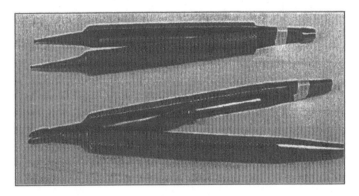

Fig. 2.2. Examples of microsurgical instruments suitable for vascular access. Figure 2.2A shows two types of nonlocking needle drivers.

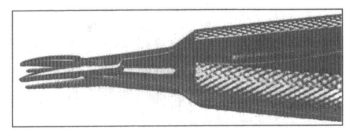

Fig. 2.2B. Depiction of one of these needle drivers in more detail.

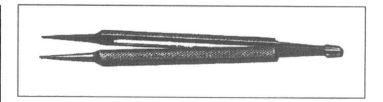

Fig. 2.2C. Example of a pair of microforceps with small indented eyes.

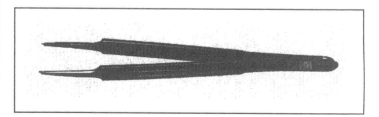

Fig. 2.2D. Representation of the sharp pointed microforceps.

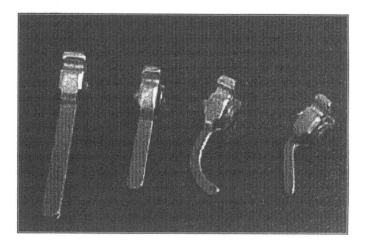

Fig. 2.3. Examples of small vascular clamps of different sizes and shapes used for occluding arteries and veins during suturing. Two strength clamps are available. The gold colored have less clamping power and therefore more suited for veins, while the stronger silver colored are used for arteries (color-coding does not show in these black and white images).

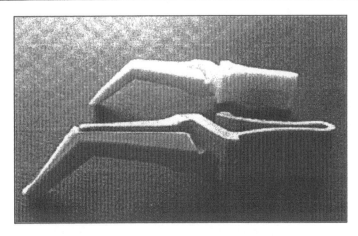

Fig. 2.4. This type of felt-covered disposable clamps are appropriate when the Heifets are too weak or too small. The authors find the yellow (smaller) clamp most usable.

For primary AV fistulas, the author uses polypropylene 7-0 suture on a CV-1 needle (8-0 sutures are also acceptable but are more technically challenging). The Gore-Tex® suture CV-6 on a TT-9 needle is also an alternative. The size of the Gore-Tex® TT-9 and TT-12 needles are the same as the suture thread which essentially eliminates bleeding from the needle holes at the anastomosis site. TT-12 suture is ideal for grafts. Gore-Tex® sutures allow approximation after the second knot since the suture can easily can be tightened after two or even three knots. The Gore-Tex® suture is very strong, and unlike polypropylene, normal use never results in breakage. It does, however, require 6-8 square knots to prevent the suture from loosening (Table 2.1). This light, bright suture is easy to see in the operating field and shows well in photographs and slides (Fig. 2.5).

Vascular dilators (Fig. 2.6) can be helpful to enlarge blood vessels in spasm in conjunction with topical 1-2% lidocaine. Great care should be taken not to force these into vessels which causes internal rupture and subsequent thrombosis. Self-retaining retractors (Fig. 2.7) are used by many surgeons routinely. Bear in mind that the part of the retractor above skin level adds to the depth of the wound, making suturing more difficult. A smaller and preferred retractor for wrist AV fistulae and antecubital graft placement is shown in Figure 2.7B.

The authors prefer the bipolar electrocautery for hemostasis. Generally speaking meticulous atraumatic technique will impact the surgeons outcome statistics.

Table 2.1. The Gore-Tex® CV-6 TT-9 or TT-12 suture characteristics

- Needle same size as thread
- Unique tying qualities
- Very strong
- White color shows well
- Requires 6-8 square knots

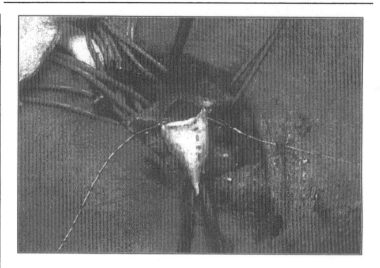

Fig. 2.5. The advantages of the Gore-Tex® suture shown in this picture are also outlined in Table 2.1.

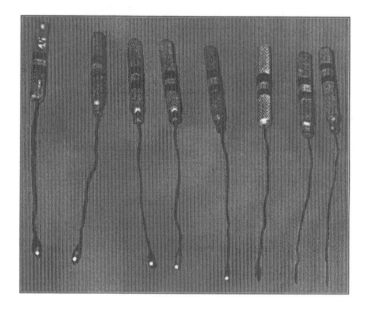

Fig. 2.6. Vascular dilators are available in sizes up to 5 mm in diameter.

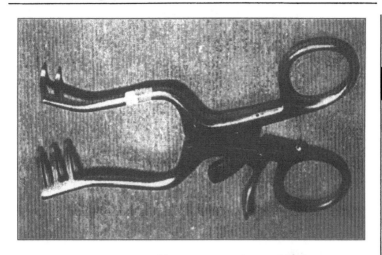

Fig. 2.7A. Self-retaining retractors of this type are sometimes useful in larger upper arm surgery in obese patients. The smaller (Alm) retractor is more suited in vascular access surgery (Fig. 2.7B).

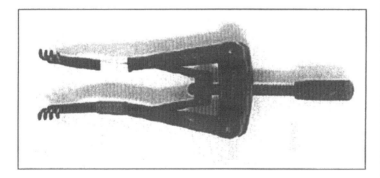

Fig. 2.7B.

Rough tissue handling, especially the use of regular electrocautery in the small operating field typical for vascular access surgery, is prone to have disastrous effects (Fig. 2.8).

Useful Tools/Instruments

There are several other tools and instruments the authors use. Some of these most useful will be mentioned in the next few paragraphs and images. First, the angled, smooth Christmas tree is a quite helpful tool when attached to a syringe to flush small vessels or grafts with heparinized saline (Fig. 2.9). The straight, rugged christmas tree is very traumatic and must not be used in connection with any living tissue.

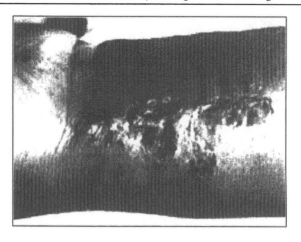

2

Fig. 2.8. Rough handling of tissues, including the use of regular monopolar electrocautery, lends itself to disasterous results.

Thrombectomy Catheters

There are several embolectomy catheters on the market (Table 2.2). The most recent ones are significantly stronger than previous generations. The author is also using the Latis graft cleaning catheter, where the balloon is encaged in a mesh. This device removes semi-resistant or adherent material (Fig. 2.10).

Mechanical Declotting Devices

There are some newer devices aimed at breaking up clots and removing wall adherent material. These are either battery driven rotational devices or jet-powered spray (Table 2.3).

Described in detail elsewhere are two important instruments or tools, but they pertinent to mention in this context. First, the Site-Rite®, portable ultrasound device (Chapter 5, Fig. 5.8A-B) and then the micropuncture set (Chapter 5, Table 5.4, Fig. 5.2 A-B). Both these devices will markedly improve the outcomes and patient safety in vascular access procedures, especially central vein catheter placements.

Table 2.2. Commercially available thrombectomy catheters

Model	Mfg	Size (F)	Length (cm)	Max Vol (cc)	Balloon Dia (mm)
Latis Graft Cleaning	App Med	4	40, 80	0.75	9
Embolectomy (Spring Tip)	App Med	2-7	40, 60, 80, 100	0.08-2.5	4-14
Embolectomy (Regular Tip)	App Med	3-7	40, 80	0.2-2.5	6-14
Thrombectomy/Arterial plug	App Med	3-5	40, 80	0.2-1.5	6-11
Graft Thrombectomy	Fogarty	5-6	50	n/a	n/a
Venous Thrombectomy	Fogarty	6, 8	80		12, 13, 19
Adherent Clot	Fogarty	4-6	80		6, 8, 10
Thru Lumen Embolectomy	Fogarty	3-7	40, 80	0.15-1.6	5-14
Arterial Embolectomy	Fogarty	2-7	40, 60, 80, 100	0.15-1.6	5-14

Applied Medical Embolectomy—Regular tip. Removal of emboli and thrombi from the arterial system. Latex balloon 0.75 ml. Cath available in 3-7 F. Also comes in spring tip with same balloon, tip minimizes intimal damage. A nonlatex version is strong, maintains shape. Balloon is nonfragmenting, with symmetric inflate/deflate system to maintain shaft in center of vessel.

Applied Medical Arterial plug—Removal of thrombi from venous system and vascular graft. The balloon is latex, with 0.75 ml fill volume, rated 90% stronger than traditional embolectomy catheters. Cath is durable—removes arterial plug. Spring tip minimizes intimal damage.

Applied Medical Latis Graft Cleaning—Removes thrombi from vascular grafts. Balloon is latex free, encased in mesh; engages clot, protects balloon from rupture. 0.75 ml fill volume. Cath is durable—removes arterial plug.

Fogarty Graft Thrombectomy—Flexible wire coil for removal of mature thrombus from synthetic grafts. The spiral wire retracts for maneuverability.

Fogarty Venous Thrombectomy—Removal of thrombi from the venous system and vascular grafts. The 4F cath has a latex balloon with 0.75 ml fill volume. The tip is flexible with a steel spring in the cath body for increased torsion strength and direction control.

Fogarty Adherent Clot—For arterial plug or resistant thrombus. The spiral shaped latex covered stainless steel cable in a corkscrew configuration retracts into the protective membrane, making the cath safe for synthetic grafts and native vessels.

Fogarty Thru Lumen Embolectomy—Has a second lumen for introduction of guidewire, injection, blood sampling or temporary occlusion. The balloon has stainless steel bushings on each side for fluoro visualization. Each size is rated for acceptable pull force.

Fogarty Arterial Embolectomy—Designed for removal of embolus and thrombus from the arterial system. 4F has 0.75 cc balloon, breaks easily. Wide variety of sizes and packaging configurations.

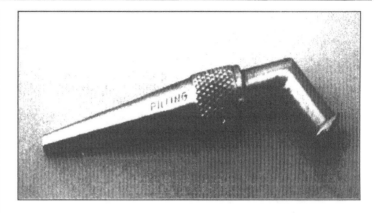

Fig. 2.9. The smooth angled "Christmas tree" is a valuable tool for flushing vessels and grafts. (Pilling, 420 Delaware Drive, Fort Washington, PA 19034, phone: (215) 643-2600. www.pillingsurgical.com). Plastic, rough "christmas trees" must not be used in native vessels.

Fig. 2.10A. The Latis catheter (Applied Medical, 22872 Avenida Empresa, Rancho Santa Margarita, CA 92688-2650, phone: (949)713-8000. www.appliedmed.com) has a plastic mesh around the balloon that conforms and moves around the balloon (Fig. 2.10B) when inflated.

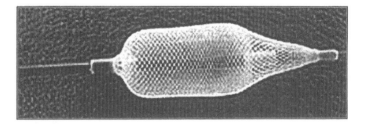

Fig. 2.10B.

Table 2.3. Commercially available mechanical thrombectomy devices

Model	Mfg	Size (F)	Type
Amplatz Thrombectomy Device	Helix	7	H
Oasis	MediTech	7	H
AngioJet	Possis	6	H
Hydrolyser	Cordis	7	H
Solera	Bacchus	7	A
Treratola Thrombectomy Device	Arrow	5	WC
Castaneda Thrombolytic Brush	Micro Therapeutics	6	WC

H=hydrodynamic; A=aspiration; WC=wall contact mode of throbectomy

Selected References

1. Collins AJ, Hanson G, Ument A et al. Changing risk factor demographics in end-stage renal disease patients entering hemodialysis and the impact on long-term mortality. Amer J Kid Dis 1990; 5(5):422-432.
2. Surratt RS, Picus D, Hicks ME et al. The importance of preoperative evaluation of the subclavian vein in dialysis access planning. AJR 1991; 156:623-625.
3. Walters GK, Jones CE. Color duplex evaluation of potential hemodialysis graft failure. J Vasc Tech 1992; 16(3):140-142.
4. Hodde LA, Sandroni S. Emergency department evaluation and management of dialysis patient complications. J Emer Med Vol 1992; 10:317-334.
5. Windus D. Permanent vascular access: A nephrologist's view. In-depth review. Amer J Kid Dis 1993; 21:5.
6. Diskin CJ, Stokes TJ, Jr., Pennell AT. Pharmacologic intervention to prevent hemodialysis vascular access thrombosis. Nephron 1993; 64:1-26.
7. Sreedhara R, Himmelfarb J, Lazarus JM et al. Anti-platelet therapy in graft thrombosis: Results of a prosepctive, randomized, double-blind study. Kid Inter 1994; 45:1477-1483.

CHAPTER 3

Primary Arterio-Venous (Native) Fistulas (PAVF)

Ingemar J. A. Davidson, Illustrations: Jonathan P. Young

Detailed Surgical Procedure

The most practical and common anastomosis site for PAVF is between the radial artery and the cephalic vein in the distal forearm. This end-of-vein to side-of-artery site is used in the following detailed surgical outline. All illustrations in this chapter are of the left upper extremity.

First, mark the course of the radial artery and cephalic vein in the distal half of the forearm and mark the proposed longitudinal skin incision midway between the two vessels (Fig. 3.1). A transverse incision is less optimal since it cannot be extended should it become necessary to move more proximally on either vessel. In the author's opinion, there is also a greater risk of kinking or angling the short segment of vein that can be exposed through a transverse incision.

If present, identify and isolate the common dorsal branch of the cephalic vein (Fig. 3.1). The use of this bifurcating site provides a "patch" anastomosis which is optimal from a hemodynamic standpoint and also technically easier to perform (see below). Make sure that the cephalic vein is of adequate caliber, (i.e., palpable) along the forearm to the antecubital fossa. The presence of multiple small mid-arm venous collaterals around the main cephalic vein usually indicates that the main branch has been thrombosed or fibrosed from previous needle punctures and therefore is not usable. Anastomosing to any vein but the main branch of the cephalic vein is doomed to failure since a single small collateral branch will not dilate sufficiently to form a usable venous limb.

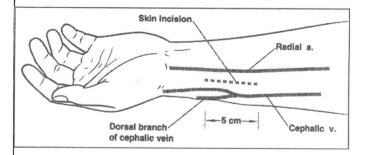

Fig. 3.1. The skin incision (the dotted line) is made to facilitate using the dorsal branch of the cephalic vein in a "patch" anastomosis.

Access for Dialysis: Surgical and Radiologic Procedures, 2nd ed., edited by Ingemar J.A. Davidson. ©2002 Landes Bioscience.

Supplement with local anesthesia (1% lidocaine) as needed. DO NOT use epinephrine since it causes vasospasm. Topical lidocaine is also used repeatedly during the procedure to help dilate both the radial artery and cephalic vein which uniformly develop various degrees of spasm.

Using a #15 blade, an incision 5-6 cm in length midway between the artery and vein is adequate (Fig. 3.1). The distal limit of the skin incision should be proximal to the radial styloid process. The level of skin incision should be dictated by the venous anatomy, i.e., location of the dorsal branch in order to choose the optimal anastomotic site for the vein.

Dissect through subcutaneous tissue using blunt (mosquito hemostat) and sharp knife techniques (Fig. 3.2). Hemostasis is obtained with bipolar electrocautery. Regular electrocautery, because of the much greater potential for tissue damage (burn injury), is not recommended. Bipolar (micro type) electrocautery on the other hand, correctly used, is less or equally traumatic as tying. Larger vessels are ligated with 5-0 polygalactin (Vicryl®) suture. The cephalic vein is mobilized by dividing the tissue on top of the vein as shown in Figure 3.2. Using this technique enables the connective tissue planes to fall on either side with minimal bleeding and tissue damage.

It is best to avoid grasping the vein proper. Only hold the adventitia with DeBakey or fine eye forceps. Place vessel loop(s) around the vein(s). Ideally, a dorsal branch, if present, of the cephalic vein is also dissected in order to use the bifurcation as a "patch" for anastomosis to the artery.

Isolate the artery for a distance of 4-5 cm. Dissect straight down to the artery, using the fine hemostats or Blue Darter forceps to identify the plane adjacent to the artery (Fig. 3.3). This technique is identical to that utilized for the vein above (Fig. 3.2). There is a thin layer of periarterial tissue that needs to be divided (Fig. 3.4A). The concomitant veins run parallel to the artery on either side. The final step will separate the two small concomitant veins from the artery in a loose plane of connective tissue. No dissection should or needs to be done on the sides of the artery, since this will lead to damage of the small concomitant vein, one on each side, and artery branches and cause unnecessary bleeding. Uniformly there are no arterial branches from the anterior (volar) aspect of the artery. However, there are usually several paired arterial branches leaving the radial artery on each side somewhat in the posterior direction (Figs. 3.4A-C).

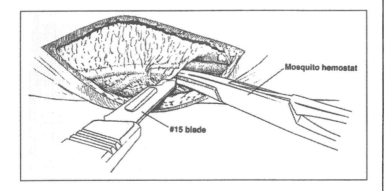

Fig. 3.2. The dissecting technique to isolate vessels is similar for arteries and veins.

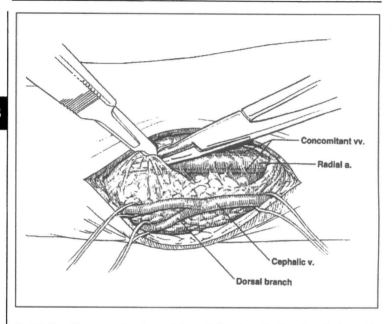

Fig. 3.3. Vessel loops are placed around the cephalic vein and its branches. The fascia-like structure on the top of the radial artery is being divided.

As the fine periarterial tissue is divided, (Fig. 3.4A) the concomitant veins are separated away from the artery (Figs. 3.4A-C), exposing the three to four small, often paired arterial branches on either side of the radial artery. At this point, two vessel loops are easily slipped behind and around the artery. Care should be taken not to tear the small arterial branches off the radial artery, as it causes bothersome bleeding. Figures 3.4B-C illustrate how dividing the periarterial tissue will expose and separate the concomitant veins and expose the small arterial branches. The side branches are ligated and divided with nonabsorbable 5-0 ligatures. The tie is placed slightly away from the radial artery, or about 2 mm out on the small branch to avoid forming a "waist" when the artery later dilates (Fig. 3.5).

At this point, the vessels are adequately freed and mobilized to create the fistula (Fig. 3.6). If severe venospasm is present, which is common especially in younger patients, 1-2% lidocaine is topically applied. Next, hemostat(s) are applied distal to the veins (Fig. 3.7). Cut the vein(s) partially (about two-thirds) with a #11 blade, dilate the opening of the venous branch and place a corner stitch using 7-0 polypropylene (Prolene®) with double armed CV-1 needles or Gore-Tex® CV-7 on TT-9 needles. Once the corner stitch is placed, the remaining vein is transected. The dorsal branch is divided in the same manner after placing the second corner stitch. A segment of a few millimeters longer is left on the dorsal branch. Rubber-shod clamps are placed on the sutures to maintain orientation of the vein at all times and to avoid rotating which, if undetected, would result in thrombosis within a few hours after creation of the anastomosis.

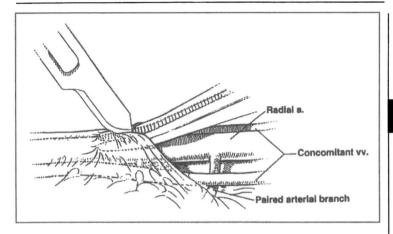

Fig. 3.4A. By dividing the thin periarterial adventitial layer, the artery is exposed without damaging the concomitant small veins.

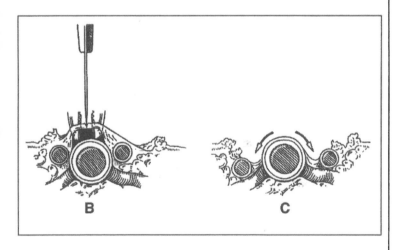

Fig. 3.4B-C. By dividing this structure on top of the artery, the concomitant veins are moved away from the artery, exposing the small paired arterial branches.

Open the vein patch by cutting between the open ends of the veins, 180 degrees from the corner stitches (Fig. 3.8). Uneven or excess vein can now be carefully trimmed, using fine scissors. Any valves that may be present, near to, or in the opened vein patch, should be excised to avoid technical problems if they are inadvertently sutured into the anastomosis. The cephalic vein is prevented from twisting or rotating by keeping the rubber-shods attached and oriented at all times to the respective sutures. This is even more important when a standard (nonpatch) end-of-vein to side-of-artery anastomosis is performed. By leaving the back wall of

3

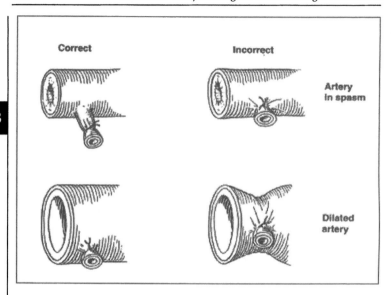

Fig. 3.5. Technique of ligating the small paired arterial branches, avoiding "waist" formation when the artery dilates.

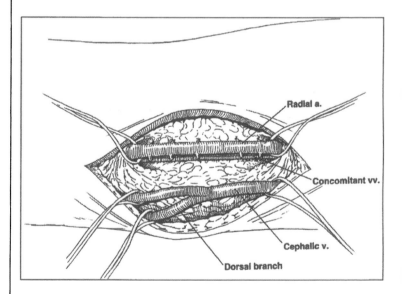

Fig. 3.6. The radial artery and the cephalic vein, with its dorsal branch have been adequately dissected.

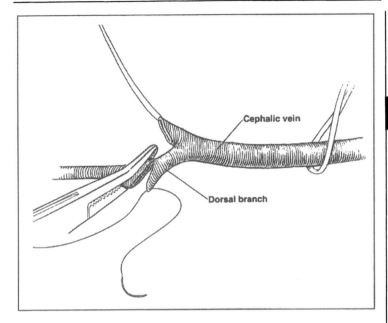

Fig. 3.7. The dorsal branch is left slightly longer to avoid kinking of the cephalic vein after completion of the anastomosis.

the dorsal branch vein intact, one can more easily divide, (Fig. 3.8A) dilate, (Fig. 3.8B) and flush (Fig. 3.8C) the vein with heparinized saline and trim the edges (Fig. 3.8D). The vein is also stabilized by having the two corner sutures under slight tension (Fig. 3.9).

Flush the proximal vein with heparinized saline (5 ml of 1000 units/cc heparin in 500 ml of saline) using a 20 ml syringe with an 18 ga angiocatheter (Fig. 3.8C). The vein may be gently dilated with vascular dilators (Fig. 3.8B). Stop at the slightest resistance. A Heifet's vascular clamp is sometimes needed on the vein to prevent back flow.

Double ligate or suture ligate the distal vein(s) with 5-0 polyglactin ligature (Fig. 3.9).

Apply two Heifet's clamps to the artery, one proximal and one distal to the proposed arteriotomy site (Fig. 3.9). Place these in such a direction that the sutures will not become entangled in them. The clamps can usually be "hidden" under the skin edges, away from sutures and instruments (Fig. 3.10A).

Make a small, 1-2 mm arteriotomy, with a #11 blade, over the radial side of the artery facing the vein (Fig. 3.10A). Using a snugly fitting #18 angiocatheter, the artery can be locally heparinized by injecting heparinized saline sequentially while opening and closing the vascular clamps in both directions (Fig. 3.10B). Systemic heparin is not necessary nor desired because of its propensity to cause postoperative wound hematomas. An artery in spasm can be dilated by heparinized saline under moderate pressure or by using vascular dilators. Over dilatation, however, will cause

3

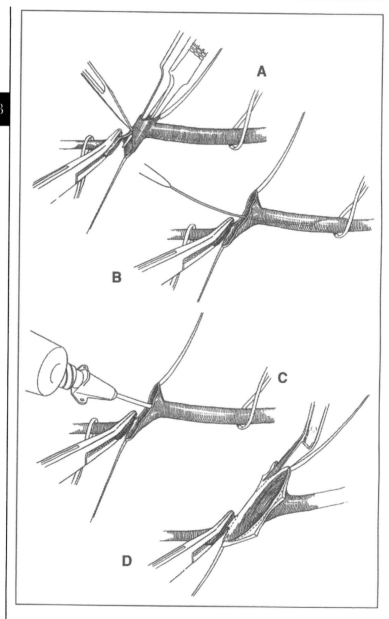

Fig. 3.8A-D. Techniques outlining the creation of the venous patch and preparation of the vein.

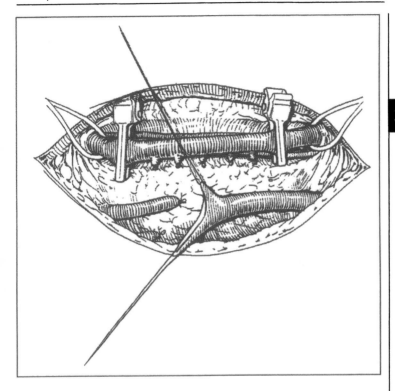

Fig. 3.9. The two corner stitches are used to keep the vein and the patch oriented at all times.

intimal rupture and subsequent thrombosis. Frequent use of 1% topical lidocaine will help to prevent and reverse vasospasm.

Extend the arteriotomy to the appropriate length (8-10 mm), matching the vein using fine Dietrich scissors (Fig. 3.10C).

Take the appropriate needle of the previously placed proximal double-armed 7-0 polypropylene corner suture and suture inside out to the proximal arterial corner. Have the assistant dilate the artery using a Blue Darter forceps for exact suture placement and to keep from catching the back wall with the suture (Fig. 3.11). The corner bites should be small (approximately 1 mm). Tie the suture in three square knots. Correctly placed, the knot is on the outside of the vessel (Fig. 3.12A).

Place the second double-armed polypropylene suture in similar fashion to the distal arterial corner, but do not tie the suture at this time (Fig. 3.11). This helps to expose the back wall while placing the first 2-3 proximal sutures (Fig. 3.12).

An artery or vein may be closest to the surgeon, depending on which side of the arm the surgeon prefers to sit. The running suture is begun at the back wall in the proximal corner. The very first stitch is placed from outside-in on the vessel closest to the surgeon (Fig. 3.12A). The purpose of this first stitch is just to get inside the vessel with the needle passing as close to the corner knot as possible. Alternatively,

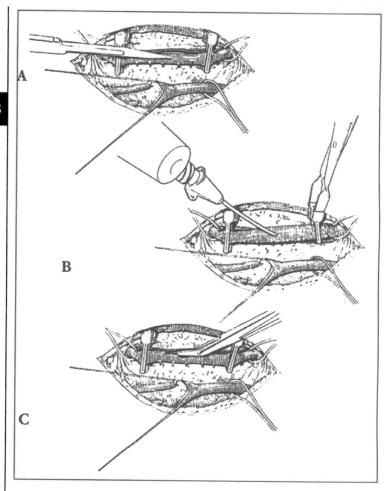

Fig. 3.10A, B. Techniques for dilating and local heparinization of the radial artery. C:
The arteriotomy is extended to match the size of venous patch.

one may pass the suture (the needle) between the back walls and then place the first
stitch out-in on the vessel next to the surgeon (Fig. 3.12B).

The first back wall suture and all subsequent back wall sutures go inside-out on
the opposite vessel and outside-in on the vessel nearest the surgeon (Fig. 3.12C). In
these illustrations, the surgeon is on the ulnar site closest to the radial artery.

The stitches closest to each corner are placed taking small bites (1 mm) of the
vessels with minimal travel. This will maximize blood flow and preserve lumen size
by preventing a purse-string effect. Large bites in the corners jeopardize the very
survival of the fistula. No rough handling of the vessels is permissible. Forceps may

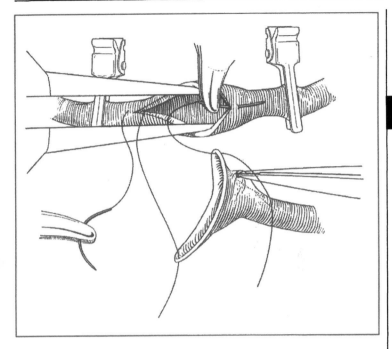

Fig. 3.11. By using the sharp microforceps to dilate the artery, exact stitching is facilitated.

be used to push and direct vessels during suturing, but not to grasp. The intima should never be picked up by forceps. The only acceptable grabbing is of the perivascular loose connective tissue using fine forceps.

After the first 2-3 back wall stitches have been placed, the distal arterial corner stitch is tied, or this suture can be left untied under slight tension until the back wall is completed. In either case, this suture is attached to a rubber-shod clamp hanging over the patient's hand. This gives the appropriate tension and lines up the back walls nicely for precise suturing. Every stitch is strategically placed to maximize the fistula size and resultant blood flow (Fig. 3.13A).

The last stitch of the back wall goes inside-out on the vessel away from the surgeon (Fig. 3.13B).

If not done before the distal corner stitch is now tied in three knots (sutures a & b) (Fig. 3.14A). One end is rubber-shod and the other end is used to tie to the back wall suture (suture c) using 6-7 square knots (Fig. 3.14B). These last two tied sutures are then cut.

Start in either corner and run the anterior wall (Fig. 3.15A). The first 2-3 stitches should be double bites while the assistant gently dilates the vessels with a Blue Darter forceps. When the anterior anastomosis is halfway complete, this suture is rubber-shod. Then start the suture from the other corner (Fig. 3.15B), to meet midway on the anterior wall. If the surgeon runs both sutures toward himself, the very last stitch may be reversed so that the knot can be tied across the anterior wall (mainly an aesthetic point).

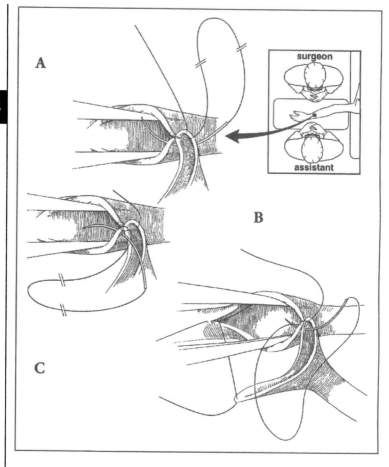

Fig. 3.12. A) The first stitch of the back wall anastomosis goes outside-in on the vessel closest to the surgeon. B) Alternatively, this stitch can be placed after passing underneath the vein patch. C) All subsequent stitches are single bites through the back wall anastomosis.

If a Heifet's clamp was placed on the vein, it is removed now. The distal arterial Heifet's clip is removed, and then the proximal arterial clip is released.

There is always slight bleeding from the suture lines at this time. Even if the bleeding seems significant, simply apply gentle pressure for a few minutes. Unless there is a technical mishap along the suture line, the bleeding will stop. A serious mistake often made immediately after removal of the vascular clamps is to start placing extra sutures to stop small bleeding points from needle holes.

Figure 3.16 shows an overview of a "patch" cephalic vein PAVF. Often, however, there is no suitable dorsal branch. Figure 3.17 illustrates the steps for an end-of-vein to side-of-artery without "patch." The principal technique is identical to the patch

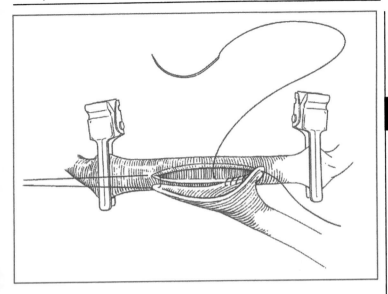

Fig. 3.13A. The back wall running suture.

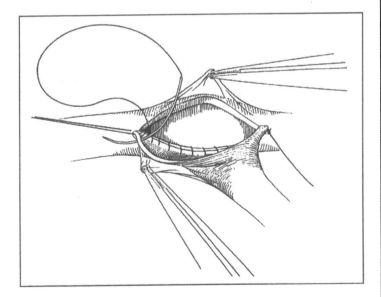

Fig. 3.13B. The very last stitch of the back wall anastomosis.

3

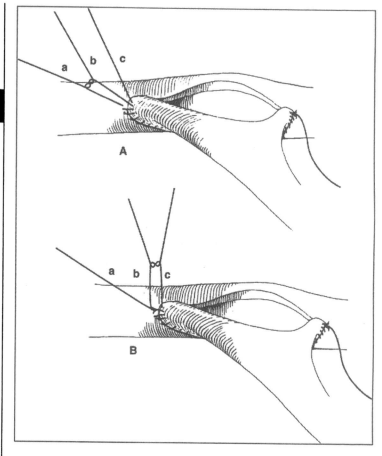

Fig. 3.14. Technique for tying the distal corner sutures.

steps. However, because of the absence of the patch, exact suturing technique becomes even more important. Also, placing the corner stitches before dividing the vein to keep orientation is imperative (Figs. 3.17C-D). All suturing techniques described above for the patch PAVF apply.

When the bleeding has stopped, the vessels should be examined for strictures (from fibrous bands or vasospasm). Topical 1-2% lidocaine and judicious cutting of fibrous bands with microscissors will resolve these problems. There is often pronounced spasm in the vein at the level immediately beyond the point where the dissection stops. Be sure that the vein makes a smooth curve, and then gently spread or cut along the vein for another 1-2 cm. A sponge soaked in lidocaine in contact with the vein for a few minutes will usually relieve the vasospasm. Finally, make sure the entire wound is absolutely dry before skin closure.

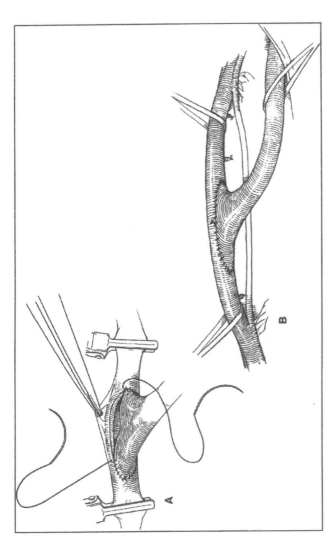

Fig. 3.15. A: Completion of the front wall anastomosis. B: The proximal and distal sutures are tied midway.

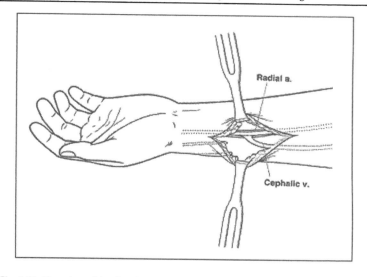

Fig. 3.16. Overview of the "patch" primary AV fistula technique.

Two or three subcutaneous sutures may be placed, avoiding suturing over the vein. The authors prefer subcuticular (5-0 polyglactin) skin closure with steri-strips and a loose dressing applied. A circular or even semi-circular tightly taped wound covering may obstruct fistula flow and cause hand edema. The patient is encouraged to elevate the arm resting on pillows and to make fists over a soft ball to prevent swelling.

Complications of Primary AV Fistulas

Early problems after PAVF placement are often related to surgical/technical factors and include thrombosis, postoperative bleeding, infection, hand ischemia ("steal") and paresthesia from peripheral nerve injury during anesthesia or surgery.

Late complications are usually related to dialysis practice and needle puncture technique. The most common are vascular stenosis at various levels, thrombosis, usually starting at a stenosis site, infection/inflammation usually in association with thrombosis, false aneurysm at the anastomosis site, infiltrating hematoma after dialysis needle puncture, true aneurysm along the vein and venous hypertension in the hand.

Early Complications

Thrombosis is the most common early complication. The incidence depends on the criteria (i.e., the quality of vessels, usually the vein) used for placement of PAVF. One should, however, always suspect a technical problem such as a kinked or twisted vein, problem with suturing, compressing hematoma, a too tight subcutaneous closure with edema, preexisting unrecognized proximal venous occlusion or a dressing that is too tight. Sometimes, thrombosis occurs at the anastomosis within a few days of operation. Most often, the vein is patent proximal to the clotted anastomosis. The fistula should be explored since the problem, if found, can usually be corrected. However, one may alternately find thickened, inflamed vessels which place the

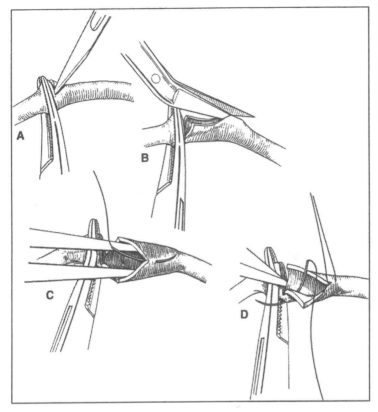

Fig. 3.17. Techniques for preparing the cephalic vein without patch technique. The length of the venotomy is determined by the local anatomy, including the angle between the vein and artery. The suturing technique is identical to that described for the patch anastomosis.

fistula at risk for rethrombosis after repair. Moving the anastomosis 2-3 cm proximally in order to use unaffected tissue and vessels is recommended.

Kinks and rotations of the vein must be corrected. This requires taking down the anastomosis, properly aligning the vessels and carefully re-anastomosing the vessels. Again, if the vessels show signs of inflammation, one is better off creating a new anastomosis.

A more proximal and unrecognized, preexisting venous stenosis (usually from previous needle punctures) can be assessed with a Fogarty catheter (#3-5) or smooth dilators (#3-5) if they will reach. Some surgeons routinely pass a Fogarty catheter during the initial surgery to ensure an adequate vein all the way to the antecubital fossa. If there is no adequate passage, and assuming the artery is of size and quality deemed likely to result in a successful fistula, another form of vascular access should be considered such as a straight polytetrafluoroethylene (PTFE) graft from the distal radial artery (already exposed). If the adequacy of the artery is in question, it is wiser to place a loop PTFE AV graft with both anastomoses in the antecubital fossa (Chapter 4).

Postoperative bleeding is uncommon and requires exploration if continuous or in the case of expanding hematoma. A small anastomotic bleed will usually require a carefully placed 7-0 polypropylene suture. Exact suturing is facilitated with a neuro-suction held by the surgeon's left hand, while the suture is placed with the needle driver in the right hand. Only when the surgeon can see the bleeding hole in the suture line can an exact stitch be placed. This may be obvious only for a fraction of a second, and the coordination between the surgeon's left (suction) hand and right (suture) hand is critical. A larger bleed from an anastomotic defect will require clamping of the artery.

Bleeding from other sites is addressed accordingly. A sloppy ligature on the distal vein(s) may produce profound acute bleeding requiring compression and exploration. Minor bleeding or oozing can be stopped by a bipolar electrocautery. Hematoma formation causing compression of the vein may result in thrombosis.

Infection is and should be extremely uncommon. Every attempt should be made to save a well-functioning fistula using common surgical principles and judgment. Late infections along the vein are uniformly related to dialysis needle puncture technique.

Hand ischemia ("steal") is caused by reversal of blood flow through the radial artery away from the hand. This complication is less common with primary AV fistulas than with PTFE grafts (Chapter 4). However, the treatment is simple and consists of suture ligating the radial artery distal to the AV anastomosis using permanent sutures (Fig. 3.18). This increases pressure and thereby flow to the palmar arch of the hand from the ulnar artery. The diagnosis of arterial steal is made from clinical examination, but should also include duplex Doppler ultrasonography to obtain flow determinations and finger pressures before and after manual occlusion of the artery distal to the anastomosis. Based on duplex Doppler measurements and finger pressures, there will always be evidence of some arterial steal with any type of AV fistula. However, clinical symptoms such as pain, coolness and tingling are quite uncommon after PAVF. Differential diagnoses include nerve damage (from radial nerve compression during surgery or related to axillary block anesthesia), distal embolization and carpal tunnel syndrome.

Late Complications

Vascular stenosis can occur at any level. Often, it is seen in the cephalic vein 1-2 cm from the anastomosis. Even though this can be corrected with a vein or PTFE patch angioplasty, it is most appropriate to create an entirely new anastomosis a few centimeters up the artery as illustrated in Figure 3.19A. A stenosis further up the vein can be corrected by a patch angioplasty (Fig. 3.19B) or a graft interposition (Fig. 3.19C), depending on length, severity and other anatomical considerations, as well as the surgeon's preference.

Thrombosis can also occur at any level along the vein. In fact, it often starts at a stenosis site. Therefore, the development of a venous stenosis often precedes thrombosis formation. If the thrombosis occurs at the anastomosis and the proximal vein is still open, treatment consists of re-anastomosing the fistula a few centimeters up the artery as described in Figure 3.19A. A thrombosed primary AV fistula may also be declotted and the stenosis corrected with a patch or interposition graft. Another option is to utilize radiographic interventional techniques with t-PA and balloon

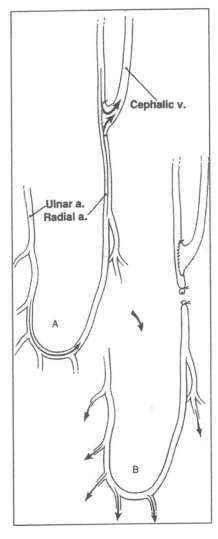

Fig. 3.18. Hand ischemia from arterial steal (A) is treated by ligating the radial artery distal to the anastomosis (B).

angioplasty (see Chapter 7). In cases of marked inflammation along the thrombosis, the likelihood of successful declotting is decreased.

False aneurysm at the anastomosis site results from bleeding between sutures. Small aneurysms can be watched. If they are cosmetically bothersome or if the skin becomes shiny (atrophic) the aneurysm needs to be excised (Fig. 3.20). These procedures are sometimes technically challenging. Generally speaking, the artery needs to

3

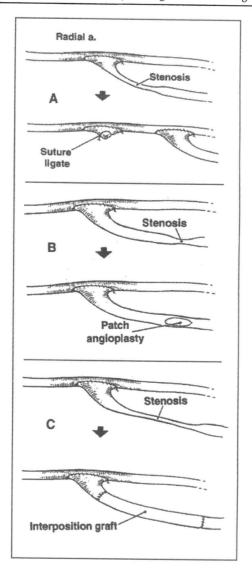

Fig. 3.19. Three different ways of managing a stenosis of a primary AV fistula. A) Creating a new anastomosis. B) Placing a patch angioplasty. C) Placing an interposition graft.

be isolated on both sides of the anastomosis, i.e., above and below the aneurysm. After obtaining control of the artery with Heifet's clips, the aneurysm can be opened and the small opening at the anastomosis site closed with 7-0 polypropylene suture.

Again, should technical difficulties prevent resection and repair, the option of creating a new anastomosis (as shown in Fig. 3.19A) more proximal remains an alternative that might have been chosen in the first place.

False aneurysm from multiple dialysis needle punctures may be hard to differentiate from true aneurysms along the vein. However, treatment and management of these is the same. When such an aneurysm becomes disturbingly big or the skin becomes atrophic, correction is recommended (Fig. 3.20). This can be done by either excising part of the wall and thereby narrowing the venous lumen or by placing an interposition vein or PTFE graft. Many of these aneurysms occur with repeated needle punctures at the same site and can be avoided by rotating dialysis needle punctures sites. This is true for PTFE grafts as well.

Venous hypertension to the hand occurs more often (15-20%) with a side-to-side AV fistula than with venous end to arterial side-type fistula. The author exclusively performs the vein end to artery side-type primary AV fistulas. Even with these, occasional venous hypertension occurs from back flow through a dorsal branch to the hand, especially if there is a more proximal stenosis in the cephalic vein. This situation needs correction only when the patient develops pain and/or ischemia. The treatment consists of dividing and suture ligating the venous branch going to the hand, usually affecting the thumb and index fingers (Fig. 3.21).

AV fistulas, both primary and grafts, may develop such high blood flow that congestive heart failure develops. This is perhaps more likely to occur with nontapered PTFE grafts. The brachial artery to cephalic vein fistula shown in Figure 3.22 had an estimated blood flow of 5-6 l/min. The blood flow is decreased by some sort of "banding" procedure; in this case, a 2 cm segment of a 6 mm PTFE graft was sutured around the vein to partially occlude the vein close to the anastomosis. The patient's cardiac status has permanently improved (8 months) after corrective banding.

Chronic complications as described here with primary AV fistulas are fairly common. Many of these, however, do not need correction but rather should be followed carefully and corrected if and when significant symptoms develop. It must be remembered that an AV fistula is the patient's lifeline and any surgical intervention may potentially result in fistula failure requiring further access procedures and acute placement of dual lumen catheters or PTFE grafts. One should exercise great judgment and err toward the conservative side.

Summary Steps in Primary AVF Creation

1. Mark radial artery and cephalic vein and dorsal branch.
2. Skin incision between the radial artery and cephalic vein.
3. Dissect cephalic vein and the dorsal branch. Place vessel loops.
4. Cut through the fascia on top of the radial artery. (DO NOT dissect on either side of the radial artery to avoid bleeding from the concomitant veins).
5. Expose the radial artery from the top, where there are no branches.
6. Gently push the peri-adventitial tissue sideways exposing the paired small arterial branches. The concomitant veins will also move away from the artery with this technique. The same technique illustrates the mechanism by which the concomitant veins move away from the artery exposing the paired side branches.
7. By tying the side branches 1-2 mm away from the artery the waist formation is avoided.
8. At this point, the radial artery is mobilized for 3-4 cm, side branches tied and two vessel loops placed around the artery.
9. The cephalic vein is freed with its dorsal branch surrounded with vessel loops.
10. Lidocaine 1% may be sprayed on the vessels to decrease vasospasm.
11. Mosquito hemostats are placed on the bifurcating veins.
12. The distal (dorsal branch) is left slightly longer, the shorter cephalic branch prevents a sharp angling of the vein that otherwise would occur as the cephalic vein is turned toward the radial artery.
13. The branches are cut partially with a #11 blade; the venotomy widened with the micro forceps or mosquito hemostats.
14. Corner stitches (Prolene® 7.0, BV-1) are placed, attached to rubber shods.
15. The two branch openings are connected using a #11 blade or fine scissors, with a micro forceps inserted into the branches, connecting the openings.
16. The cephalic vein is now gently dilated or preferentially flushed with heparinized saline, using an angiocatheter or the angled, smooth christmas tree. By leaving the distal dorsal branch backwall still attached, these maneuvers are easier to perform while the cephalic vein is stabilized.
17. Finally, the "patch" is trimmed to remove excess vein in the corners using diethrich fine scissors.
18. The patch is now ready to be sewn to the radial artery. Note that the proximal side of the patch is slightly shorter to avoid kinking of the cephalic vein.
19. Heifets clips are applied on the radial artery in a way that they can be "hidden" under the skin edges, thereby avoiding sutures catching to its parts.
20. Use precise suturing technique. Do not purse-string suture line. Do not grab vessel intima. Be exceedingly atraumatic—use magnifying glasses.
21. Do not place extra sutures on minor bleeds. Wait!
22. Close skin with subcuticular running suture, i.e., 5-0 PDS®or Vicryl® on an RB needle.
23. Steri-strips on skin, cover with gauze dressing.
24. Elevate arm and hand. Make fists around soft ball postoperatively to decrease edema. Exercise hand fists against 30-40 mm Hg blood pressure cuff after 10-14 days postoperatively to enlarge cephalic vein.

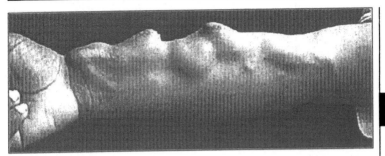

Fig. 3.20. Bothersome aneurysmatic dilatation of a forearm radial-cephalic primary AV fistula. There are no proximal venous obstructions. A new AV fistula was placed in the contralateral arm.

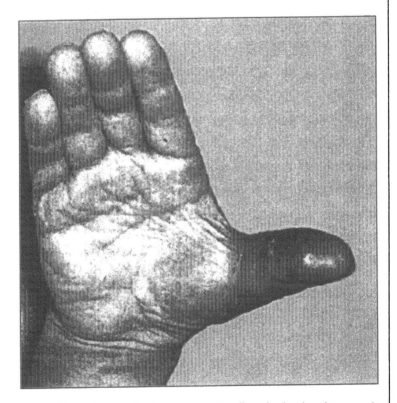

Fig. 3.21. Venous hypertension from PAVF usually affects the thumb and causes pain, bluish discoloration and eventually ulceration.

3

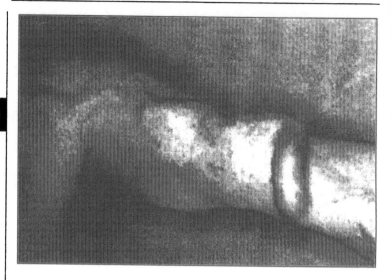

Fig. 3.22. This upper arm PAVF had an estimated 6 L/min blood flow before "banding."

Selected References

1. Humphries Jr AL, Nesbit Jr RR, Caruana RJ et al. Thirty-six recommendations for vascular access operations: Lessons learned from our first thousand operations. Amer Surg 1981; 47:4.
2. Burger H, Kluchert BA, Koostra G et al. Survival of arteriovenous fistulas and shunts for haemodialysis. Eur J Surg 1995; 161:327-334.
3. Katsumata T, Ihashi K, Nakano H et al. An alternative technique to create end-of-vein to side-of-artery fistula for angioaccess. J Amer Coll Surg 1996; 182:69-70.

CHAPTER 4

PTFE Bridge Grafts

Ingemar J.A. Davidson, Illustrations: Stephen T. Brown

Preoperative Considerations

The majority of ESRD patients requiring PTFE AV grafts are elderly, obese and anemic (hematocrit 20-25%) and up to 50% are diabetic. Intraoperative problems including respiratory arrests from anesthetics and sedatives may occur if these agents are routinely dosed on a body weight basis. General anesthesia should be avoided especially in overweight individuals, even though adequate regional anesthesia is more difficult to achieve in these patients.

The general surgical considerations for PTFE grafts are outlined in Chapter 2. The correct, atraumatic surgical technique is the key for short- and long-term graft patency. Early graft failure (thrombosis and infection) before the graft has been used, is likely the result of poor surgical technique and debilitating patient circumstances, i.e., HIV infection, diabetes, preexisting infectious conditions, intravenous drug abuse and obesity. Lack of adequate or usable vessels is unusual and more likely the result of rough surgical technique resulting in severe vessel spasm.

In the preoperative evaluation, the patients must be seen by the surgeon and evaluated for type of access. A primary AV fistula should always be considered because of lower postoperative morbidity. However, only about 50% of patients in the US will currently be candidates for a primary AV fistula. For a first time access placement the nondominant hand is preferred. However, the arm that provides the best chance for long-term access function should be chosen and the reason clearly communicated to the patient. For example, a successful primary AV fistula in the dominant arm is preferred over a PTFE graft in the nondominant arm.

It is not acceptable to examine the patient for the first time in the operating room and decide the type and site of access. This is "ghost surgery," and does not represent an acceptable basic level of care, and will lead to patient dissatisfaction with potential legal consequences. Routine examination includes palpating the patient's arms, and identifying veins with a tourniquet or with a blood pressure cuff applied at 40-50 mm Hg while the patient makes a few fists. This is done with the patient in a warm comfortable room. If the patient is cold, an adequate cephalic vein to create a primary AV fistula may be masked. Sometimes, to the surgeon's surprise, a large cephalic vein is found in the operating room after a successful axillary block, which tends to dilate peripheral vessels, facilitating not only the choice of access but the surgery itself.

Color duplex Doppler sonography is a useful preoperative screening tool especially when searching for veins in arms with multiple previous access surgeries (Chapter 8, Fig. 8.2, Table 8.1). The venous site is usually the limiting factor in repeat access cases. If the stenosis is located proximally behind the clavicle, the duplex Doppler may not detect unexpected subclavian vein stenosis or occlusion. If a subclavian stenosis is demonstrated or suspected, the patient should undergo a venogram to

Access for Dialysis: Surgical and Radiologic Procedures, 2nd ed., edited by Ingemar J.A. Davidson. ©2002 Landes Bioscience.

determine the extent of the process, as well as its suitability for balloon angioplasty (which should be done in the same setting) or surgical repair (bypassing the stenosis).

The surgeon must be present during the sonographic examination to guide the examination and determine the optimal sites for repeat access surgery. The most distal site on the arm should be chosen in order to save future sites since the patient will likely be back within months to years with still another failed access. It should be kept in mind that lack of adequate vascular access is a major contributing factor in up to 25% of all ESRD patients who die annually in the U.S.

Venograms or arteriograms are not indicated for routine or first time access unless special circumstances prevail.

Arterial steal is a common postoperative problem in diabetics and elderly patients. A 4-7 mm tapered or stepped graft from the proximal radial artery may diminish this risk, but no prospective controlled studies are available to support this statement.

In cases of bacterial infection, graft placement should be delayed and dialysis managed by temporary means.

Detailed Surgical Procedure

After induction of adequate axillary block anesthesia, once again confirm that the patient is not a candidate for a primary AV fistula. The antecubital fossa vascular anatomy is detailed in Figure 4.1. Mark the skin over the artery (palpated) and if visible, the superficial antecubital veins. The intended skin incision should be about 1-2 cm distal to the antecubital fold (Fig. 4.2). If the cephalic and median antecubital veins are missing, check for the basilic vein, which is the second best choice before using a deep concomitant vein. By using superficial veins, later revisions are technically easier since 20-30% of PTFE grafts will return for declotting procedures within one year, requiring venous outflow obstruction reconstruction. A first time PTFE graft should not pass the elbow joint.

Skin incision is made with a #15 blade, and hemostasis obtained using bipolar electrocautery and 5-0 absorbable material. Silk must not be used because of increased risks of infection and suture granuloma formation. Regular electrocautery creates excessive tissue damage (burn) in the small operating field (Chapter 2, Fig. 2.8). Tying bleeders is achieved fastest with the needle driver or a mosquito hemostat, especially in a deeper wound. The free end of the ligature should be kept very short (1-1.5 cm) facilitating this very time saving technique (Fig. 4.3). Also, having the assistant leave the hemostat (Fig. 4.3A) or the "eye"-forceps (Fig. 4.3B) on the bleeder during tying increases holding strength and diminishes the chance of tearing the tie off of friable tissue.

Static retractors are sometimes useful when working in a deep wound (Fig. 4.4A). It should be remembered, however, that the portion of the retractor that remains above skin level adds to the depth of the wound and makes suturing harder. Often, the entire procedure may be performed without the aid of static retractors, but, instead by using forceps to gently move tissue planes as needed. The small Alm retractor (Fig. 4.4B) is an excellent retractor for most forearm access surgeries.

The vein is dissected first. Again, the most commonly employed veins are the superficial cephalic, median antecubital and the diving anastomotic veins (Fig. 4.1). Typically and ideally these veins provide two or three branches at the site of anastomosis. Usually the median antecubital vein connects to the basilic vein and, consistently, on the deep side of the cephalic vein, v anastomotica (Fig. 4.1) is diving and

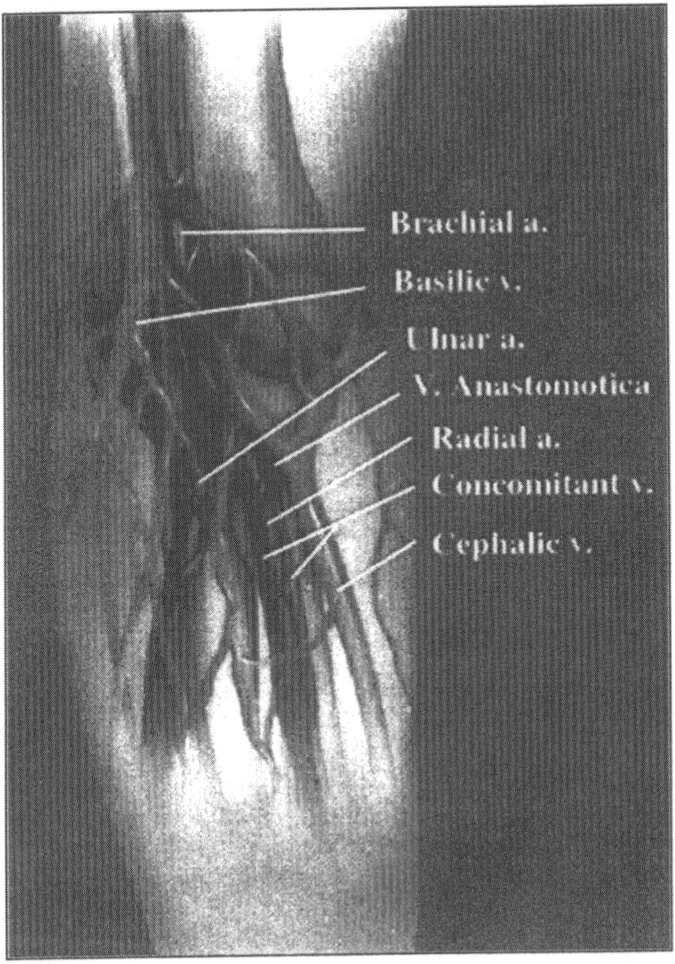

Fig. 4.1. The antecubital fossa anatomy, as pertaining to vascular access.

connecting to the deeper concomitant veins. Even though the anatomy is fairly uniform, there is considerable variation with surprises. The rule of thumb is not to divide any vein branches and sacrifice venous outflow until the venous anastomosis site has been decided upon. In fact, almost never does a venous branch need to be divided.

The vein is dissected free for about 3-4 cm and each branch surrounded with a vessel loop. In the process of dissecting, the surgeon should use a mosquito hemostat along the vein and have the assistant cut with a knife or fine scissors (Fig. 4.5). If there is no assistant, a similar technique is used with a fine forceps and scissors. This technique for dissection is identical for both arteries and veins, and was

4

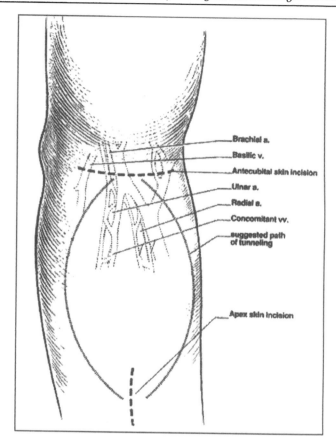

Fig. 4.2. The skin incisions and intended loop configuration in relation to the forearm anatomy.

emphasized in the creation of the primary AV fistula as well (Chapter 3, Figs. 3.2 and 3.3). Regardless of the surgeon's style and technique, atraumatic technique is an absolute necessity for short- and long-term success. For example, grabbing vessels with large forceps or pulling hard with vessel loops, or using heavy silk double looped around vessels for occlusion are unacceptable techniques. Gentle traction by grabbing periadventitial tissue, or pushing (without grabbing) will achieve the same exposure without damage to vital structures. Another basic principle which applies to all surgery is not to divide any structures unless absolutely certain about their nature.

Next, the artery is addressed. At this point the subcutaneous tissue has been divided down to the biceps aponeurosis. The fibrous aponeurosis is sharply divided with knife with the surgeon guiding with a mosquito hemostat underneath. Usually this structure also needs proximal and distal division in a cruciate formation for optimal exposure of the arteries and concomitant veins. At this point the vascular sheath is visible with the concomitant veins, one on each side of the artery,

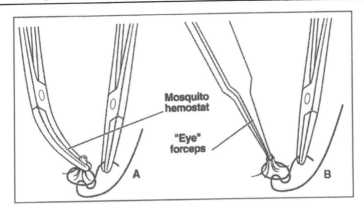

Fig. 4.3. Useful tying techniques, with mosquito hemostat (A) or "eye" forceps (B) using a needle driver.

connected by several short venous branches across the arteries (Fig. 4.6). Dissection is now carried out along the concomitant veins. Any venous branches must not be divided until the intended anastomosis sites have been determined. Even though, as in most cases, the superficial cephalic vein is used, venous outflow is partially directed via the vena anastomotica to the deep concomitant veins. Furthermore, any ligated branches may act as a nidus for thrombus formation. The concomitant veins must also be saved for future use when the superficial veins become occluded by hyperplastic processes. Usually the brachial artery and its radial and ulnar bifurcation can be dissected and exposed appropriately to facilitate creation of a technically adequate arterial anastomosis without sacrificing any veins. As was the case with the venous dissection, one should stay right on top of the artery, use a (mosquito) hemostat and identify the plane next to the artery and cut with a knife or fine scissors (Fig. 4.5). Loose connective tissue around the vessels will retract as divided and the vessels will be exposed without grabbing and potentially injuring or causing severe vasospasm. The brachial artery is freed for about 3 cm, the radial artery for 1 cm. There is often a crossing vein at this point. This crossing vein usually does not need to be divided unless the proximal radial artery is to be used for anastomosis. The ulnar artery is also freed for approximately 1 cm. All three arteries are captured with vessel loops. Retractors of the type shown in Figure 4.4 are sometimes helpful to better expose the vessels. Retractors, however, will add further depth to the wound, making parts of the operation, especially suturing, more difficult. Using forceps and a manual retractor for short moments as needed will often suffice and speed up the procedure.

Next, the graft is placed in the subcutaneous tunnel. This step is of great importance. The details of this part depend upon the type of loop, vessel anatomy, the amount of subcutaneous fat, type of tunneler available, type of graft and finally, the surgeon's personal preference.

The authors are familiar with and have used three types of tunneling devices (Fig. 4.7). First, is the so-called Noon tunneler (top of Fig. 4.7), which is essentially a 6 mm dilator at either end of a 25 cm long flexible steel rod. The graft is tied to one end and pulled through in a semi-circular movement.

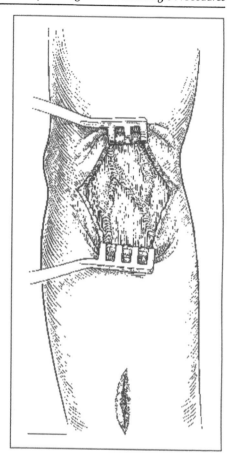

Fig. 4.4A. Static retractors of this type are sometimes needed.

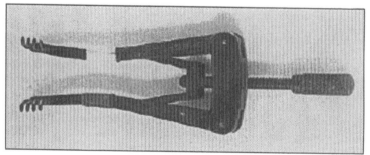

Fig. 4.4B. The Alm retractor is the author's usual choice in small wounds.

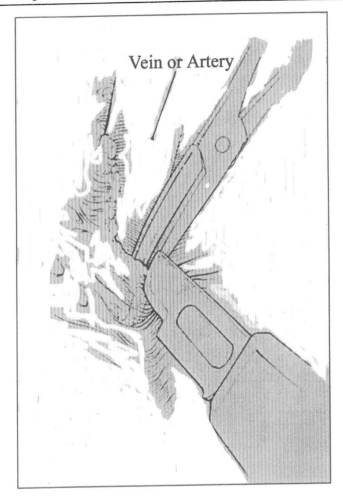

Vein or Artery

Fig. 4.5. The technique dissecting arteries or veins is similar. See also Figures 3.2-3.4 in Chapter 3.

The second type is the Kelly-Weck tunneler (middle, Fig. 4.7), which comes in various degrees of curvatures and head sizes. For the PTFE vascular access placement, the authors have used the semicircular type with head size #6 to which the graft is tied and pulled through. The authors have abandoned both these tunnel devices since these require the graft to be pulled through with potential injury to the graft itself, as well as more tissue damage.

The third type, which the authors currently use and prefer, is a sheath tunneler. Again, for the loop access placement the semicircular type is preferred (bottom, Fig. 4.7). This tunnel device consists of three parts, a semicircular sheath containing a rod to which a bullet or head is screwed (attached) during the subcutaneous tunnel

4

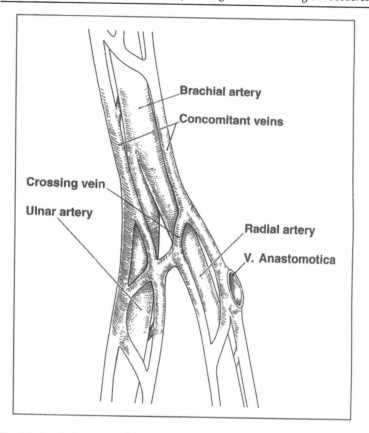

Brachial artery

Concomitant veins

Crossing vein

Ulnar artery

Radial artery

V. Anastomotica

Fig. 4.6. Detailed anatomy of the brachial artery bifurcating into the radial and ulnar artery surrounded by the concomitant veins.

penetration (Fig. 4.8). The directions and technique for use of this tunnel insertion is shown in Figures 4.9A-D. The bullet is then removed, the graft tied (sutured) to the rod and then pulled through. Finally, the sheath is pulled out leaving the graft in the subcutaneous tunnel. A longitudinal counter incision at the apex of the loop is required for all of these tunnel devices. This procedure is then repeated for the second half of the semicircular loop toward the venous anastomosis site (Fig. 4.9D).

As a variation of this procedure, instead of suturing to the rod, the authors feed the PTFE graft into the sheath tunneler and, while holding the graft, the sheath is pulled out. This makes the step less complicated and also saves time (Fig. 4.9C). Feeding the graft is also done for the second portion of the loop also, even though there is a small risk that the graft will not quite reach out from the subcutaneous tunnel (Fig. 4.9D). When placing the second half of the loop, a pair of forceps, pushing the graft into the sheath helps to ensure that the PTFE graft reaches to the venous site (Fig. 4.9D). The use of the sheath tunneler essentially eliminates the risk of the graft twisting or kinking and produces a uniform and smooth subcutaneous

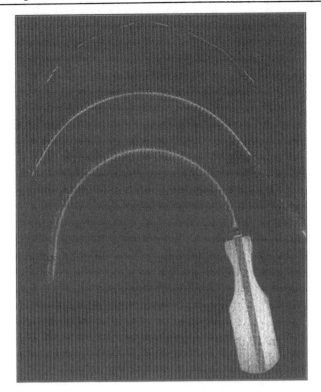

Fig. 4.7. Three types of tunneling devices for placement of PTFE grafts.

tunnel. Also, the sheath eliminates the dragging of the graft through the tissues which may potentially damage the graft and cause kinking and rotation.

Regardless of the tunneling device used, the main objective is an atraumatic procedure, placing the graft at an appropriate depth while avoiding kinking and other mechanical complications. The blue line indicator on the grafts helps to align the graft between the first and second portion of the loop to avoid twisting. Grafts placed too deep in abundant subcutaneous fat often fail due to perigraft hematomas sustained during dialysis procedures as a result of the difficulty cannulating the graft (Fig. 4.10B). Grafts placed too superficially tend to cause skin ischemia, swelling, redness and a greater risk of infection (Fig. 4.10C). To find and place the graft/ tunneler at the appropriate depth is something that the surgeon must learn from personal experience (and from talking to dialysis nurses and technicians) (Fig. 4.10A). As a guide, the surgeon may use his/her hand to hold the skin in a fold around the sheath. Also, the tunneler should be directed by proper angling to obtain the proper subcutaneous plane.

If a tapered or stepped graft is used, the tunneling step ideally starts at the arterial anastomosis side (Fig. 4.9A). In other words, the wider portion of the graft (usually 7 mm) is pulled in first. In the majority of cases, this is going to be on the ulnar side

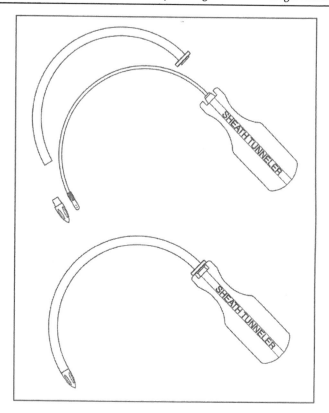

Fig. 4.8. The sheath tunneler consists of three parts.

of the forearm (little finger), since the cephalic vein is on the radial (thumb) side. Figure 4.9A shows the direction in which the sheath tunneler should be inserted, because of the collar or flange device of the current design. Even in cases where one of the deep concomitant veins is used, the arterial anastomosis is usually more favorably placed with the graft arterial anastomosis toward the ulnar side. The reverse situation, i.e., the arterial ("pull") side on the radial (thumb) side would exist if the basilic vein was clearly the optimal venous site. The anatomical location of the arterial and venous anastomoses must be clearly indicated in the medical record, as well as communicated to the dialysis personnel. Perhaps, more importantly, the patient should be informed and educated about AV graft function and use. Occasionally patients are dialyzed "backwards," making dialysis very ineffective because of recirculation. Should the direction of flow be unknown, it is very easy to compress the graft midway at the apex of the loop to determine on which side the pulse disappears and the direction of flow. When the graft is clotted it is harder to determine the anastomosis anatomy without prior knowledge. The patient will usually know which side is arterial (pull) or venous (return).

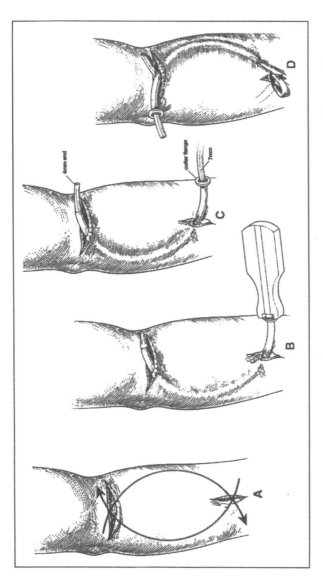

Fig. 4.9. The direction of tunneler insertion (A and B), the tapered vascular graft insertion direction (C) and the second portion of the loop placement (D).

4

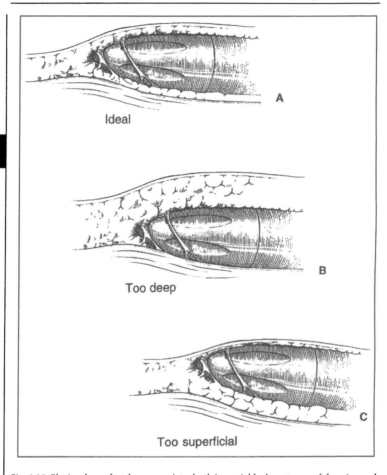

Ideal

Too deep

Too superficial

Fig. 4.10. Placing the graft at the appropriate depth is crucial for long-term graft function and survival. By directing the tunneler during insertion, the ideal level is obtained.

Once the graft is in place, it is the practice of many surgeons to expand the graft with saline to assure there are no kinks, and to palpate the subcutaneous graft ensuring its smooth course. However, the use of saline under pressure in the graft will saturate the PTFE material and possibly predispose it for continued "sweating", i.e., leaking of serous fluid through the graft, resulting in subcutaneous seroma formation. For these reasons, the authors have abandoned this technique. It is the author's impression that less postoperative swelling and inflammation (redness) occur if the graft was not expanded with saline prior to opening to blood flow. Also, the use of the sheath tunneling device causes less tissue injury and swelling (hematoma) and eliminates the graft kinking.

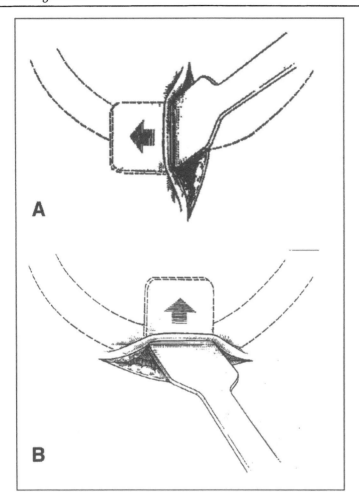

Fig. 4.11. A subcutaneous space is created at the apex to place the graft slightly away from the skin incision. In recent years the author has changed to a longitudinal incision (A), which may provide better wound healing.

By using a small retractor, the subcutaneous bands at the apex skin incision are loosened in order to place the graft somewhat away from the skin incision (Fig. 4.11). This also removes the possibility of kinking at the apex site. A longitudinal skin incision at the apex is perhaps more appropriate and is easier to close (Fig. 4.11A).

A mosquito hemostat is placed at the very tip of the 4 mm end to keep this portion of the graft away from the operating field.

4

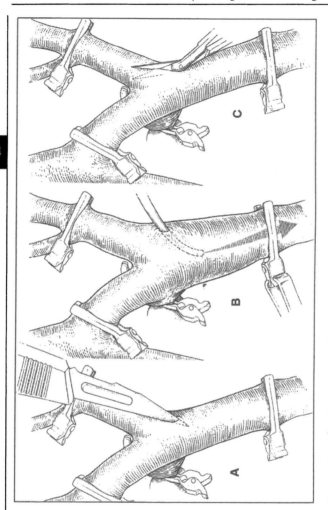

Fig. 4.12. Vascular clamps are placed on all venous branches and a small venotomy is created (A). The vein is gently dilated and heparinized locally (B) and the venotomy extended with microscissors to match the graft (C).

The venous anastomosis is completed first. At this point, one may place a small static retractor if needed. Often, however, especially if the superficial cephalic vein is used, one is better off without retractors since retractors will make the wound deeper and make it more difficult to achieve the appropriate needle driver angle during suturing. First, place Heifet's clamps on all venous branches (Fig. 4.12). Every attempt should be made to place these devices in such a way that sutures cannot be caught in the clips. This is achieved by placing the clips away from the anastomosis (Fig. 4.12). Also, angled or curved clamps can be hidden under the vein and the anastomosis.

Next, by using a #11 blade, a small incision is made at the intended site of the anastomosis (Fig. 4.12A). Using a #18 or #20 angiocath attached to a 10-20 ml

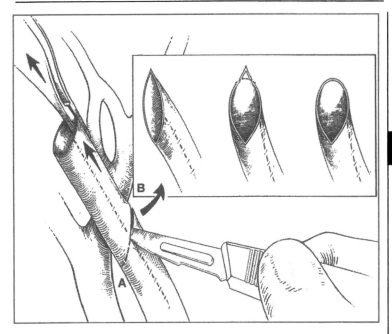

4

Fig. 4.13. The PTFE graft is divided at a level and at an angle matching the venotomy (A). The tip of the PTFE graft is rounded to maximize the venous anastomosis size (B).

syringe filled with heparinized saline inserted into this small venotomy, the vein and the anastomosis site are slightly distended under pressure (Fig. 4.12B). Also, by opening each and every Heifet's clamp the veins are locally heparinized. This technique is identical to that used and described for the creation of a native AV fistula (Chapter 3, Fig. 3.10B). The author used this technique for both arteries and veins to gently dilate vessels in spasm and also to locally heparinize. The use of systemic heparinization is not necessary and is, in fact, discouraged because of the risk of postoperative bleeding from suture lines and throughout the subcutaneous tunnel. To relieve vasospasm, locally applied 1-2% lidocaine is also helpful. Sometimes dilators may be used to ensure patency of veins.

The next step is to extend the venotomy with Dietrich's scissors to a length appropriate to match the intended anastomosis (Fig. 4.12C). This is estimated by holding the graft and estimating the angle at which the graft will be cut. The proximal end of the venotomy is more important since this will be where suturing the anastomosis begins. Should the initial venotomy be too short, it can be extended slightly at the distal end at a later point.

Ideally, one should attempt to make the venotomy right over one or more side branches which will optimize the venous outflow. Also the venotomy should be placed on the radial side of the vein from where the graft comes. A common mistake is to make the venotomy directly on top of the vein while the graft is directed toward the side of the vein. This would create a 90 degree turn, i.e., twist (rotate) of the vein 90 degrees at the anastomosis site.

4

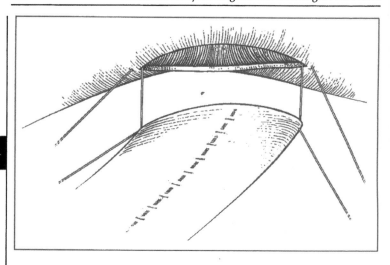

Fig. 4.14. Corner stitches through the graft and venotomy are placed.

The graft is next cut at the appropriate angle (Fig. 4.13A). This is best achieved with a #11 blade while the PTFE graft is held to match the vein. In cases of the Gore-Tex® stretch graft, the graft should be stretched and held under slight tension. The graft will then retract somewhat. Next, the most distal portion of the cut graft should be rounded in order to maximize the anastomosis opening (Fig. 4.13B). Rounding off of the graft will have a similar effect on the anastomosis as a patch angioplasty described later in this chapter in Figure 4.21.

The first stitch is ideally placed in the proximal corner. The authors prefer the Gore-Tex® CV-6 suture on a TT-9 or TT-12 needle since the needle diameter of the sutures is the same as that of its thread, thereby eliminating anastomotic bleeding from needle holes. Secondly this suture has the ability to slide 1 or 2 knots, which makes tying exceedingly convenient and easy. Also, the Gore Tex® suture is very strong and will not break with normal use (Table 2.1). Finally, this white suture material is easily seen in the operating field and reproduces well in photographs and slides (Fig. 2.5).

There is no right or preferable way in which the corner stitch is placed (Fig. 4.14). One may use both needles and place the corners from inside out or one may use one needle and go outside in through the graft and then inside-out through the vein. The order in which this is performed depends on the surgeon's preference and the anatomic factors in each specific case. The distal corner suture may also be placed at this time. The corner stitches may now be tied with 3-4 knots. The authors usually leave the distal corner untied, which makes suturing of the back wall easier.

The back wall is usually best sutured from the inside. Again, this should start at the proximal corner. Depending on which side the surgeon sits, the very first stitch goes outside in on the vessel (or graft) closest to the surgeon (Fig. 4.15). One should take great care not to take big bites in the corner since this will use up lumen and compromise venous outflow. The suture technique described here is in principle the same as that described for primary AV fistulae (Chapter 3, Figs. 3.11-3.15).

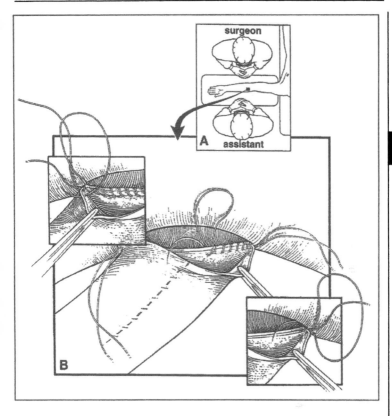

Figs. 4.15A,B. A) The direction of the suturing depends on the postion of the surgeon. B) The back wall suturing is detailed in.

Sometimes placing a temporary retention suture midway in the back wall anastomosis will improve exposure and facilitate suturing. The back wall is run continuously taking about 1-2 mm bites until the distal corner is reached (Fig. 4.15C). The very last stitch goes inside-out on the vessel/graft, that is, away from the surgeon, assuming he has been sewing toward himself (Fig. 4.15A). If not tied before, the distal corner stitch is now tied with three knots and the back wall suture tied with seven square knots to one of the corner sutures. This last (seven knot) final corner suture is then cut. This is exactly the same technique described for the distal corner with primary AV fistula in Chapter 3, Fig. 3.14. The front wall is now sutured, preferably starting again in the proximal corner stopping midway at the anterior anastomosis. The distal suture is then run midway and tied to the proximal suture (Fig. 4.16). The suturing style may vary because of anatomical considerations and surgeon's preferences.

The graft is now occluded using a soft vascular clamp, usually at the apex, to prevent venous back flow. The Heifet's clips are removed and blood is allowed to flow back through the venous anastomosis into the graft (Fig. 4.17). The graft is

4

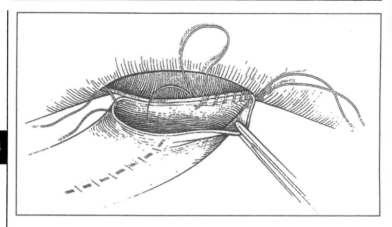

Fig. 4.15C. By leaving the distal corner suture untied, the back wall suturing is sometimes facilitated, especially when approaching the distal corner. This suturing technique is identical to that described for primary AV fistulas in Chapter 3, Figure 3.12.

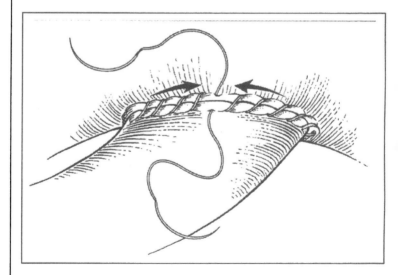

Fig. 4.16. The front wall suturing completed.

then flushed with heparinized saline (20 cc syringe with an attached blunt "Christmas tree" adapter) and the soft vascular clamp applied at the apex across the graft (Fig. 4.17). Any minor bleeding from the venous anastomosis will stop within a few minutes. Unless there is a suturing error (mishap), no extra sutures should be needed. Again, a common mistake at this point is to hasten to place sutures to stop minor bleeding which may in fact come from needle punctures.

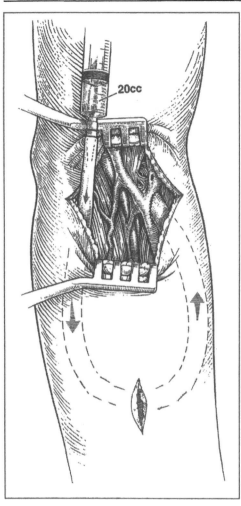

Fig. 4.17. The graft is flushed with heparinized saline.

20cc

Attention is next directed to the arterial anastomosis (Fig. 4.18). A small static retractor may be needed to obtain adequate exposure (Fig. 4.4). First, Heifet's clamps are placed on the brachial artery. The type, location and direction of the Heifet's on the artery depends on the specific anatomical situation. Most commonly a curved Heifet's directed from underneath the brachial artery is ideal to keep the clamp out of the operating field and to avoid catching sutures. The next Heifet's clamp is placed on the ulnar artery from underneath and directed medially or in the ulnar direction. Finally, an angled, or short straight Heifet's is placed on the radial artery (Fig. 4.18). When larger or atherosclerotic arteries are encountered, different clamps are necessary, i.e., a felt covered disposable yellow (Chapter 2, Fig. 2.4). In this case the anastomosis is placed on the distal brachial artery extending slightly into the radial artery (incision type b in Fig. 4.18), a strategy that often is preferable. The

4

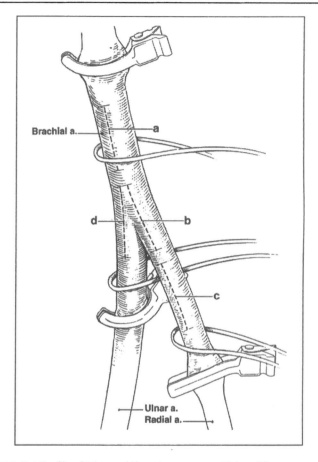

Fig. 4.18. Details of brachial artery bifurcation anatomy with four different anastomosis sites indicated.

arteriotomy is made usually at a 45 degree angle in the ulnar direction toward the PTFE graft. This arteriotomy with a #11 blade is made just large enough to snugly fit an angiocatheter. The angiocatheter is used to slightly expand the artery in spasm and also to locally irrigate vessels in all directions by opening the Heifet's clamps while injecting heparinized saline. This technique is identical to that described for the vein except that the arterial blood pressure requires synchronization between the assistant and the surgeon (Chapter 3, Fig. 4.12B and Figs. 3.10A,B). Ideally, the surgeon does both the saline flushing and the opening of the Heifet's clips. The critical step is to make the arteriotomy small enough to prevent blood from leaking out. In cases where the arteriotomy is too big for an angiocatheter, this type of local heparinization of the blood vessels can be made with a smooth "Christmas tree" adaptor (Chapter 2, Fig. 2.9A). The arteriotomy is next extended using Dietrich's scissors to a length of approximately 6 mm. The 4 mm end of the graft will be cut at a slight angle, thereby matching approximately the 6 mm arteriotomy (Fig. 4.18). The exact location of the arteriotomy may vary with local anatomy. The most common is probably at the distal brachial artery with slight extension into the radial artery (Fig. 4.18, incision type b) or less often into the ulnar artery. Many surgeons prefer to be completely on the brachial artery (Fig. 4.18, incision type a). Sometimes the entire anastomosis is made on the proximal radial artery, especially if this artery is judged to be of adequate diameter (Fig. 4.18, incision type c). Occasionally (in cases of anomaly) one may place the entire anastomosis on the ulnar artery (Fig. 4.18, incision type d). Again, to maximize the anastomotic opening, the proximal corner of the cut graft should be rounded similar to that used on the venous anastomosis site (Fig. 4.13B). The authors prefer to suture the arterial anastomosis using only one suture which may begin at the proximal corner or start midway back wall (the author's favorite style). In case of either corner as the starting point, the very first stitch goes outside-in on the corner of the arteriotomy, while the artery is held open with micro forceps. Then the same stitch is placed inside out on the graft and this corner stitch is tied with four square knots. To approximate the graft and the artery, the assistant may gently pull the graft toward the artery while the surgeon is tying.

If starting at the proximal corner, the back wall is sutured first and the very first stitch is placed outside-in on the back wall of the artery (assuming the surgeon is sitting on the graft side), taking a very small bite of the artery. The back wall is then run in a similar manner as the venous anastomosis (Fig. 4.15). The difference being that no distal corner stitch is placed on the artery. This will in fact increase exposure and facilitate the ability of the surgeon to see and be able to exactly place each stitch. As the distal corner is approached, separate bites through the graft and artery must be placed. If a distal corner stitch is used it should be left untied until the back wall is completed to increase exposure enabling exact suture placement. On the front wall anastomosis, after two or three stitches, one should switch to the proximal corner and run the anterior anastomosis, then finally tie to the previous suture by seven square knots, as described for the venous anastomosis (Figs. 4.15 and 4.16) and that for the PAVF (Figs. 3.12-3.15 in Chapter 3).

At this point, the vascular clamp at the apex is removed. The Heifet's clamps are also removed starting with the radial and ulnar. Lastly the brachial artery is released. There should now be brisk flow, often felt as a thrill through the graft and the vascular anastomosis sites. Except for a slight temporary ooze, there should be no significant bleeding from either anastomosis. Placing a surgical gauze on top of the operating field for 3-5 minutes uniformly stops all oozing. Only persistent bleeding, especially between sutures, will require an extra stitch. Bleeding from needle holes will stop. Constant manipulation of the anastomoses, looking for bleeding points, will only prolong bleeding. This is a common and potentially dangerous mistake, since unnecessary extra stitching at this time can compromise an otherwise perfect anastomosis. The final PTFE graft anastomotic sites, now complete, are shown in relationship to the antecubital anatomy in Figure 4.19.

The wound is closed using 4-0 or 5-0 polyglactin interrupted subcutaneous sutures, the skin is closed with a 5-0 subcuticular suture. Only a loose gauze is applied. No tight circular or semicircular dressings or tapes should be used. The patient is encouraged to use the hand and the arm freely and to elevate the arm while at rest, preferably on pillows. The arm or hand should NOT be suspended in a sling for any prolonged periods.

Fig. 4.19. The finished product. The 4-7 mm
PTFE graft in relation to vascular anatomy.

Complications from PTFE Grafts

Early Complications

Early thrombosis of the PTFE graft occurs about 5% of the time. It is probably
more common when the graft has been used for dialysis early (< 14 days), in patients
with preexisting or debilitating factors such as suboptimal vascular anatomy, a history of IV drug abuse, long illness with multiple venopunctures, diabetes and advanced age. An early clotted PTFE graft should be explored through the original
incision, while in cases of late thrombosis, a longitudinal skin incision should be

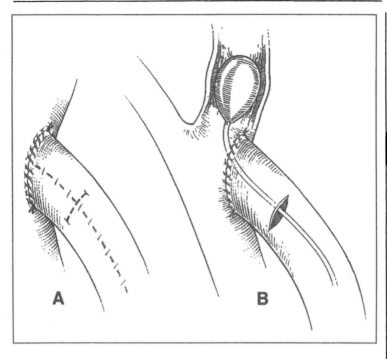

Fig. 4.20. For early thrombosis in PTFE grafts, a transverse incision on the graft is preferred (A). The venous anastomosis is examined using dilators or balloon catheters (B).

made along the graft onto the vein. The management of late thrombosis is further described below.

Early thrombosis is approached through a small transverse incision made approximately 2 cm from the venous anastomosis through the graft (Fig. 4.20A). (For late clotting, a longitudinal incision close to the anastomosis is preferred since most of these will require a patch angioplasty procedure) (see Fig. 4.21). The venous anastomosis is carefully examined using balloon catheters and vascular dilators of increasing diameter (Fig. 4.20B). Unless a #4 to #4.5 can be passed, there is likely an anastomotic problem which needs to be addressed. The exact nature of this procedure must be assessed based on the local anatomy, i.e., the degree of any stenosis, and the extent and cause of thrombosis. If the vein is a superficial cephalic vein, one may attempt to place a PTFE patch (Fig. 4.21). If the vein is thickened and narrowed from inflammation, a new venous outflow must be found. Inflamed, irritated veins can result from rough handling during the initial surgery. The first option would be to determine whether the deep concomitant venous system is accessible, possibly without using an extension graft (Fig. 4.22). Another venous outflow option is a large basilic vein which will require the placement of an extension graft over the arterial anastomosis to the ulnar side, or to any suitable antecubital vein (Fig. 4.23). The variations are multiple and the success rate depends on the surgeon's intuition and skill.

4

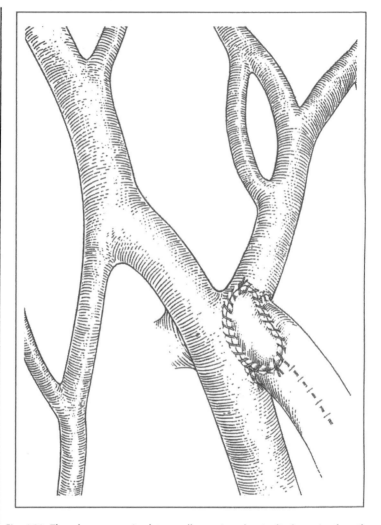

Fig. 4.21. Thromboses occurring late usually require a longitudinal opening from the graft onto the vein with a patch angioplasty.

As the last resort, an extension graft across the elbow joint to the upper arm cephalic, basilic or brachial veins may be performed. In this case, the extension graft should be of the ringed enforced type and pass the elbow on either side of the elbow joint to avoid kinking of the graft with arm movement (Fig. 4.24). A recent graft with intrawall rings seems to be optimal for these antecubital crossings (Intering, W.L. Gore, Flagstaff, Arizona). The authors have used this extension graft on several occasions with excellent outcome (graft patency). The intrawall rings are made of PTFE and can be cut if necessary and included in the suture line.

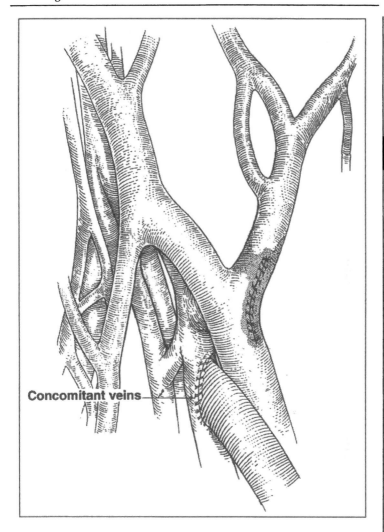

Concomitant veins

Fig. 4.22. Creation of a new venous outflow to the deeper concomitant vein without the need for an interposition graft.

A special variation of venous outflow reconstruction is shown in Figure 4.25 where an extension graft is placed between the loop, the cephalic vein and the diving branch (v anastomotica) in order to maximize venous outflow. If a suitable vein is found, the graft is declotted using a Fogarty balloon catheter (#5 or #6) (Fig. 4.26). Thrombectomy of the graft should be performed step-wise, 5-10 cm at a time. The appropriate length of catheter needed to reach the arterial anastomosis can be estimated by measuring the loop with the catheter placed on the skin on top of the loop.

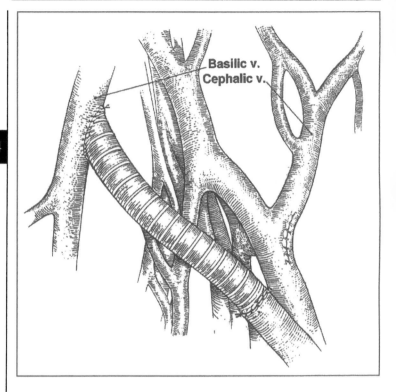

Fig. 4.23. The basilic vein is used to create venous outflow using an interposition ringed PTFE graft.

The graft may further be cleaned/stripped of wall attached hyperplastic material using the Latis® Thrombectomy Catheter (Applied-Medical) or the Fogarty® Adherent Clot Catheter (Baxter) or the Solera® declotting device (Table 4.2, Chapter 2, Tables 2.2 and 2.3). (These and other new devices, although moderately effective, must be used correctly in order not to injure native vessels). The final step involves removal of the arterial plug by pushing the Fogarty catheter beyond the arterial anastomosis into the artery.

The arterial plug has a typical whitish appearance with a concave surface toward the arterial lumen. The shape reflects the PTFE graft configuration. Figure 4.27A shows the shape of a plug from a 4-7 mm tapered graft, Figure 4.27B represents a thrombosis from a 6 mm graft and Figure 4.27C is from a bovine graft. Estimating the force which the arterial anastomosis will tolerate requires great skill (experience) and also depends on the age of the graft. Rough handling may rupture or injure the artery. Unless the arterial plug is retrieved, declotting is not likely to be long lasting. Occasionally, the author has used the Latis® Catheter to retrieve the plug in resistant cases. The catheter balloon is then placed at the anastomosis, and the balloon inflated under manual control when pulled. Adequate arterial inflow can be judged

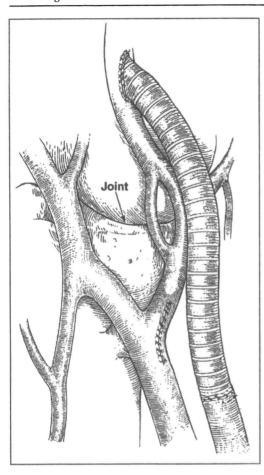

Fig. 4.24. A ringed graft is used to bypass the elbow when no antecubital veins are found. See also Appendix I, Case #23.

from the briskness by which blood flushes out. Also, arterial inflow can be estimated by assessing whether a 20 ml syringe (with a "Christmas tree" adapter) can be filled with blood in 4 seconds, which translates to 300 ml of blood flow per minute. Likewise, venous return can be tested by injecting 20 ml of saline into the venous system in 4 seconds. If this is possible without significant resistance, the venous outflow can at least accept a flow rate of 300 cc per minute, which is sufficient in most dialysis situations. If the Fogarty® catheter can be passed and pulled smoothly when inflated through the entire length of the graft, there is unlikely to be any obstruction of the graft. Should there be a mechanical problem, this is most likely to occur at the apex of the loop in a recently placed graft under the incision. Should there be a suspected resistance or a kink at the apex, the apical incision is opened, the graft exposed and subcutaneous bands are released. If a twist exists, the graft should be divided, realigned and reanastomosed end-to-end. In "older" grafts, stenosis is likely to be present at repeat puncture sites or at pseudoaneurysm formation sites.

Fig. 4.25. One of many possible variations is illustrated, where an interposition graft is also anastomosed to the vena anastomotica connecting to the deep concomitant venous system.

Postoperative swelling around the incision site may be caused by bleeding or excessive sweating from the PTFE graft with seroma formation. Sometimes there is redness and increased skin temperature along the graft; a situation that is very suggestive of infection. However, if the patient has no fever or other signs of infection, the author usually advocates surveillance and preventive treatment with antibiotics, i.e., cephalexin or vancomycin. In more alarming cases, the patient is hospitalized and the arm elevated with clinical assessments every 4-6 hours. In the author's experience, these situations usually turn out to be a noninfectious inflammatory process which dissipates. The necessity to explore grafts for postoperative swelling or bleeding is quite rare. It is the author's impression that some grafts are explored unnecessarily only to increase morbidity, mainly infection and poor wound healing. It is also the author's opinion that perigraft swelling and reaction is lessened by using the sheath tunneler, avoiding saline expansion of the graft and employing atraumatic surgical technique, postoperative arm elevation and ice packs for 6-12 hours over the graft area.

Early Infections

Early infections (<30 days) with proper surgical technique and prophylactic perioperative antibiotics, (i.e., 1 g of vancomycin) are uncommon. In the author's recent experience only 3 (1.1%) of 270 consecutive PTFE grafts developed infection within 30 days. When the infection involves the entire graft, it has to be removed. In such cases, the venous anastomosis may either be oversewn or the vein ligated, but the artery has to be carefully repaired. When the arterial anastomosis is completely on the proximal radial artery, a segment of this artery can be resected and each side suture ligated (Fig. 4.28A). There are three options for managing the brachial artery anastomosis. First, one may attempt to suture the artery, as shown in

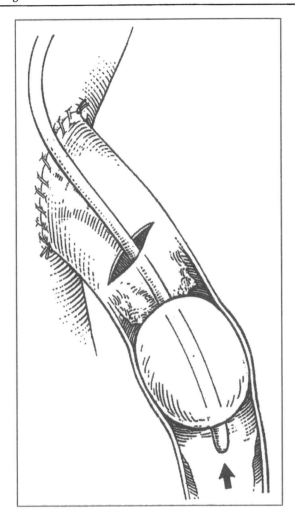

Fig. 4.26. First, when adequate venous outflow has been secured, clots inside the graft are removed using a balloon catheter.

Figure 4.28B. Usually, this will result in a severe stenosis. A vein patch angioplasty is a safer alternative (Fig. 4.28C). Most surgeons, including the authors, would leave a short segment of the PTFE graft and oversew the graft with a running suture (Fig. 4.28D). In the author's experience, occasionally this leads to a chronic infectious process, requiring further surgery (Appendix I, Case #24). In such challenging instances, a vein patch angioplasty is the treatment option. Late infections are discussed next.

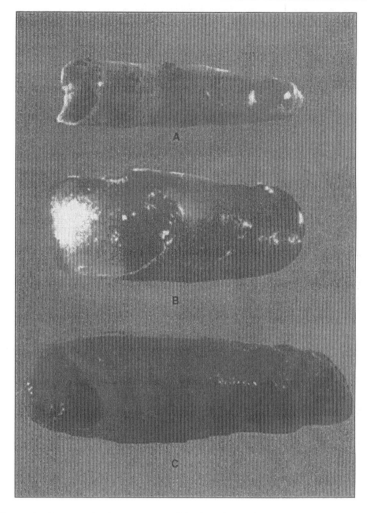

Fig. 4.27. The somewhat harder portion of the thrombus at the arterial anastomosis has a typical appearance reflecting the size of the 4 mm arterial side of the PTFE graft (A), the 6 mm graft (B) or bovine graft (C).

Arterial Steal

Arterial steal is becoming an increasingly common complication because of the higher incidence of elderly and diabetic patients admitted for dialysis treatment. The symptoms are distinct and consist of coolness of the hand, tingling and pain. When symptoms are severe with a cool and bluish hand, urgent correction is warranted. The diagnosis is confirmed if the patient's hand becomes warm and symptoms disappear or improve when the graft is partially manually occluded. The

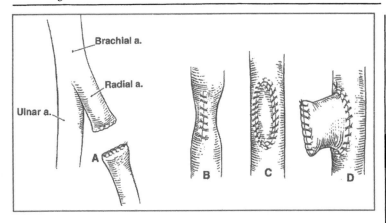

Fig. 4.28. Options for managing the arterial anastomosis in cases of peri-graft infection, requiring removal.

diagnosis is further confirmed by finger pressure measurements before and after near complete manual occlusion of the graft (Table 4.1).

Many different techniques have been used to correct this problem. It is the author's view that "banding" must be performed at the apex of the graft to ensure good inflow, as well as venous return during the dialysis procedures. For several years in the operating room, the author has used a large hemoclip that is gradually tightened over the apex while our vascular laboratory technician measures finger pressures (Table 4.1). Preoperatively, finger pressures in cases of arterial steal are typically below 20 mm Hg, i.e., not measurable. One should ideally improve finger pressures for at least 60-80 mm Hg to relieve the patient's symptoms. This usually requires at least a 50% occlusion (Fig. 4.29). In the author's experience, it is difficult or impossible to achieve this type of exact gradual or partial occlusion with a tie using material such as polypropylene or umbilical tape. Others have described gradual occlusion by suturing the graft and thereby gradually occluding the lumen or by placing a 4 mm interposition graft.

The author has used the hemoclip technique with good outcome in more than 30 cases without complication, except for late graft clottings. In several such instances, after graft thrombosis and declotting, weeks or months after banding, the hemo-clips have been removed without a return of ischemic symptoms. In our more recent experience from Medical City Hospital in Dallas between 9/98 - 11/1/01, 13 of 15 (87%) consecutive bandings stayed open for up to a year (Appendix V, Fig. 5B). Therefore, banding is a worthwhile procedure rather than more extensive surgery, such as the distal levascularization and internal ligation (DLIL), or abandoning the graft by ligation.

Late Thrombosis

Thrombosis is the most common late complication and cause of graft failure. In many series, only 50-60% of all PTFE grafts are patent at 2 years. In the author's experience with 811 first time forearm access placement during the last ten years, graft survival has decreased from 92% to 76% at one year (Appendix V). Two

Table 4.1. Finger pressure measurements in suspected steal

Arterial steal Finger Pressure (mm Hg)

	Right (normal)	Left (ischemic)	Left (AV outflow occluded)	Intra-op (left) (right)	Post-op (left)	Post-op (right)
Thumb	58	< 20	44	80	64	72
2nd	52	< 20	52	72	58	60
3rd	56	< 20	46	68	62	64
4th	72	< 20	36	136	80	86
5th	70	< 20	52	112	64	82

factors seem to be at play. Patients are 15 years older, and we now place native fistulae in 47% of cases versus 17% in the early 90's, leaving the older, high risk patients for PTFE grafts (Appendix V, Fig. 3A-B). Although the author strongly supports placing primary AV fistulae whenever suitable, (as evidenced by current 47% of first time access), it is somewhat surprising that the outcome (graft survival) is the same or better with PTFE placement (Appendix V, Figs. 3A and 3.6). However, this comes at a higher cost and higher severity of patient morbidity mainly declotting and revision procedures (Fig. 4.30).

In the vast majority, or more than 90% of cases, late thrombosis is associated with intimal hyperplasia at the venous anastomosis. Cellular debris is deposited, which gradually obstructs venous outflow at the anastomosis site, eventually leading to thrombosis. Ideally prophylactic intervention with surgical revision or radiologic balloon angioplasty is instituted return when venous pressures exceed target values at specific blood flow rates. The clotting event can often be predicted by increasing venous (return) pressure during dialysis treatments. In cases of late thrombosis, the venous anastomosis is explored through a skin incision parallel to the graft, extending to and slightly across the venous anastomosis. After surrounding the graft with a vessel loop, the anastomosis and the veins are carefully isolated. Depending on the direction of the venous outflow system, the skin incision may be extended as needed. After the venous branches have been carefully dissected free and surrounded with vessel loops, and Heifet's clamps have been placed on the veins, a longitudinal incision is made into the graft across the anastomosis into the vein. Clots and debris are removed. Sometimes the hyperplastic material can be removed by a procedure analogous to endartectomy. The venous outflow is tested with a Fogarty® catheter and flushed with 20 ml of heparinized saline using the metallic smooth so-called "Christmas tree" (Fig. 2.9). By passing a Fogarty® catheter to or beyond the shoulder region and then pulling back with inflated balloon, clots are removed and a venous stenosis is confirmed or excluded. Also, if 20 cc of saline can be injected in 4 seconds with no resistance, the venous system is adequate for fistula flow.

The rest of the graft is then cleaned of thrombosis using a Fogarty® catheter. As described above (Fig. 4.26), this is made step-wise to make sure the entire graft is free from clot. This maneuver is also used to assess the quality of the remaining graft, that is the presence of stenoses from repeated dialysis neddle punctures. Sometimes a 6 mm dilator can be inserted to dilate stenoses in the graft from fibrosis and multiple needle punctures. Recently developed devices may further remove wall adherent material not removed with standard Fogarty® balloon catheter (Table 4.2).

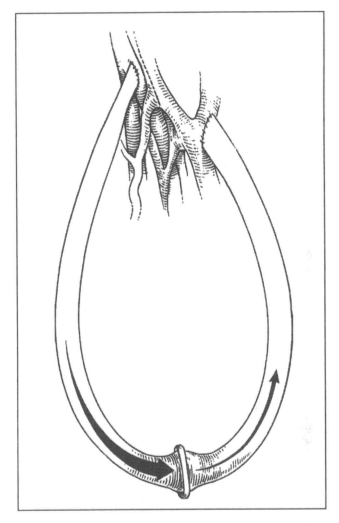

Fig. 4.29. Arterial steal with PTFE grafts requires partial "banding" at the apex.

Their use is determined by effectiveness, user friendliness and cost. Care should be taken not to use these devices in the native vessels, avoiding intimal injury.

Finally, the arterial plug is removed by passing the Fogarty® catheter up into the artery; the plug can be retrieved in the majority of cases. If the arterial plug with its typical appearance (Fig. 4.27) cannot be obtained, an extra skin incision is made on the arterial side. The graft is then exposed, surrounded with vessel loops, and a small transverse incision is made. Sometimes, the balloon can now be passed. Also, dilators can now be inserted into the artery. Arterial plugs can often be loosened and

Table 4.2. Thrombectomy and embolectomy devices

Device	Company
Latis	Applied Medical
Fogarty Adherent Clot Catheter	Baxter
Solera	Bacchus Vascular
AngioJet	Possis Medical

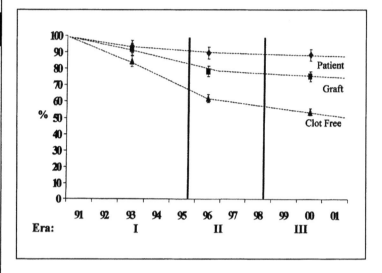

Fig. 4.30. Patient, graft and clot-free survival in 811 consecutive first time forearm loop Gore-Tex® Stretch Grafts over a ten year period 1991-2001 (Appendix V).

retrieved after dilating the graft up to 4 mm. Alternatively, a small mosquito hemostat can be inserted to grab hard material stuck to the graft and to the arterial anastomosis.

Exploration and Corrective Measures

Depending on the venous outflow anatomy and the surgeon's experience and preference, various types of patches and interposition grafts may be utilized. These measures are in principle the same as those for early thrombosis. The most common variations are shown in Figures 4.21-4.25. A patch should extend about 0.5-1.0 cm beyond the end of the stenotic area (Fig. 4.21 and Appendix I, Case #20 and 21). Since the patch crosses the previous anastomosis, the old cut sutures may be tied to both sides of the patch suture as it passes. Tying to the previous sutures is more important in a fairly recently placed graft. Often, various types of interposition or extension grafts are needed, because the old graft will not reach to the intended venous outflow site. Figure 4.25 shows a variation where an interposition graft is placed to the cephalic vein and also anastomosed to a diving branch (v anastomotica) to improve outflow. The principle is to utilize every possible outflow to prolong the access survival time.

Should the superficial vein be inadequate or severely fibrosed, a deep concomitant vein should be sought as described in Figure 4.22. Often, the deep concomitant veins along the brachial artery are dilated from existing connections between the failed superficial system and the deep veins. This dissection is sometimes technically challenging and time consuming but often rewarding. If successful, the patient can immediately go back to hemodialysis using the old graft. Also, finding new venous outflow for established grafts preserves access sites for future use and also avoids temporary central vein catheters (Fig. 4.23). Also the authors in increasing frequency cross the elbow and place an interposition "ringed" graft to the upper arm, basilic, cephalic or brachial veins (Fig. 4.24). For this purpose a new Intering graft has been very effective to prevent kinking (Appendix I, Case #18 and 22). Occasionally, when the graft has multiple stenoses and/or pseudoaneurysms from repeated needle punctures which are beyond salvage for any extended time, one may place a new PTFE graft around the old graft and anastomose the new graft to the old graft 2-3 cm from the arterial anastomosis, and then on the venous side to the most optimal venous site available; this may in fact be the graft itself, assuming the venous anastomosis is widely patent. The many possible variations for revising failing or thrombosed grafts leave opportunities for the open-minded surgeon, keeping in mind that graft survival also means patient survival.

Late Infections

Late-occurring infections are uniformly associated with or caused by needle punctures during dialysis. In fact, most infections start from such needle punctures. Infection may progress from a small subcutaneous abscess to involve the entire graft, surrounding it with pus. Management depends on the extent of infection, the general status of the patient and the surgeon's experience.

Small localized infections should be drained without exposing the graft itself. Localized but more involved infections, often associated with hematoma formation usually require more extensive debridement. Even though many surgeons would remove the entire graft, the authors have on numerous occasions bypassed the involved area with an interposition graft. The excluded, infected portion of the graft is excised. This should be attempted, especially if the patient is otherwise unaffected, without symptoms of generalized sepsis. Antibiotic coverage with vancomycin should be instituted as soon as infection is diagnosed or suspected, and changed appropriately based on cultures.

Infection involving the entire graft with redness, fever and septic symptoms, requires immediate attention. Under IV antibiotic coverage, i.e., Cephalexin or Vaniomycin, the entire graft is removed. Incisions are made over the venous, as well as the arterial side and also at the apex of the loop. The venous side is usually addressed first. The vein, proximally and distally, and any branches suture ligated with absorbable sutures, are best removed with the graft. Usually the graft can easily be pulled out of its tract. The management of the arterial anastomosis requires some thoughtful considerations, as described in Figures 4.28A-D. Should the infection clearly surround the arterial anastomosis, especially if there is bleeding from this area, the entire graft needs to be excised from the artery. To safely do this, an upper arm tourniquet may be applied and the artery exposed under a bloodless field. The treatment options are described in Figures 4.28A-D.

The infected PTFE graft can usually be removed by passing a 6 mm dilator inside the graft and pulling (stripping) it out through the apex site after tying it to this dilator or tunneling device. The infected tunnel is rinsed with saline containing antibiotics, i.e., gentamicin, and drained using Penrose drains for 1 or 2 days and thereafter packed with gauze. Again, these seemingly aggressive infections will usually heal quickly after graft removal. The dialysis has to be managed through other means, usually through dual lumen dialysis catheters. Ideally one should wait 1 or 2 days before placing these catheters, to give the patient time to clear the blood stream from bacteria, usually Staphylococcus species. The author suggests placing a percutaneous femoral catheter at the time of next needed dialysis, and after 3-4 days place an internal jugular vein cuffed tunneled catheter (Chapter 5).

Graft Aneurysm

Aneurysms can occur at any time after graft placement. First, an aneurysm can occur at the anastomosis sites as a result of a suturing defect. Secondly, aneurysms may result from needle punctures, which connect the graft to the aneurysms. The aneurysm wall consists of a pseudo-membrane formation from surrounding subcutaneous tissue (Fig. 4.31A). When these aneurysms become enlarged and especially when the skin becomes shiny and atrophic, surgical correction is warranted. Often, when these areas are exposed, one will find that large portions of the graft have been totally destroyed by repeated needle punctures and repair is impossible. Under these circumstances one may choose to place an interposition graft, replacing the destroyed PTFE graft (Fig. 4.31B). Another option is to bypass the affected area, as in the case with localized infections. If the area does not have signs of infection, one may not need to remove the old graft since this is often difficult and traumatic because

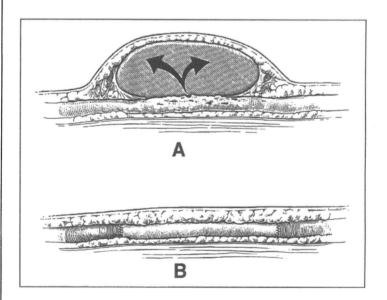

Fig. 4.31. Large aneurysms are best managed by excision (A) and reconstructed with an interpostition graft (B).

of firm incorporation. The bypassed graft then will rest under new, unaffected skin. Occasionally an aneurysm has one tiny hole into the graft requiring only a stitch.

Arterial steal may occur late (especially after correction of a flow restricting stenosis), but is more common early after placement. The management was described earlier in this chapter and in Figure 4.29 as well as Table 4.1.

Selected References

1. Taucher LA. Immediate, safe hemodialysis into arterio-venous fistulas created with a new tunneler. An 11-year experience. Amer J Surgery 1985; 150.
2. Gifford RRM. Improved positioning of the upper arm graft fistula for hemodialysis. Amer J Surg 1986; 151.
3. Curl GR, Jakubowski JA, Deykin D et al. Beneficial effect of aspirin in maintaining the patency of small-caliber prosthetic grafts after thrombolysis with urokinase or tissue-type plasminogen activator. Circulation 1986; 74(I).
4. McKenna PJ, Leadbetter MG. Salvage of chronically exposed Gore Tex® vascular access grafts in the hemodialysis patient. Plas and Recons Surg, 1988.
5. Mattson WJ. Recognition and treatment of vascular steal secondary to hemodialysis prostheses. Amer J Surg 1987; 154.
6. Bell DB, Rosenthal JJ. Arterio-venous graft life in chronic hemodialysis. Arch Surg 1988; 123.
7. Schwab SJ, Raymond JR, Saeed M et al. Prevention of hemodialysis fistula thrombosis. Early detection of venous stenoses. Inter Soc Nephrol 1989; 36:707-711.
8. Windus DW, Audrain J, Vanderson R et al. Optimization of high-efficiency hemodialysis by detection and correction of fistula dysfunction. Inter Soc Nephrol 1990; 38:337-341.
9. Odland MD, Kelly PH, Ney AL et al. Management of dialysis-associated steal syndrome complicating upper extremity arterio-venous fistulas: Use of intraoperative digital photoplethysmography. Surgery 1991; 100:4.
10. McMullen K, Hayes D, Hussey JL et al. Salvage of hemodialysis access in infected arterio-venous fistulas. Arch Surg Pct 1991; 126.
11. Kumpe DA, Cohen MA. Angioplasty/thrombolytic treatment of ailing and failed hemodialysis access sites: comparison with surgical treatment. CardioVasc Dis 1992; XXXIV(4):263-278.
12. Rivers SP, Scher LA, Veith FJ. Correction of steal syndrome secondary to hemodialysis access fistulas: a simplified quantitative technique. Surgery 1992; 112:3.
13. Levy SS, Sherman RA, Nosher JL. Value of clinical screening for detection of asymptomatic hemodialysis vascular access stenoses. Angiology J of Vasc Dis 1992.
14. Mehta S. Statistical summary of clinical results of vascular access procedures for hemodialysis. In: Sommer II H, ed. Vascular Access for Hemodialysis. Precept Press, 1993.
15. Jain KM, Simoni EJ, Munn JS. A new technique to correct vascular steal secondary to hemodialysis grafts. Surgery, Gyn & OB 1992; 175.
16. Nolph KD. Access problems plague both peritoneal dialysis and hemodialysis. Kid Inter 1993; (43)40:S81-S84.
17. Beathard GA. Mechanical versus pharmacomechanical thrombolysis for the treatment of thrombosed dialysis access grafts. Kid Inter 1994; 45:1401-1406.
18. Padberg FT, Smith SM, Eng RH. Accuracy of disincorporation of identification of vascular graft infection. Arch Surg 1995; 130:183-188.
19. Dawidson IJA, Ar'Rajab A, Melone LD et al. Early use of the Gore Tex® stretch graft. Vasc Access for Hemodialysis-IV. 1995; 109-117.
20. Davidson ISA, Smith BL, Nichols D et al. Vascular access survival following banding for hand ischemia. Presented to Vascular Access for Hemodialysis VIII, Ranuto Mirage, CA. May, 2002.

CHAPTER 5

Dual Lumen Catheters for Dialysis

Ingemar J.A. Davidson, W. Perry Arnold and Frank Rivera

Introduction

While central vein hemodialysis catheters are often life saving, there is a remarkable variation in their indications and frequency between dialysis units. For example, the average use of catheters for chronic use in the state of Texas was 17% (Fig. 5.1A) but varied between 2% to 40% for dialysis centers (Fig. 5.1B). The national average catheter use in the US is estimated to be 20%. The DOQI guidelines aim for less than 10%. It is the authors' opinion that the appropriate use of catheters for chronic use can be 5% or less. In sharp contrast to these idealistic numbers stands the fact that 40% of all patients initiating dialysis in the US do so with a temporary dual lumen catheter. Changing these statistics will take concentrated educational efforts of the dialysis unit personnel, surgeons, nephrologists, radiologists and the patients. These efforts for improvement initiatives in dialysis access in general are badly needed and long overdue. Organizational and fiscal support currently is not well defined. Since ESRD programs are Federally funded, the ESRD networks are the appropriate administrative body to be charged with implementation and outcome documentation of such efforts.

Indication for Dual Lumen Catheters (Table 5.1)

When to Place

1. **Emergent need for dialysis**: Uremic patients with fluid overload, shortness of breath or hyperkalemia and, therefore in emergent need for dialysis are best served with a percutaneous, preferably femoral dual lumen catheter. After two or three dialysis treatments when patient is more stable, a cuffed dual lumen catheter is placed, optimally in the right internal jugular vein. Consecutively or later pending patient status and clinical circumstances, a permanent access may be placed, such as a primary AV fistula, PTFE AV graft or a PD Tenckhoff catheter.

2. **Urgent need for access**: Stable patients with no other access needing dialysis ithin 1 or 2 days may have a cuffed dual lumen catheter placed in the operating room or the angiographic suite. At the same time or later, depending on clinical situations, a permanent access may be placed.

3. **Maturing access**: In situations where a primary AV fistula is not ready for use, a cuffed dialysis catheter in the internal jugular vein may be placed.

4. **Thrombosed central veins**: Consult with interventional radiology regarding the possibility of re-establishing venous vascular patency. When both internal jugular veins are thrombosed, the authors prefer the femoral veins for cuffed dual len catheters, rather than using the subclavian vein. The groin

Access for Dialysis: Surgical and Radiologic Procedures, 2nd ed.,
edited by Ingemar J.A. Davidson. ©2002 Landes Bioscience.

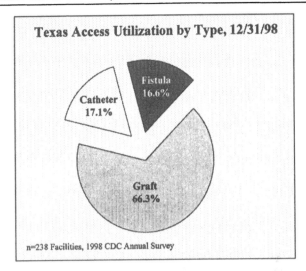

Fig. 5.1A. The low utilization of the native AV fistulae and common use of dialysis catheters for chronic dialysis reflect many ESRD background system problem issues. (ESRD Network 14 data, used with permission)

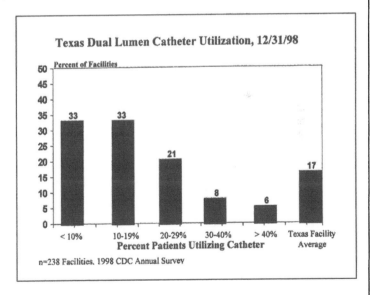

Fig. 5.1B. The system problem is dramatically exemplified by the sharp differences between catheter use at different dialysis units. Thirty five percent of all centers utilize catheters as dialysis access in more than 20% of the patients. A few centers utilize catheters for chronic dialysis in more than 50% of their patients.

Table 5.1. When to place dual lumen hemodialysis catheters

1. **Emergent: Need Dialysis Now**
   1. Place a percutaneous femoral dual lumen catheter
   2. When patient is stable, place a cuffed (right) internal jugular vein catheter (i.e., Tesio or split ash)
   3. Consecutively or later place permanent access (primary AV fistula, PTFE or Tenckhoff catheter)
2. **Urgent: Need Dialysis within 24-48 Hours**
   1. Place a cuffed dual lumen catheter (right) internal jugular vein
   2. At the same time or later place permanent access (primary AV fistula, PTFE or Tenckhoff)
3. **Maturing Access: i.e., Primary AV fistula not ready for use**
   1. Place a cuffed dual lumen catheter (right) internal jugular vein
4. **Unfavorable Upper Body Central Vein Anatomy**
   1. When both internal jugular veins are thrombosed, use femoral vein for cuffed dual lumen catheter. Tunnel these to an appropriate exit site.
5. **The Only Choice for Permanent Access**
   1. Elderly patients—to evaluate dialysis improvement in quality of life
   2. Exhausted all other access options (RARE!)
   3. Imminent kidney transplant, i.e., LRD (< 4 weeks)

catheter is tunneled up onto the lower abdomen or an appropriate exit site based on local anatomy (fat pendulum).

5. **The only choice for access**: Dialysis catheter as the only possible alternative for chronic dialysis access is rare. However, with the increase in elderly ESRD patients admitted for dialysis, the authors rely more heavily on catheters, at least initially. Should an elderly patient recover, a permanent access is considred. Exhausted access sites other than catheter is a rare situation, and may represent less than 1% of all dialysis patients.

Patients expecting an imminent kidney transplant (within 2-4 weeks) may be a candidates for a short term dual lumen cuffed catheter. This is often the situation with living donor transplant situations.

When to Remove or Exchange (Table 5.2)

Successful placement and use of permanent access should result in immediate removal of dialysis catheters. Likewise, a kidney transplant that functions properly mandates prompt removal of catheter. In such instances the author uses the catheter for post transplant polyclonal antibody (i.e., thymoglobulin) treatment, and the catheter is removed about day 5 post operatively. A nonfunctioning, thrombosed catheter that failed declotting also must be removed or exchanged. An infected catheter, as evidenced by tunnel and lumen infections with positive blood culture, fever and chills or drainage at the exit site, also mandates removal.

When to exchange: Malfunctioning catheters and those with low flow characteristics may be exchanged over a guidewire (Figs. 5.27A-E and 5.28C-E). It is advisable especially in cases positive blood cultures to establish a new skin exit site. Catheters with positive blood cultures may be intraluminally infected and often can be successfully changed to new exit site under antibiotic (i.e., vancomycin) coverage.

Table 5.2. When to remove or exchange dual lumen hemodialysis catheters

1. Successful use (x 2-3) of new permanent access
2. Following kidney transplant
3. Nonfunctioning catheter, failed declot (i.e., t-PA or radiologic stripping)
4. Infected catheter: tract infection, positive blood cultures, fever, chills
5. When exchanging catheter through same venotomy site, use new exit site

The fibrin sheath surrounding a malfunctioning catheter may be dealt with in different ways. The interventional radiologist may choose to "strip" or remove this material through a femoral vein approach. When a new catheter is inserted, the fibrin sheath tract may be expanded with balloon angioplasty. Also, a newly inserted catheter should be advanced beyond the level of the malfunctioning, removed catheter.

Don'ts in Dialysis Catheters (Table 5.3)

Do not ever use subclavian veins for dual lumen dialysis catheters. The thrombosis/radiologic stenosis rate approaches 50%. Pulmonary emboli have been reported in up to 10%. Also, that extremity is ruined for future access. Pneumo-/hemothorax and mortality are also increased.

The authors strongly advise against placing percutaneous temporary catheters in the right internal jugular vein since this is the most successful site for cuffed catheters, should this be needed for a prolonged period. Left side dual lumen cuffed catheters have a high incidence of malfunction. Percutaneous catheters MUST not be placed for physician or surgeon convenience or to accommodate dialysis unit scheduling.

Percutaneous femoral vein catheters should not be left in place for more than 48 hours. When percutaneous catheters are replaced with cuffed, tunneled catheters over a guidewire, a new exit site must be used. When placing any type of percutaneous catheters, NEVER leave catheter unsecured without sutures (tape is not adequate). Accidental dislodgment may cause rapid exsanguination.

Preoperative Considerations

Anesthesia

Uncomplicated placement of dual lumen cuffed catheter does not require general anesthesia. Lidocaine 1% with NaHCO$_3$ 4% (Neut®) (5:1 ratio) is preferred (bicarbonate alleviates burning pain during injection). If done in the OR setting with anesthesiologist, light IV sedation may be used. The authors caution against overuse of sedation because it can make an awake patient an uncooperative individual jeopardizing the procedure. By creating a calm, trusting environment patients can be "talked" through the "steps". A snoring patient with an obstructed airway induces negative intrathoracic pressure on inhalation, increasing the risk for air embolism. Also, the large internal jugular vein present before the case "disappears" on each forced inspiration.

Often catheter placement access occurs concomitant with permanent access placement, i.e., primary AV or PTFE, ideal for regional block anesthesia.

Preoperative antibiotics may be given, i.e., cephalosporins (Ancef®) 1g IV. This is modified in cases of history of infection and comorbidity.

Table 5.3. Don'ts in temporary dual lumen hemodialysis catheters

1. Do not EVER use subclavian veins
   - Increased incidence of pneumo-/hemothorax
   - Causes thrombosis / stenosis / occlusion in 50%
   - Causes pulmonary emboli in 9-12 %
   - Ruins extremity for future access
2. Avoid placing percutaneous catheters in the right internal jugular vein
   - Increases morbidity for future cuffed catheters
   - Left side "long term" internal jugular catheters have a higher incidence of malfunction
3. Do not place percutaneous catheters for convenience unless failed declotting & revision.
4. Do not leave percutaneous femoral vein catheters for > 48 hours.
5. Do not exchange a percutaneous catheter over a wire with a cuffed tunneled dialysis catheter, using the same exit site.
6. Do not leave percutaneous catheters unsecured (always suture in place).

5

The Site Rite® Ultrasound Device

The Site Rite® ultrasound device is an extremely valuable tool when assessing and accessing central veins for hemodialysis. Not only can the operator determine the exact anatomy, but also see and guide the needle for placement of lines. Its use for this purpose is in fact recommended by the DOQI guidelines. Uses for the Site Rite® ultrasound device are outlined in Chapter 1 (Table 1.1).

Recently, a newer and updated version (Site Rite 3) came on the market, which represents a significant improvement over the older version (Site Rite 2). Site Rite 3 features a larger, flicker-free screen, a third and deeper (10-18 cm) tissue penetration probe, as well as several other technical improvements (Fig. 5.8A-B). The recommended technique of Site Rite® ultrasound use is outlined in Chapter 1, Table 1.1. Again, the right internal jugular vein is the preferred site for dual lumen, cuffed, tunneled catheter insertion. With increased experience, the use of disposable needle guides may be omitted or bypassed, which saves time and simplifies insertion. The authors strongly recommend the use of Site Rite® for all large vein access procedures. Strange anatomy is sometimes encountered.

Duplex Doppler is another excellent screening tool to assure central vein (internal jugular, subclavian vein) patency, as well as extremity vein mapping for vessel availability and optimal anastomotic sites. Preoperative duplex Doppler examination is indicated whenever there is a history of previous central vein catheter placement (Chapter 1, Table 1.1, Chapter 8, Table 8.1, Chapter 10).

The Micropuncture Set

For all central vein catheter placements the authors strongly recommend the use of the micropuncture set (Fig. 5.2). There are currently four on the market (Table 5.4). The micropuncture set consists of a 0.018 micro guidewire, puncture needle (21 gauge), a double catheter 4.0 or 5.0 French introducer and an inner sheath (Fig. 5.2A). The main benefit of the micropuncture set is the increased safety margin with the smaller 21 gauge needle. It is the authors' impression that the micropuncture use prevents perivascular bleeds, and thereby decreases post placement vascular compression and stenoses (Figs. 5.9 and 5.11).

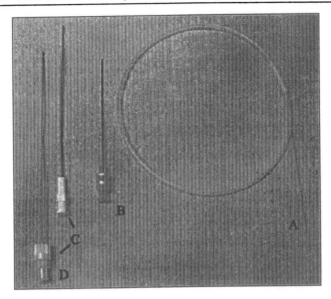

Fig. 5.2A. Detailed view of the Cook micropuncture set components: A. Microguidewire; B. Puncture needle (21 g); C. Double catheter (4.0 F) introducer D. Inner catheter.

Detailed Insertion Technique

Type of Catheters

There are many commercially available dual lumen hemodialysis catheters, and more are under development or revision. The reader is encouraged to become familiar with more than one type. The author has used the Tesio brand in the past, currently the split ash catheter is popular. Most catheters come in a shorter (i.e., 28 cm) length for the right side and a longer (32 cm) left side. The most recent device on the market is the Lifesite dual implantable port system (Vasca, Inc. 3 Highwood Drive, Tewksbury, MA 01876, phone (888) 827-2203, fax (978) 863-4469. www.vasca.com). Another single port system, the Dialock (Biolink, 136 Longwater Drive, Norwell, MA 02061, phone (781) 871-9353, fax (781) 871-4530. www.biolink.com) is under development but at the time of this publication printing not FDA approved (Table 5.5). The indications for implantable ports in dialysis patients are not well outlined but will be more clearly defined as the access function rate and morbidity becomes available.

Central Vein Anatomy

When considering central vein anatomy, it is easy to understand why dialysis catheters are better fitted to be placed on the right side through the internal jugular vein (Fig. 5.3). The left sided catheters need to be longer (32 cm), make two turns and are more likely to interfere with the cava wall, causing obstruction if too short. Ideally, the tip of the catheter is around the level of the right main bronchus or at the junction of the right atrium and superior vena cava (Fig. 5.3). Some operators prefer

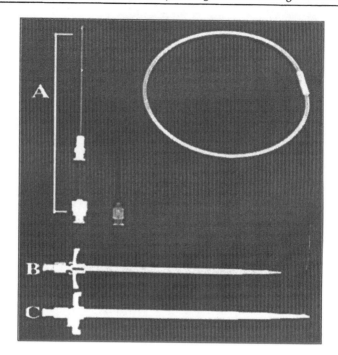

Fig. 5.2B. The micropuncture set (A) (Cook, P.O. Box 489, Bloomington, IN 47402) is highly recommended when placing central lines or dual lumen catheters. The main benefit of the micropuncture set is the increased safety margin with the smaller 21g needle. Also in this picture for comparison of the size difference between the peel away introducer sets for the Tesio catheter (B) and a dual lumen cuffed catheter (C) is depicted.

Table 5.4. Micropuncture sets currently on the U.S. market

1. **Cook Critical Care**, P.O. Box 489, Bloomington, IN 47402. Phone (800) 457-4500, Fax (800) 554-8335. www.cooksurgical.com/vascular access
2. **Bard Access Systems**, 5425 Amelia Earhart Drive, Salt Lake City, Utah 84116. Phone (800) 545-0890, Fax (801) 595-5975. www.bardaccess.com
3. **AngioDynamics**, 603 Queensbury Avenue, Queensbury, NY 12804, Phone (800) 772-6446 Fax (518) 798-1360. www.angiodynamics.com
4. **Boston Scientific**, 500 Commander Shea Boulevard, Quincy, MA 02171, Phone 800-225-3238 Fax 888-272-9444.www.bsci.com

---

both ports be in the cephalad portion of the right atrium in order to reduce the incidence of fibrin sheath formation.

Commercially Available Dual Lumen Hemodialysis Catheters

There are numerous catheters on the market, and more are under way (Table 5.5). The reader is encouraged to become familiar with more than one type. The most recent device on the market are the implantable port systems. Catheter type,

Table 5.5.

I. Non-cuffed, percutaneous catheters

Company	Product	Model	Shape	Lumen Diameter	Insertion Length	Flow Rate
Bard	Vas Cath	Flexxicon II	c/s	11.5	12.5,15,20,24	200-300
	VasCath	Niagara	c/s	12.5	12,15,20,24	200-300
Medcomp	Dual Flow	Internal Jugular	c	11.5	12,15,20,	200-300
	Dual Flow	XTP	s	9,11.5	12,15,20	200-300
	Raulerson/DuoFlow	Internal Jugular	c	9,11.5	12,15,20	200-300
	Soft Line		s	7,9,11.5	10,12,15,20	200-300
	Soft Line	Internal Jugular	c	11.5	12,15,20,	200-300
	Hemo Cath	Silicone	s	8,11.5	12,15,20,24	200-300
	Schon	XL	s	14	15,20	300-400
Arrow	Dual Lumen		s	12	13,15,20	200-300
	Arrow g+ard Blue		s	12	13,15,20	200-300
HMP	Neostar		s	11.5,13.5	15,18,19,20,23,28	300-400
Quinton	Mahurkar		c	10,11.5	13.5,15,16,19.5	200-300
	Mahurkar		s	10,11.5	12,13.5,15,16,19.5,24	200-300

continued on next page

5

Table 5.5., continued.

II Cuffed, dual lumen catheters

Company	Product	Model	Shape c/s	Lumen Diameter	Insertion Length	Flow Rate
Bard	Hickman	Dialysis/Apheresis	s	13.5	11,15,19,23,28,33,38,42,51	300-400
	Opti-Flow	Opti-Flow PC	c/s	14.5	19,23	300-400
	Vas Cath	Soft-Cell	c/s	12.5	12,15,17,19,23	300-400
Boston Scientific	Vaxcel	Chronic Dialysis	s	16-14	19, 23, 28	>400
Medcomp	Hemo Cath	Silicone	s	8,12.5	18,24,28,32	200-300
	Ash	Split Cath	s	14	28,32	300-400
	Ash	Split Cath II	s	14	28,32	300-400
	Tesio	Modified	s	10	7,10	300-400
	BioFlex	CS	s	10	52,72	300-400
HMP	Neostar		s	11.5,13.5	15,18,19,20,23,28	300-400
	Lifejet	multiple	c/s	15.4	19,23,28	300-400+
Quinton	Perm-Cath		s	13.5	28,36,40	300-400

III. Subcutaneous port implants

Company	Product	Model	Shape c/s	Lumen Diameter	Insertion Length	Flow Rate
Vasca	Lifesite			13	60	>400
Biolink	Dialock					300-400

C = Curved, S = Straight

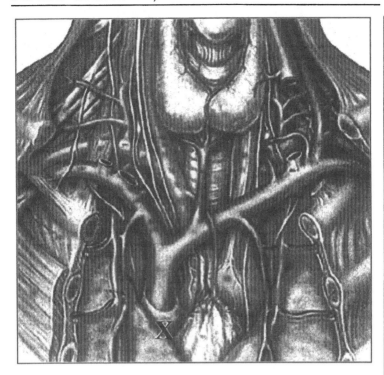

Fig. 5.3. The central vein anatomy. It is easy to understand why dialysis catheters are better fitted to be placed on the right side through the internal jugular vein; the left sided catheters need to be longer, and are more likely to interfere with the cava wall causing obstruction (if too short). Ideally, the tip of the catheter is around the right main bronchus (X in picture), or lower. Again the right internal jugular vein is the ideal site for temporary dialysis catheter placement.

brand and length are chosen based on insertion site (left or right), body size, anatomical variation and operator experience.

Dual Lumen Cuffed Catheters (Split Ash)

The lateral approach is preferred for internal jugular vein puncture, allowing a smooth wide curvature of the catheter. The patients head is turned slightly to the opposite side. The right internal jugular vein is the preferred site, as shown in the picture (Fig. 5.4).

Prior to insertion, the internal jugular vein outline is marked (Fig. 5.5) using the Site Rite® or duplex Doppler (Fig. 5.4). The intended subcutaneous tunnel is also marked as well as the exit site (Fig. 5.5). The appropriate catheter length is chosen based on the patients size, exit site level, and side right or left. Carefully select the exit site for patient comfort, especially in heavy females where the breast will pull the catheter when standing. In the operating room fluoroscopy is necessary to identify proper guidewire and catheter insertion positions.

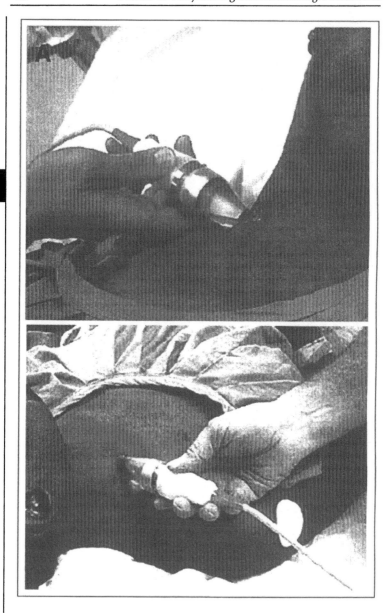

Fig. 5.4. A) The "lateral" approach is preferred for internal jugular (IJ) vein puncture, allowing a smooth wide curvature of the catheter. Patient's head is turned slightly to the opposite side. The right IJ is the preferred site, as shown in this picture. B) Same situation as previous image, indicating the lateral approach more clearly.

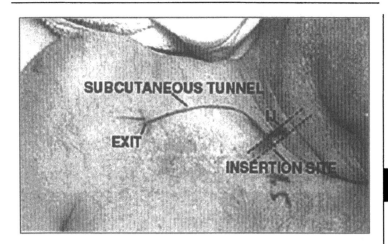

Fig. 5.5. Prior to insertion the internal jugular vein is outlined and the intended subcutaneous tunnel and the exit are marked. Based on patient size and exit site location, the appropriate catheter length is chosen. The micro guidewire can be used effectively to measure (estimate) the distance from the mid right atrium to the exit site position. The Cook micro guidewire is exactly the same length as the 32 cm split ash catheter. Note the lateral smooth intended catheter tract and low insertion site on the neck, both of which are aimed to minimize the propensity for catheter kinking.

The split ash catheter kit has only eight loose parts (Figs. 5.6A and 5.7A), while the Tesio kit has more than 40 loose parts (Fig. 5.18). Prior to insertion, the author lines up all parts in the order in which they will be used for safety and convenience, since during the procedure the surgeons' left hand must remain at the patient's neck part of the time. Presenting the contents of the catheter kit as shown in Fig. 5.6B indicates lack of knowledge and poor attitude. For the procedure to proceed safely, the assisting nurse or OR technician must be familiar with each step for proper and safe catheter placement (Fig. 5.6A).

After choosing the appropriate length of catheter (Table 5.5), usually the longer version (32 cm) for left and the shorter version (28 cm) for the right side, the catheter is flushed with heparinized saline (10 units/ml) or plain saline.

The tunneler is attached to the venous (return) end of the catheter, which is the longer of the two ports (Fig. 5.7A-C). The tunneler can be bent to facilitate a smooth subcutaneous curve (Fig. 5.7D). When inserted from the upper chest, appropriate depth will minimize cosmetic concerns and also make future removal or exchange of the catheter technically easier. Before pulling through the subcutaneous tunnel, the plastic sheath is pulled over the catheter covering both the venous and arterial ports (Fig. 5.7D).

The portable Site Rite® ultrasound device is used to identify the internal jugular vein and used to guide the needle puncture (Figs. 5.4 and 5.8).

5

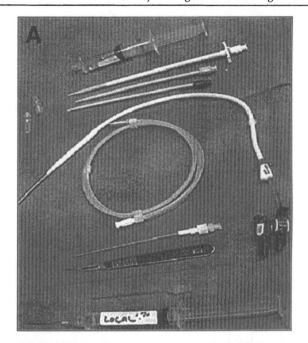

Fig. 5.6. Prior to starting the catheter insertion procedure, the many items lined up in the order they will be used (A) greatly increases safety and eliminates frustration. B) This way of "delivery" the catheter items demonstrates lack of knowledge and poor attitude.

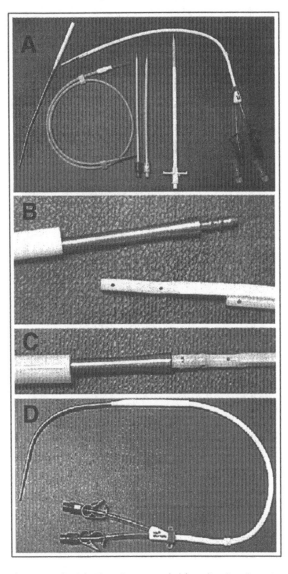

5

Fig. 5.7. A) There are only eight "loose" parts needed for split ash catheter insertion. The blue plastic skin suture fixation device can be removed by sliding it off the catheter. The author does not suture this device to the skin. B) The subcutaneous tunneller is connected to the venous (return) end of the catheter. C) The blunt end is inserted into the catheter end. D) The tunneller is bent to facilitate the smooth subcutaneous curve. Appropriate depth will minimize cosmetic concerns and also make future removal or exchange of the catheter technically easier. The plastic sheath is pulled over the catheter with the red port catheter also pushed into this plastic sheath.

5

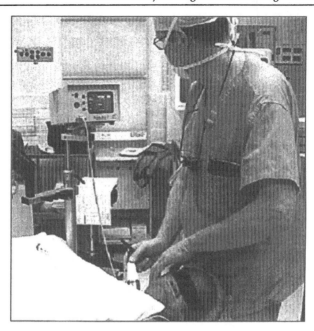

Fig. 5.8A. The Site-Rite® ultrasound device is a valuable tool in central vein dialysis catheter access; (in fact, recommended in DOQI guidelines) it can be used to guide needle puncture or as in this case identifying the vein and marking its location and depth.

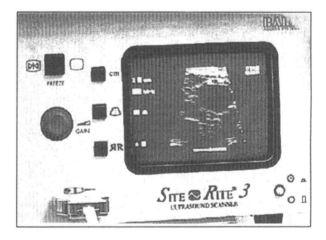

Fig. 5.8B. The Site-Rite® screen identifying the large internal jugular (IJ) vein and the carotid artery below and medial to the vein. The center of the vein in this case is at 1.5 cm depth under skin level. The IJ can be punctured under direct Site-Rite® guidance or after marking the skin and depth memory. (Also, see Table 1.1 for details on the Site-Rite®).

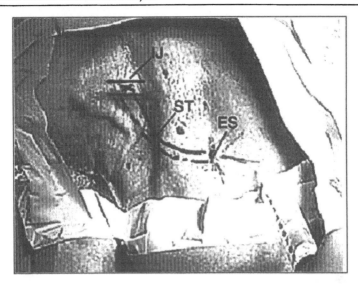

Fig. 5.9. The exit site and tunnel-tract are anesthetized with 1% lidocaine, which also elevates the skin, facillitating the tunnelling in passing over the collar bone, especially in thin individuals. (IJ=internal jugular vein; ST=subcutaneous tunnel; ES=exit site)

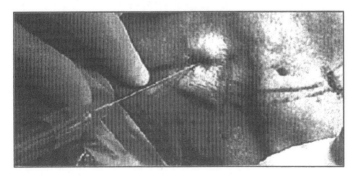

Fig. 5.10A. Using a 10 cc syringe on the micropuncture (21g) needle, the IJ is punctured. Note the lateral approach. Some lidocaine solution may be left in the syringe should more anesthesia be needed as the vein is entered.

Lidocaine 1% is used to infiltrate skin and the subcutaneous tissues and the tunnel from the upper chest crossing over the clavicle (Fig. 5.9).

Using a 10 cc syringe on the micropuncture (21 gauge) needle the internal jugular vein is punctured, again using the lateral approach (Fig. 5.10A).

With aspiration, blood will suddenly enter the syringe. If there is any suspicion of pulsating blood, or blood being "too red", the needle-syringe must be removed and the vein repunctured. With the use of the Site Rite® ultrasound, inadvertent carotid artery puncture is quite unlikely.

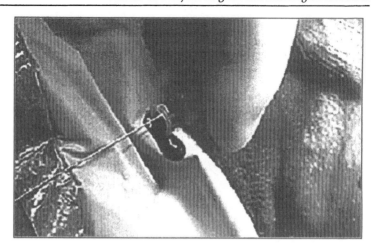

Fig. 5.10B. The dark blood assures the correct position in the IJ vein.

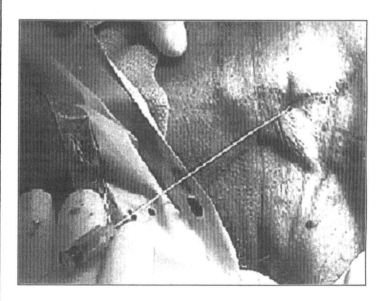

Fig. 5.10C. The micro guidewire is inserted; no resistance should occur. If the guidewire goes into the right ventricle, extra heartbeats will occur; another assurance of successful venous cannulation. The author prefers the heart rate on audible device. Multiple ventricular ectopy prompts immediate retraction of wire.

With the syringe removed and the patient in Trendelenburg or head-down position a slow drip of dark blood is seen (Fig. 5.10B). At this point the micro guidewire is inserted (Fig. 5.10B-C).

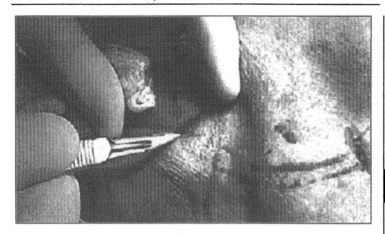

Fig. 5.10D. A small #11 blade incision at the micro guidewire exit site (the micro guidewire does not appear in this image. It is hidden behind the knife blade).

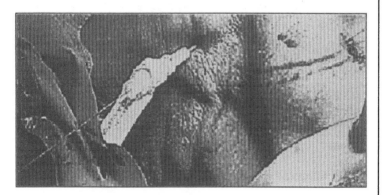

Fig. 5.10E. The double coaxial catheter, micro guidewire introducer in place with the micro guidewire inserted. At this time the distance to mid right atrium from the intended skin or cuff level is measured by making a manual kink in the 0.018 micro guidewire. This measurement will dictate the catheter tip to cuff length.

If the micro guidewire goes into the right ventricle, extra heartbeats may occur, another sign of correct venous puncture. The guidewire position is confirmed by fluoroscopy. The authors prefer the heart rate on audible device. Multiple ventricular ectopy prompts immediate retraction of any guidewire.

A small incision (#11 blade) is made at the micro guidewire skin exit (Fig. 5.10D), now allowing the insertion of the double coaxial catheter over the microguidewire (Fig. 5.10E).

Then the micro guidewire and the inner catheter are removed (Fig. 5.10F). The larger guidewire is now inserted and its position confirmed with fluoroscopy (Fig. 5.11A). Ideally, the author tries to advance the guidewire past the heart into the

5

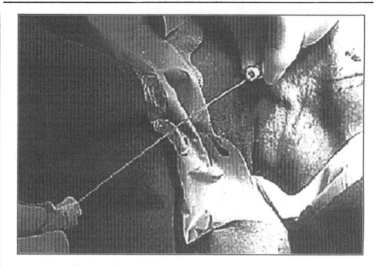

Fig. 5.10F. The micro guidewire and the inner catheter are then removed. Again, assure that dark (nonpulsating) venous blood is dripping.

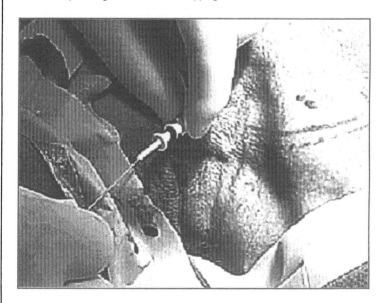

Fig. 5.11A. The large (0.038") guidewire is inserted, its position confirmed with portable C-arm fluoroscopy. Ideally the wire can be advanced into the IVC under active fluoroscopy.

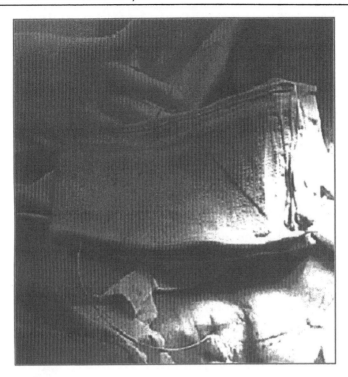

Fig. 5.11B. The guidewire may be supported (stabilized) by placing it in a folded towel to prevent accidental removal and placed out of way for the next several catheter insertion steps.

inferior vena cava. This position both decreases the chance of inadvertent pull out and also eliminates catheter induced arrhythmia. The larger guidewire is supported or stabilized in a folded towel on the patients chest (Fig. 5.11B).

The skin incision next to the guidewire exit is now extended to about 5 mm to allow the catheter and the dilator introducer sheath (Fig. 5.12A). The exit site previously marked is also incised about 4 mm (Fig. 5.12B).

The subcutaneous tunneler is now inserted into the chest incision and passed above the clavicle into the neck wound (Fig. 5.13A-B). The tunneler and the plastic sheath covering the catheter are now pulled into the neck wound. The catheter is pulled all the way, with the dacron cuff now being almost into the neck wound. The cuff will later be pulled back and adjusted to place the two ports optimally at the right atrium-superior vena cava (SVC) level (Fig. 5.13C). The arterial (red) port is placed laterally locating the pull line medially, preventing sucking against the lateral cava wall.

Next, the tract into the internal jugular vein is dilated over the guidewire (Fig. 5.14A). The split ash catheter has two sized dilators to help facilitate the passage of the sheath introducer. These dilators are introduced with rotating movements, while

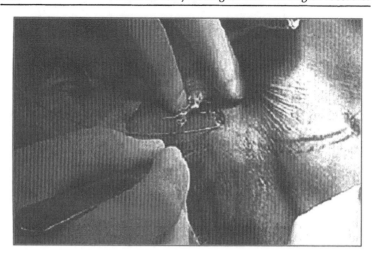

Fig. 5.12A. The skin incision in the neck is extended about 5 mm to allow the catheter and the dilator/introducer sheath.

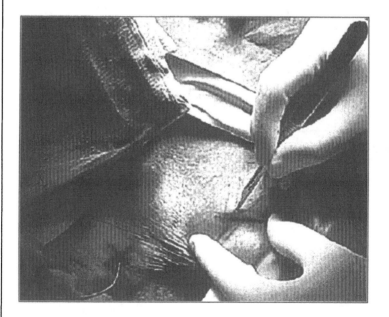

Fig. 5.12B. The catheter exit site (previously marked) is incised ~ 4-5 mm, allowing the dacron cuff to be advanced, but snugly.

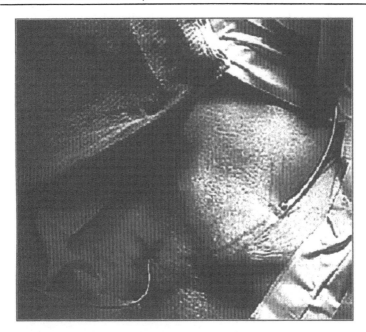

Fig. 5.13A. The subcutaneous tunneler is bent slightly to facilitate passage along the predetermined subcutaneous tract.

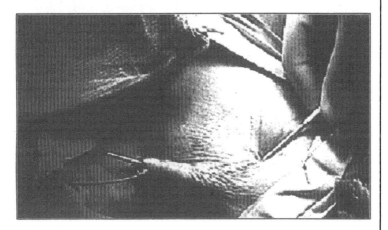

Fig. 5.13B. At this point the tip of the tunneller is coming out next to the guidewire at the neck incision site.

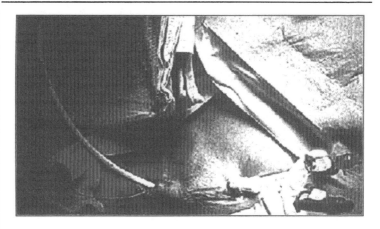

Fig. 5.13C. The split ash catheter is pulled all the way into the tunnelled tract. Based on fluoroscopy determination, it will be adjusted outward to place the tips at the right atrium/superior cava junction. Rotate catheter so that the short/red port will be directed medially to avoid sucking against the lateral vena cava wall.The external red port is located laterally

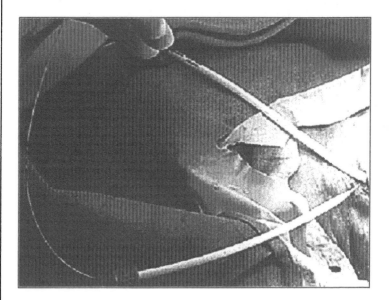

Fig. 5.14A. The split ash set has two sized dilators to help facilitate the passage of the sheath introducer. These are inserted with rotating movements, while assuring free movement of guidewire.

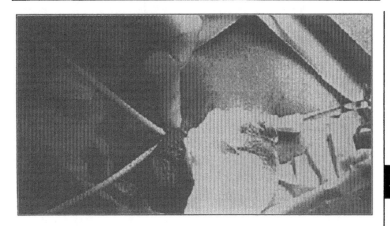

Fig. 5.14B. As the dilators are removed, finger pressure is applied to prevent backbleeding.

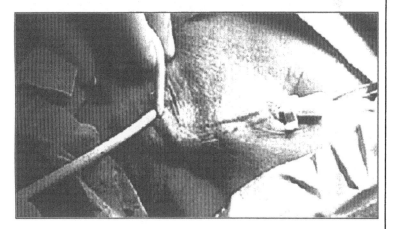

Fig. 5.14C. The second, larger dilator is introduced the same way. It is wise to look at the angle of the dilator in relation to the guidewire under fluoroscopy.

assuring free movement of the guidewire. As the dilators are removed, finger pressure prevents back bleeding (Fig. 5.14B). The second larger dilator is approximately the same size as the introducer sheath (Fig. 5.14C). Finally, the introducer sheath is inserted, again with rotating movements over the guidewire (Fig. 5.14D). The sheath is introduced half way (Fig. 5.14E). As a safety precaution, the author pushes the sheath while the stiffer dilator is kept constant to decrease the risk of vascular injury in the face of an uncertain anatomic environment (Fig. 5.14F).

Next the dilator inside the sheath is removed. At this point the patient must be in Trendelenburg position to prevent air embolism. The sheath is pinched between the surgeon's fingers and the catheter is inserted (Fig. 5.15A). Some blood loss is

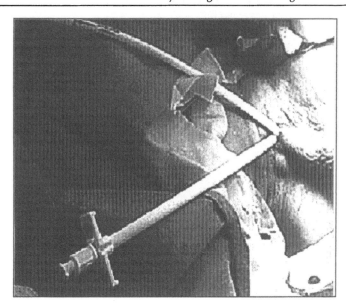

Fig. 5.14D. Finally, the even larger sheath introducer is inserted cautiously with rotating moves, again ensuring repeatedly a freely moving guidewire. Again to increase safety, the authors recommend advancing the wire into the inferior vena cava under active fluoroscopy.

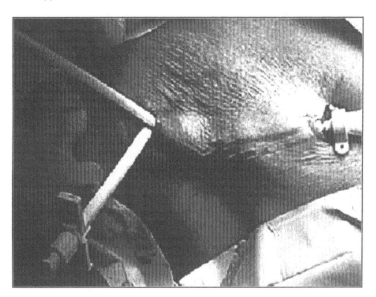

Fig. 5.14E. The sheath is introduced "half way".

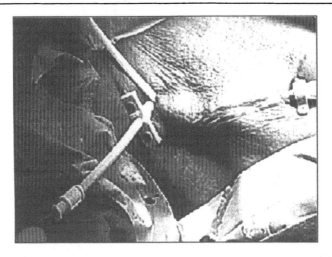

Fig. 5.14F. Then the soft sheath only is advanced while the stiff dilator is kept constant (this is to decrease the risk of vascular injury in an uncertain anatomic environment).

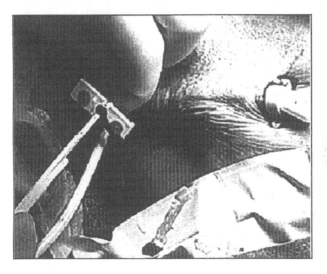

Fig. 5.15A. As the dilator is removed, (patient must be head down to prevent air embolism), the sheath is pinched between the surgeon's fingers, and the catheter is inserted. Some blood loss is always encountered during the 3-5 seconds the catheter is inserted. At the sheath introducer step the patient needs to be in the head down position; if awake, is the patient instructed to "hum" or "sing" to avoid an unwanted forceful inspiration which may cause massive air embolism (a practice that is most helpful to increase safety of this procedure). With positive ventilation this is less likely, still the Trendelenburg position (head down) is warranted. In cases of a nonflexible, OR table a safe technique involves two guidewires through each port while the catheter is inserted without use of the large sheath, providing a "sealed" system at all times.

Fig. 5.15B. Once the catheter is advanced, minimal blood loss is encountered. The two lines (arterial short and venous long) have previously been split to the black mark, shown in this image.

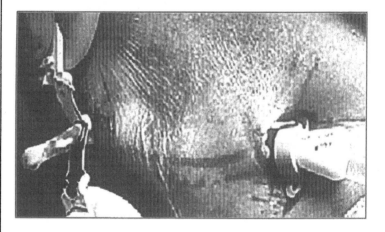

Fig. 5.15C. As the catheter is maximally introduced, the sheath is split and pulled out, while the catheter is further introduced.

always encountered during the 3-5 seconds the catheter is inserted (Fig. 5.15B-D). It cannot be overemphasized that the patient has to be head down to avoid air embolism. If awake, the patient is instructed to hum or sing to avoid an unwanted forceful inspiration, which may induce massive air embolism. With anesthesia delivering positive ventilation pressure, this is unlikely, yet still head down position is warranted.

In cases of nonflexible OR table, a safe technique involves two guidewires inserted through each port while the catheter is inserted without the use of the large

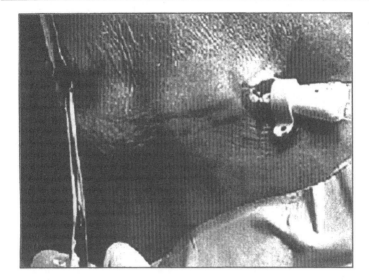

Fig. 5.15D. The final stage of sheath splitting. The dialysis catheter is now completely in the subcutaneous space.

Fig. 5.15E. When the sheath is removed and the catheter is completely inside, the bleeding will stop (reversing the head-down position will also decrease bleeding).

sheath, providing a sealed system at all times. The split ash catheter, because of design, can be inserted over one guidewire going through both ports (Fig. 5.27).

Once the catheter is fully inserted with the sheath split and removed, no further blood loss is encountered (Fig. 5.15E). Gentle pressure over the area will stop all bleeding. Also, reversing the head down position will decrease bleeding.

Based on fluoroscopy, the catheter is pulled back to assure proper position at the right atrium-SVC level (Fig. 5.24A).

Both ports are tested for rapid filling return of blood, such as pulling 20 cc in 2 seconds (Fig. 5.16A). Ideally, the red port (priming value = 2.0 cc) at the exit site is located laterally, which will place the red port medially at the cava level, decreasing the chance of sucking against the lateral SVC wall (Fig. 5.16B).

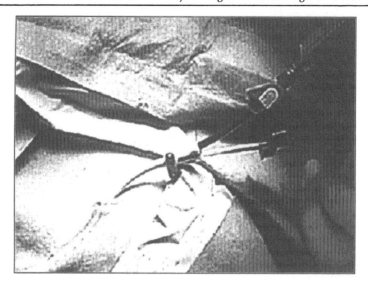

Fig. 5.16A. Based on fluoroscopy, the catheter is pulled back to assure the proper position at right atrium-superior cava level. Both ports are also tested for rapid filling-returning blood (i.e., 20cc in 2 seconds), assuring adequate dialysis machine flow rate.

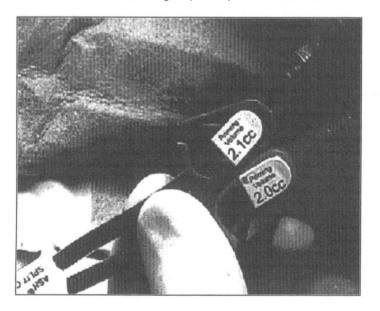

Fig. 5.16B. The split ash catheter requires priming volumes of 2.1 cc and 2.0 cc in the blue and red ports for the longer (32 cm) catheter and 1.9 and 1.8 cc for the short (28 cm) catheter, respectively.

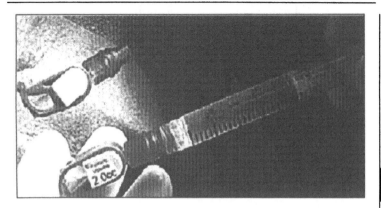

Fig. 5.16C. Using small volume syringe assures exact volume injection. In this image, the red port (32 cm catheter) is primed with 2.0 cc (of 1000 u heparin/cc).

Fig. 5.16D. The caps are placed on the ports.

Finally, the split ash requires priming volume of 2.1 and 2.0 ml in the blue and red ports respectively for the longer (32 cm) catheter and 1.9 and 1.8 cc for the right (28 cm) catheter. For accuracy, use small volume syringes (Fig. 5.16C). These "priming" volumes are clearly marked on the newer generation of most catheters (Fig. 5.16B-C). The author uses heparin 1000 u/cc for priming catheters (for hospital policy reasons). However, it is our belief that 10 u/cc is as effective but safer. Using 5000 u/cc of heparin is unnecessary and lends itself to errors and bleeding complications.

Finally the caps are placed on the ports (Fig. 5.16D).

The neck wound is closed with one 5.0 PDS subcuticular suture.

The blue collar at the exit site of the split ash catheter can be removed (Fig. 5.17A). It is the authors' opinion that suturing this wing to the skin causes unnecessary harm to the patient and often pulls the dacron cuff out of the exit site. This is more common in overweight females with heavy breasts, especially if the exit site is placed close to the breast tissue.

The authors place a nylon suture at the skin level to tighten the exit site, tied several times around the catheter and finally into the groove where the blue collar or wing was placed (Fig. 5.17B). This suture is removed after 7-10 days at the dialysis unit.

The neck incision is covered with a 1/2" steri strip (Fig. 5.17C), the catheter exit site is surrounded with a 2x2 gauze and covered with a bio-occlusive (i.e., Tegaderm®) tape (Fig. 5.17D). The catheter is ready for use.

Two Line Catheters (Tesio)

At the present time, the most popular two line dialysis catheter is the Tesio (Medcomp, 1499 Delp Drive, Harleyville, PA, 19438 (215) 256-4201). This section outlines only the significant differences between the split ash and Tesio catheter placements.

1. The Tesio set contains 48 loose parts, compared to 8 loose parts for the split ash (Fig. 5.18).

2. Because of the two lines, the same procedure is repeated twice. Each catheter lumen is smaller, or 10 F, compared to 14 F for the split ash (Table 5.5, Fig. 5.20).

3. The introducer sheath is also smaller, and no dilators are used prior to the sheath insertion.

4. The insertion requires two sticks, one for each line insertion. It may also require two skin incisions in the neck, depending on surgeon's preference.

5. The subcutaneous tract and catheter placement goes from the neck incision to the exit site on the chest, again two subcutaneous lines are inserted, which requires a small stab wound at each of the exit sites.

6. In order to place a dacron cuff in the subcutaneous space, two dilators are needed, with hubs that are pulled at the appropriate length to place the catheter tip as guided by fluoroscopy.

7. Difficulties exist in placing the catheter tip in the chest because of the dacron cuff placement subcutaneously, and the two subcutaneous line curves describing different length.

8. One other disadvantage with the Tesio is the difficulty obtaining adequate prolonged dialysis blood flow within the first 24 hours. This well known and common phenomenon is, as of now, unexplained. An initial 24 hour blood flow disturbance is not seen with the split ash catheter.

9. With the Tesio there is a small risk for the untrained eye to pull the wrong end of the catheter to the exit site. The external port assembly steps are shown in Fig. 5.19A-F.

10. The catheters are marked and can be cut before the external kit is attached. The catheter priming volume depends on the remaining length of catheter (Fig. 5.20A) and is clearly marked on the catheter. The uncut catheter priming volume is 2.3 cc, including the attachment kit (Fig. 5.20B). The complete Tesio external sites assembled shown in Fig. 5.21.

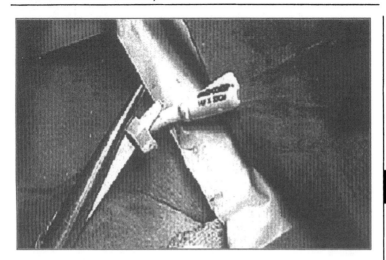

Fig. 5.17A. Remove the blue wing designed to be sutured to the skin (a practice we advise against). The wing can be cut or removed by sliding off the catheter prior to insertion.

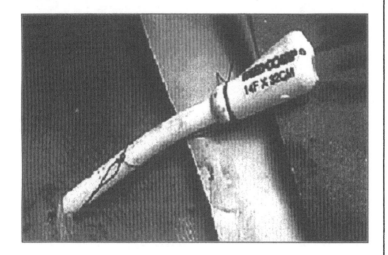

Fig. 5.17B. The catheter is attached to the exit site with a nylon suture tied around the catheter and into the groove previously occupied by the blue collar (the authors strongly advise against suturing the blue plastic collar to the skin). However, the suture may be passed through the "wing" (if kept) to further secure the catheter. Suturing the wing to the skin causes pain and traction on the skin (especially in obese patients or females with the exit site close to the breast). When the patient sits up the catheter may move out as much as 5 cm when the breast or fat pulls the catheter out, the fixed point being the cuff. A second "fixed point" should be avoided since it causes the catheter and cuff to migrate towards the exit site causing infection and loss of the catheter.

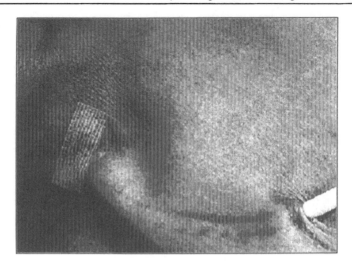

Fig. 5.17C. A steri-strip dressing is all that's needed on the neck puncture site.

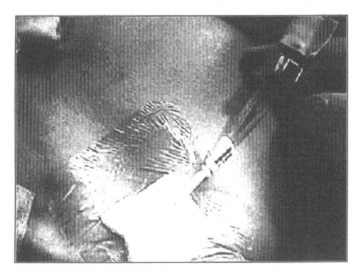

Fig. 5.17D. Surgical wound dressing may include a 2x2 gauze around the exit site covered with a bio occlusive dressing. A tape around the caps provides extra security.

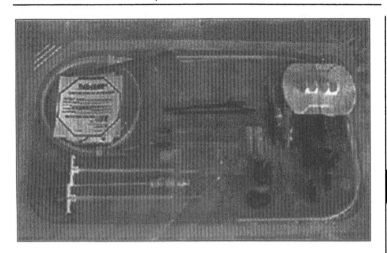

Fig. 5.18. The Tesio kit has more than 40 loose parts, including the two separate lines.

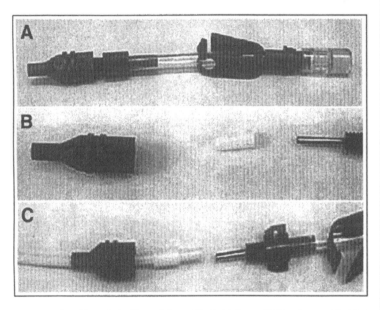

Fig. 5.19A-C. A) The exit site kit as delivered in the Tesio set. B) The disassembled Tesio external kit. C) First the color coded hub and white plastic sleeve are placed over the Tesio catheter.

5

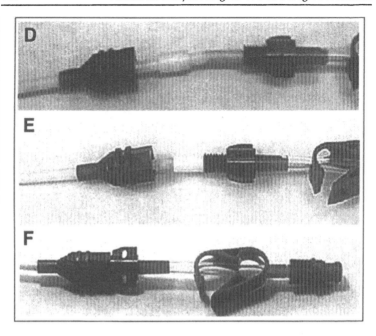

Fig. 5.19D-F. D) The metallic end is then pushed into the catheter end all the way. E) The white plastic sleeve is pushed distally and the hub screwed tightly over it. F) The assembled exit site kit.

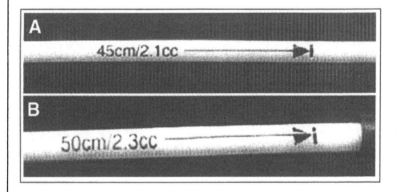

Fig. 5.20. A) The external catheter length can be shortened as needed for patient and dialysis comfort purposes. It then requires less priming volume as indicated on the catheter. B) The priming volume for the entire 50 cm catheter is 2.3 ml, including the red or blue exit site kit.

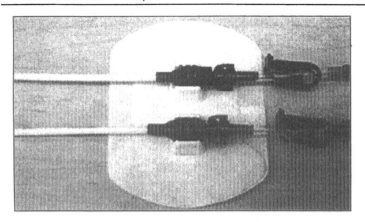

Fig. 5.21. This adhesive device supplied in the Tesio kit is taped to the patient's chest to prevent catheters from being pulled and only lasts a few days. Patients must be taught to care for their catheters to minimize complications. Catheters should not hang down freely, which causes slow migration and eventually dacron cuff migration/exposure/infection. Catheters can be taped in a loose curvilinear fashion or secured in the bra (females) to allow free movements at the exit site, avoiding pulling.

Removal of Cuffed Catheters

Removal of cuffed catheters is indicated for tract infection, sepsis, positive blood cultures and is the catheter is no longer needed.

Depending on where the cuff is located, two techniques are used.

1. The cuff is located immediately adjacent to the exit site. In this case, the exit site is anesthetized with 1% lidocaine (Fig. 5.22A). The exit site is dilated slightly with a mosquito hemostat or incised with a knife, and the catheter may be lodged out with this simple procedure (Fig. 5.22B). In some cases where the cuff is more incorporated, a small skin incision longitudinal to the catheter is used with sharp and blunt dissection around the cuff, which will release the catheter. Usually there is back bleeding which is temporarily stopped by finger pressure (Fig. 5.22C). Compression of the exit site for about 30 minutes with the patient sitting up will usually prevent further bleeding. However, the author recommends a Figure-of-eight suture with nylon or PDS to prevent backbleeding (Fig. 5.22D). The patient is asked to keep pressure on the site for the next 1-2 hours, and avoid strenuous physical activity for the next 24-48 hours (Appendix IV).

2. In cases where the cuff is located further up the subcutaneous tract, the area on top of the dacron cuff is infiltrated with 1% Lidocaine. A transverse incision is made (Fig. 5.23A). The catheter is isolated distal to the cuff, clamped with a hemostat (Fig. 5.23B) and divided distal to the hemostat (Fig. 5.23C). The catheter may now fall out from the exit site. The cuff is dissected free, exposing the catheter proximal to the cuff. Often there is a sheath of tissue surrounding the catheter at this point. A pursestring suture is placed around the catheter, grabbing the tissue around the tract (Fig. 5.23D). This pursestring suture of PDS is tied as the catheter is pulled out to prevent backbleeding.

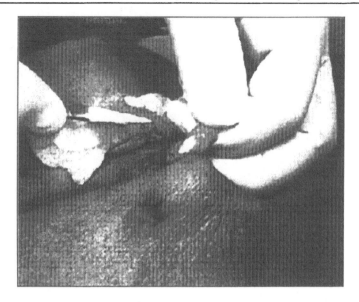

Fig. 5.22A. Removal/exchange of cuffed dual lumen catheters; in this case, the cuff is immediately adjacent to the outside. After infiltration of lidocaine 1% and removal of concentrated heparin in the ports, the cuff is dissected free with a knife, hemostat or sharp scissors.

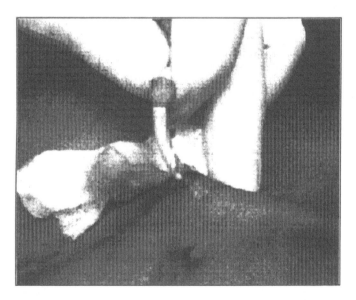

Fig. 5.22B. When the cuff is free, the catheter is pulled out in its entirety (Appendix IV).

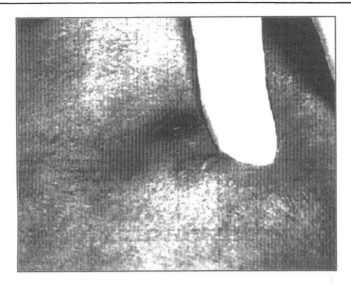

Fig. 5.22C. Usually there is back bleeding, which is temporarily stopped by finger pressure. Compression by the patient of the exit site for 30 minutes usually prevents further back bleeding.

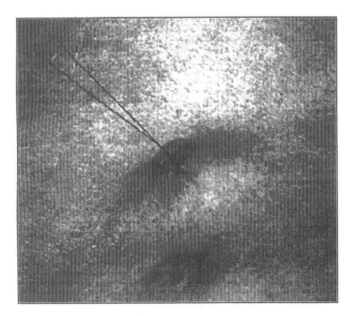

Fig. 5.22D. The author recommends a figure-of-eight suture to prevent any future back bleeding. In addition, the patient is asked to keep pressure on the site for the next 1-2 hrs and avoid strenuous physical activity for 24-48 hours.

5

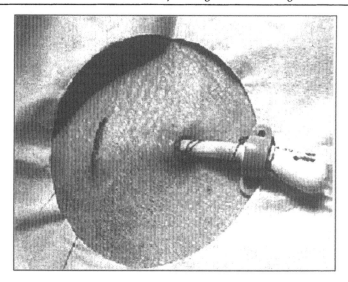

Fig. 5.23A. When the cuff is at a distance from the exit site, a cut down over the cuff is necessary. (When the cuff is at skin level blunt dissection is all that is needed around the cuff).

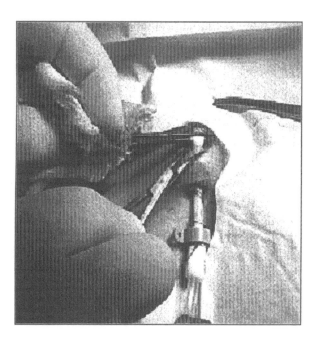

Fig. 5.23B. The catheter is clamped distal to the cuff and cut/divided.

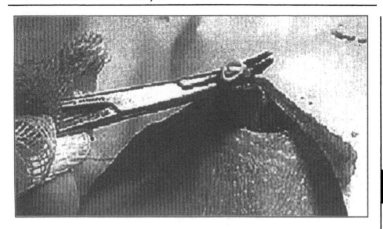

Fig. 5.23C. The cuff is dissected free to expose the catheter proximal to the cuff.

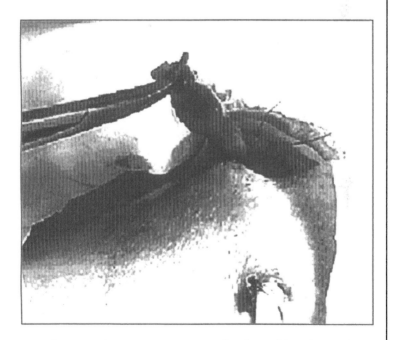

Fig. 5.23D. A pursestring suture, i.e., 4-0 PDS' is placed around the catheter and tied as the catheter is pulled to prevent back bleeding. Lastly, the suture (if present) at exit site is cut and the catheter pulled out of the exit site.

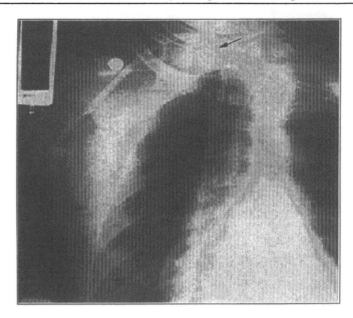

Fig. 5.24. Correctly placed, the catheter will make a smooth, lateral curve at the neck level, avoiding kinking.

The skin incision is closed with one or two subcuticular PDS sutures and covered with a 4x4 dressing and bio-occlusive dressing (Tegaderm®). The patient is asked to keep pressure for the next 1-2 hours and avoid strenuous exercise for 24-48 hours (Appendix IV).

The Ideal vs Kinked Catheter

The ideal catheter on chest X ray describes a smooth right-side, lateral curve without kinking (Fig. 5.24). The catheter lines end at the right atrium/SVC level. Kinking usually occurs at the internal jugular vein insertion site (Fig. 5.25A-F). When removed, this stiffer catheter also describes the central vein anatomy (Fig. 5.25C). Kinking may involve both ports, or as in this case only the arterial (red) port (Fig. 5.25C-D). Ways to avoid a kinked catheter include wide smooth subcutaneous tunnel achieved by bending the tunneler. Also a lower jugular approach lessens the likelihood of kinking at the vein entrance level. Rarely, despite optimal placement, kinks may occur (Fig. 5.25E-F).

Exchanging Cuffed, Tunneled Catheters

The indications for changing catheters are outlined in Table 5.2 and vary with the type of catheter. Generally, changing cuffed catheters over a guidewire using the same exit site is not recommended. The authors prefer the following technique, if technically feasible.

A skin incision is made at the neck level, close to or at the previous neck insertion site (Fig. 5.26A). The catheter is dissected free and surrounded with a vessel loop (Fig. 5.26F). The new catheter is inserted from the upper chest using a new exit

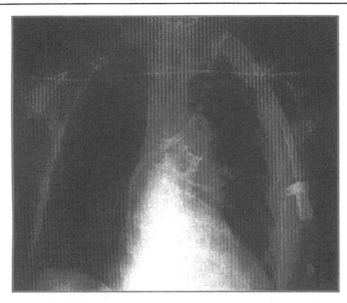

Fig. 5.25A. Classic kink of catheter with a sharp angle at the IJ exit. This is caused by too sharp a "turn" of catheter at the IJ vein.

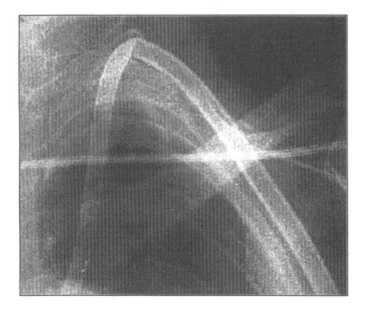

Fig. 5.25B. Close up view of same catheter.

5

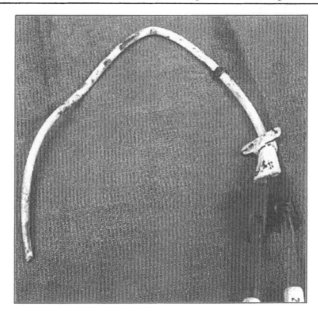

Fig. 5.25C. The removed catheter outlines the central vein anatomy.

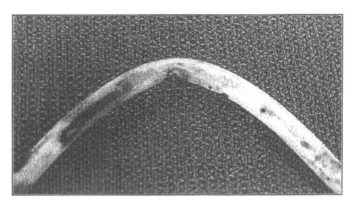

Fig. 5.25D. Close up view of the kink at the neck IJ level.

site, usually lateral to the previous one. The new catheter is tunneled and brought into the neck wound. The catheter to be removed is guidewired, either from the end by making a small incision in one of the catheter lines at the neck wound. It is not recommended to completely divide the catheter at this level since it may be lost, causing foreign body embolization (Fig. 5.26C). A guidewire is inserted, and the catheter can now be removed. Back bleeding is prevented by finger pressure. The guidewire is pulled out of the old catheter. The tract may be dilated over the guidewire,

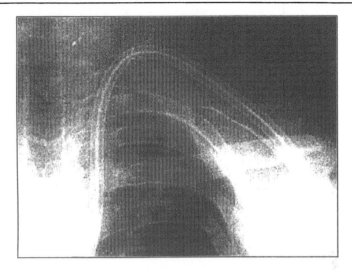

Fig. 5.25E. Despite a smooth curve of this catheter, an undulation involving one lumen was seen immediately after placement. The kink is magnified in image F. Patient underwent uneventful dialysis with >300 ml/min flow the following day. (No follow-up x-rays of this catheter available!)

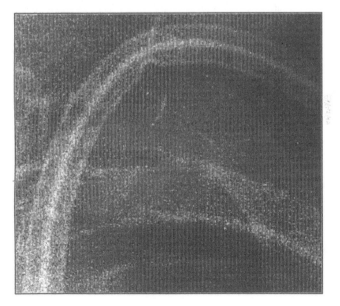

Fig. 5.25F. Close up view of kinked catheter.

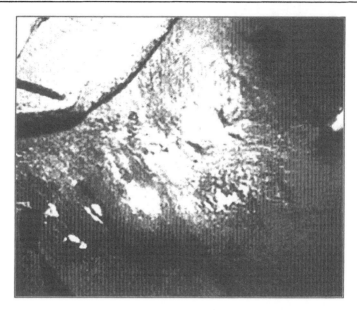

Fig. 5.26A. Make an incision at the previous neck insertion site; the catheter is mobilized.

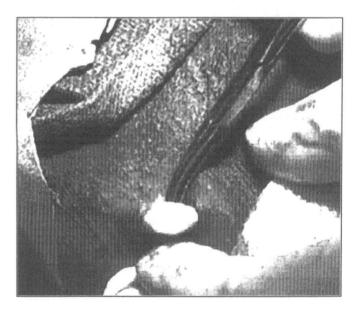

Fig. 5.26B. The catheter is pulled up. Catch the catheter with a vessel loop. At this point the catheter may be partially divided to allow guidewire insertion. This will also prevent losing the catheter into the blood stream.

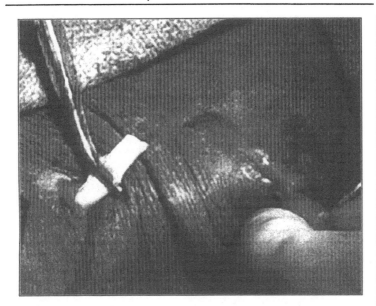

Fig. 5.26C. The old catheter with cuff is removed. No back bleeding should occur. Do not completely divide the catheter at the neck incision level as shown in this image, since the operator can lose stability and lose the catheter into the bloodstream, causing foreign body embolization.

but may not be needed. The new catheter is inserted as described above, using the introducer sheath. Since the tract is already present, the introducer sheath may not be necessary. In cases of split ash, a special technique can be used where the guidewire is inserted into the distal venous port and using the side holes, the guidewire can come out of this venous port and now go into the end of the arterial port, and the catheter can be inserted over one guidewire without using the introducer sheath. Sometimes using a stiff shaft hydrophilic wire ("glidewire") facilitates these steps, since the regular stiff guidewire produces too much friction while in the catheters. Also, as an alternative method, a rapid exchange can be achieved with dual guidewires, one in each port avoiding "hang ups" at the venous site. The old catheter is now removed, using cut downs over the dacron cuff, as described above. The neck wound is closed with subcuticular PDS, and the catheter is again sutured to the exit site as described above after having been adjusted based on fluoroscopy.

Split Ash Single Guidewire Insertion

Figure 5.27 illustrates a useful technique of insertion of a split ash catheter. The old catheter is removed over a guidewire inserted from the outside (blue) port or from the neck incision (Fig. 5.26A-B), where a small cut is made into the catheter. The author advises against completely dividing the catheter at this site (Fig. 5.26C) for risk of losing it. A wet stiff hydrophilic 0.035" (0.89 mm) glidewire is

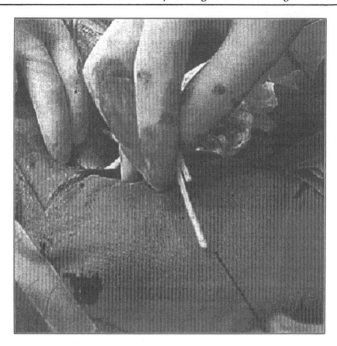

Fig. 5.27A. As the old catheter comes out, finger pressure is applied to prevent back bleeding. The guidewire remains in the IJ and preferably down in the IVC (confirmed by fluoroscopy). The end of the wire is now inserted into the blue port of the split ash.

recommended to avoid getting stuck, especially when the catheter is sharply curved from the left side insertion.

As the old catheter comes out, finger pressure is applied to prevent back bleeding. The glidewire remains in the IJ and preferably down in the IVC (confirmed by fluoroscopy).

The end of the wire is now inserted into the blue port of the split ash (Fig. 5.27A), brought out through the second side hole of this port (Fig. 5.27B), inserted into the end of the red port (Fig. 5.27C), and fed back to the external red port and secured (Fig. 5.27D). The split ash catheter is now inserted into the IJ and positioned under fluoroscopy (Fig. 5.27E). The glidewire is now removed. Both ports checked for patency and primed.

This technique allows a water tight seal and prevents air embolism, even when the patient is not in the Trendelenburg (head down) position. This one guidewire insertion could also be used in first time catheter placement after dilating the tract over the guidewire, but replacing the last step, i.e., not using the sheath (Fig. 5.14-5.15), with the single wire steps.

A two guidewire insertion technique for other catheters can be used, in which both ports will have one glidewire as the catheter is inserted.

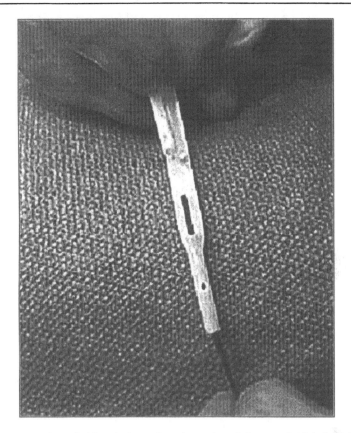

Fig. 5.27B. The end of the wire is now brought out through the second side hole of this port.

The Infected Catheter

There are different degrees and severity of catheter infections based on clinical findings.

Patients may have a perfect looking catheter, dialyze well but experience low grade fever, occasional chill, especially during dialysis. This is likely to represent an intraluminal infection that is treated with antibiotics and catheter exchange using a new exit site. Others may have the same symptoms and in addition a red, inflamed exit site or tender catheter tract. Most surgeons would permanently remove such catheters. The authors take a slightly different approach. After 24-48 hours of IV antibiotics (i.e., vancomycin and gentamicin), symptoms improve or disappear. At that point the catheter is replaced through a new exit site. Antibiotics are continued for 10-14 days. In patients with purulent drainage around catheter and clinical signs of sepsis, the catheter must urgently be removed. New access can be placed when the patient has been afebrile for >48 hours or blood cultures are negative. Dialysis is delivered through temporary femoral percutaneous catheters.

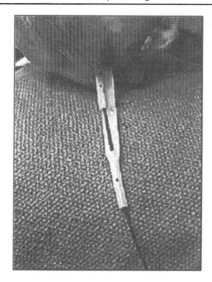

Fig. 5.27C. And inserted into the end of the red port.

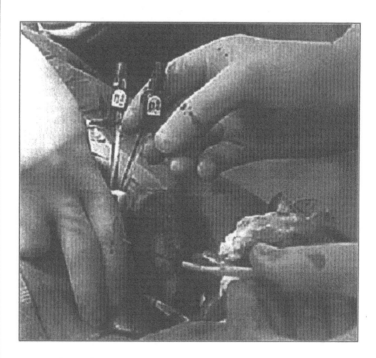

Fig. 5.27D. Then fed back to the external red port and secured.

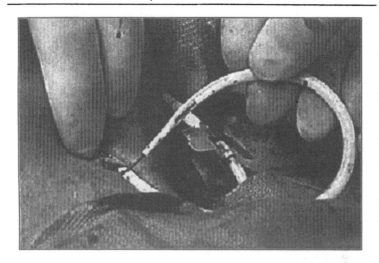

Fig. 5.27E. The split ash catheter is now inserted into the IJ and positioned under fluoroscopy.

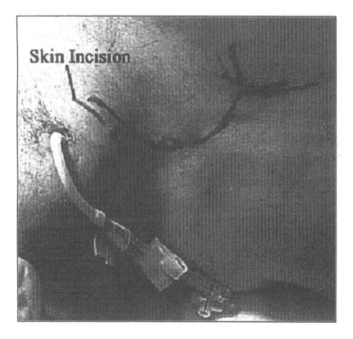

Fig. 5.28A. The skin is marked for (in this case) split ash subcutaneous tract and exit site. Skin incision is marked for catheter exposure.

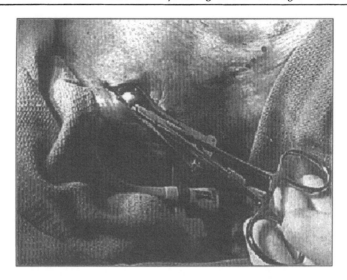

Fig. 5.28B. The catheter is mobilized and surrounded with a vessel loop.

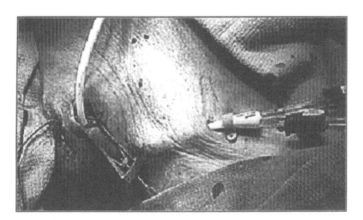

Fig. 5.28C. The split ash is now pulled into the neck wound from its exit site.

Percutaneous Catheter Conversion to Cuffed

The technique described here (Fig. 5.28) applies to the right internal jugular vein site. However, the principles apply to any site. The right neck and the entire percutaneously placed catheter are thoroughly prepared with Betadine solution. The "wing" often sutured to the skin is removed. The skin is marked for (in this case) the split ash subcutaneous tract and exit site (Fig. 5.28A). Skin incision is made and the catheter is exposed (Fig. 5.28A), the catheter is mobilized and surrounded with a vessel loop (Fig. 5.28B). The split ash is now pulled into the neck wound from its

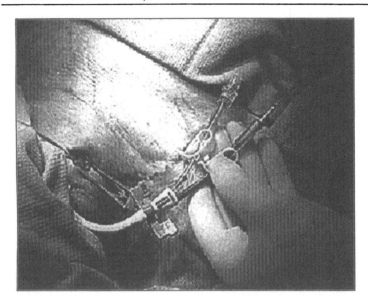

Fig. 5.28D. The blue port is used for a stiff guidewire (0.035″ or 0.89 mm, straight or angled) insertion into the percutaneous catheter that is then removed leaving the guidewire in place, confirmed by fluoroscopy. Ideally, the guidewire goes down into the inferior vena cava.

exit site (Fig. 5.28C). The blue port is used for a stiff guidewire (0.035" or 0.89 mm, straight or angled) insertion into the percutaneous catheter (Fig. 5.28D) that is then removed leaving the guidewire in place, the position confirmed with fluoroscopy. Ideally, the guidewire goes down into the inferior vena cava. The guidewire is now retrieved into the neck incision wound using the vessel loop (Fig. 5.28E). The guidewire is now used to insert the split ash using the single wire technique (Fig. 5.27) or the sheath technique as described in Fig. 5.14D.

The Thrombosed Catheter

Several options, generally in a stepwise, progressive fashion are available for dealing with a nonfunctioning catheter. One or both ports may be partially or completely occluded for either pull, return flow or both.

1. First the operator should attempt vigorous flushing with a 10 or 20 cc syringe to aspirate the clot.
2. Second step includes using brushes that rotate inside the lumen breaking loose clots. The authors have used the ureteral cytology brushes for this purpose, with questionable results.
3. Thrombolytic agents may be used, such as t-PA (tissue plasminogen activator), using indwelling or 2-4 hour infusion of 1 mg per port per hour (Table 5.6)
4. Often there is a fibrous sheath surrounding the catheter. This will require interventionalradiology to strip (snare) from a femoral vein approach. Recent studies suggest catheter exchange is superior to stripping (Table 5.6).

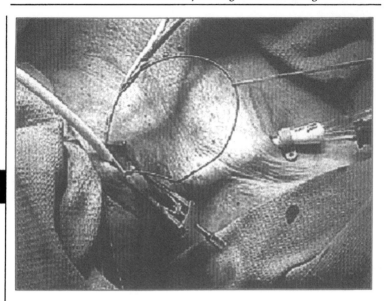

5

Fig. 5.28E. The guidewire is now retrieved into the neck incision wound using the vessel loop.

5. These procedures may be combined with thrombolytic treatment, i.e., t-PA injected as a bolus or infusion over 2-4 hours.
6. Balloon dilatation to 12-14 mm of the innominate vein, SVC and the right atrium in order to disrupt the fibrin sheath is carried out.
7. Finally, the catheter can be exchanged over guidewire to a new exit site or inserted on a new side. In either case the catheter should be inserted slightly further beyond the fibrous sheath to ensure catheter longevity. The sheath can be visualized through an SVC/venogram as the catheter is removed. If present, the sheath is subjected to 8-10 mm balloon angioplasty.

Selected References

1.   Merport M, Murphy T, Egglin T et al. Fibrin sheath stripping versus catheter exchange for the treatment of failed tunneled hemodialysis catheters: randomized clinical trial. JVIR 2000; 11:1115-1120.
2.   NKF-DOQI Clinical Practice Guidelines for Vascular Access. New York: National Kidney Foundation, 1999.

Table 5.6. Protocol for clearance of thrombosed catheters

Clotted/Thrombosed Catheters

1. Attempt vigorous saline flush/aspiration.
2. If unsuccessful, try repeated, small, frequent t-PA catheter injections (both ports) or infusions depending on local protocols.
3. Radiologic stripping of catheter (femoral vein access).
4. Exchange catheter over guidewire, using an incision at neck site exploring the catheter; always use a new tract and exit site. Inside tract may require balloon angioplasty. Insert an introducer sheath if the IJ or IJ/SVC junction is stenotic.
5. Techniques used require sound judgment and should vary depending on local availability of equipment, technical expertise and patient condition.
6. Patients with a history of fever need preop antibiotics (i.e., vancomycin and gentamicin). If no evidence of tract infection, i.e., tenderness or drainage, it is assumed to be intraluminal catheter infection, and new catheter is placed. Using the same cutaneous tract or exit site is strongly discouraged.

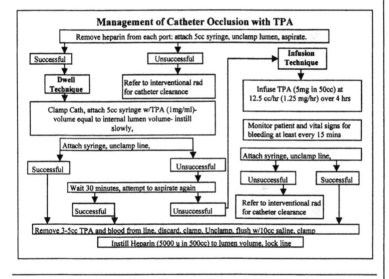

Management of Catheter Occlusion with TPA

Remove heparin from each port: attach 5cc syringe, unclamp lumen, aspirate.

Successful → Dwell Technique

Unsuccessful → Refer to interventional rad for catheter clearance

Infusion Technique → Infuse TPA (5mg in 50cc) at 12.5 cc/hr (1.25 mg/hr) over 4 hrs

Dwell Technique: Clamp Cath, attach 5cc syringe w/TPA (1mg/ml)- volume equal to internal lumen volume- instill slowly,

Attach syringe, unclamp line,

Successful / Unsuccessful

Wait 30 minutes, attempt to aspirate again

Successful / Unsuccessful

Monitor patient and vital signs for bleeding at least every 15 mins

Attach syringe, unclamp line,

Unsuccessful / Successful

Refer to interventional rad for catheter clearance

Remove 3-5cc TPA and blood from line, discard, clamp. Unclamp, flush w/10cc saline, clamp

Instill Heparin (5000 u in 500cc) to lumen volume, lock line

Chapter 6

Abdominal Catheters for Peritoneal Dialysis: A Detailed Surgical Procedure

Ingemar J.A. Davidson; Illustrations: Stephen T. Brown

Background

The percentage of end stage renal disease (ESRD) patients using peritoneal dialysis (PD) varies markedly between different countries and societies. In the US, PD patients constitute 16% while in some European countries this number is between 40-50%. Patients typically suited for PD are young professionals who have regular working hours, travel needs and require a high degree of freedom. PD can be done intermittently (IPD), continuously with several exchanges throughout the day (CAPD), or continuously during the night with a cycling machine while the patient is asleep (CCPD). The dialysis hypertonic (dextrose) solution prescribed may vary from 1.5%-4.25% to regulate body water content.

Preoperative Considerations

A number of factors are associated with the likelihood of success with peritoneal dialysis (Table 1). Patients with diabetes, impaired vision, obesity, poor compliance or understanding of technical details and lack of dexterity are less likely to benefit or do well.

The decision as to what access to place in a specific patient requires a well-informed patient and effective communications between treating team members, that is nephrology, surgery and dialysis personnel.

Detailed Operative Procedure

Marking of The Skin

With the patient standing prior to surgery, the incision, as well as the intended subcutaneous tract and exit site of the catheter are marked with an indelible pen (Fig. 6.1). This is especially important in case of abundant abdominal fat that changes position at standing (60% of the US population is overweight). An exit site away from the beltline is preferred. Possible future transplant incision sites may be taken into consideration. Therefore, in a virgin abdomen, the left side is chosen because the right iliac fossa is the preferred site for a first kidney transplant.

Choosing the Catheter

The different catheters, type, lengths, number of cuffs and manufacturers are detailed in Table 6.2. It is the author's strong opinion that only curled (pigtail) type catheters should be used. Therefore, the straight catheters have been excluded from the table. The curled catheter comes in left- and right-sided configurations. The only difference between the two is the orientation of the radiopaque line on the

Access for Dialysis: Surgical and Radiologic Procedures, 2nd ed.,
edited by Ingemar J.A. Davidson. ©2002 Landes Bioscience.

Table 6.1. Factors influencing the dialysis modality decision

Patient desire, including lifestyle, profession, age, body habitus
Patient restriction to learn technique, i.e., impaired vision, advanced age, lack of
    manual dexterity
Distance to hemodialysis unit
Socioeconomic factors
PD training facility availability
Previous abdominal surgery with adhesions
Recent or recurrent abdominal infections, i.e., diverticulitis
Overweight, hygiene issues
Heparin intolerance
Inability to establish vascular access

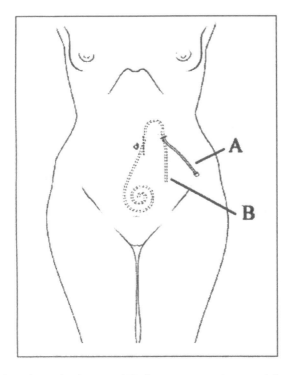

Fig. 6.1. The author prefers the paraumbilical transrectus muscle approach (level guided by patient size and catheter length). The catheter my be exteriorized at the time of placement (A) or buried subcutaneously (Moncrief method), to be externalized 4-5 weeks later (B).

Table 6.2. Commercially available curled peritoneal dialysis catheters

Company	Product	Model	Cuffs	Length Overall	Length Internal	Comment
Vas Cath		Pigtail	2	56	30	
Vas Cath		Pigtail	2	60	34	
Vas Cath	Acute	Pigtail	1	56	30	
Vas Cath	Acute	Pigtail	1	60	34	
HMP	Neostar		1	39.25, 41, 56.5, 60		
HMP	Neostar		2	56.5, 52		
Quinton	Curl Cath		1	57, 62		
Quinton	Curl Cath		2	39, 57, 60		cuffs placed at time of implant
Quinton	Curl Cath		Loose	62	variable	L/R radiopaque strip
Quinton	Curl Cath	Swan Neck	2	62.5	41	Swan Neck, L/R radiopaque, lean-average
Quinton	Curl Cath	Missouri	2	62	42.3	Swan Neck, L/R radiopaque,
Quinton	Curl Cath	Pre-Peritoneal	2	112.8		Swan Neck, L/R radiopaque
Moncrief-Popovich	Curl Cath		2	62.5	42.5	

It is the author's opinion that only "pigtail" catheters should be used for patency reasons. The swan-neck catheters come in right and left interchangeable configurations, referring to the side with the radiopaque strip.

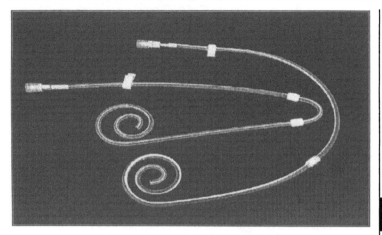

Fig. 6.2A. There are several types of Tenckhoff's catheters available, with one or two dacron cuffs. The innermost cuff is placed between the peritoneum and the posterior fascia; the outermost cuff is placed subcutaneously 1 cm from the sign. In cases of one cuff, this goes at the peritoneum site. Using a swan neck type curved exit portion, the Moncrief design calls for a 6 week period of curing the entire catheter and then externalizing the part distal to the second outermost cuff. Since the introduction of the pigtail catheter, intraabdominal obstruction by omentum has been largely eliminated. For complete listing of catheters, see Table 6.2.

catheter. This line provides easy x-ray visualization of the catheter course and may be useful in identifying possible kinks in the catheter. The direction of the curl will be left or right in the pelvis depending on which side the catheter is placed, a matter of no consequence. The number of cuffs is a matter of surgeon's choice. Properly positioned, the second cuff is placed 1-2 cm away from the skin. The author recommends the two cuff catheter. The sliding of the catheter at the exit site, as is the case with only one cuff, is likely to induce catheter tract infection. The proper position of the curled portion in the pelvic space is obtained by appropriately choosing catheter length and the incision site (Fig. 6.1). Figure 6.2A shows two types of catheters contrasting one versus two cuffs and swan neck preformed versus regular. The authors prefer the two cuffed, swan neck type catheter (Fig. 6.2B).

Anesthesia

As in all cases of hemodialysis access, patients are uremic, often with significant comorbidity, most commonly heart disease, hypertension, anemia and diabetes. Dosing of anesthesia drugs must be decreased accordingly and given with great caution to avoid serious complications such as intraoperative circulatory arrest, hypotension, jeopardize the procedure itself or inducing other complications such as strokes, cardiac events or requirements for intubation and prolonged postoperative ventilator treatment. PD catheter placement is preferably placed under local anesthesia, i.e., epidural with careful sedation; young otherwise healthy individuals may be safely given general anesthesia. HIV positive individuals, in the author's view, are candidates for epidurals or general anesthesia to protect the OR personnel.

6

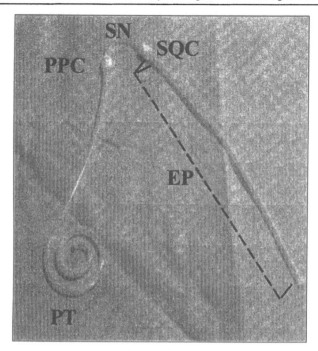

Fig. 6.2B. The swan neck (proximal), pigtail (distal), 62.5 cm double cuffed Quinton peritoneal dialysis catheter. The parts of the catheter are marked as follows: PT: Pigtail; PPC: Preperitoneal Cuff; SN: Swan Neck; SQC: Subcutaneous Cuff; EP: Externalized Portion. (Sherwood Davis & Geck, St Louis, MO 63103.(800) 962-9888).

Antibiotics

Cefazolin (Ancef) 1 gram IVPB or vancomycin 1 gram IVPB is given or started before skin incision. Both the skin incision and exit site have been marked with the patient standing. The abdomen is shaved as needed and prepped and draped as per routine. 1% lidocaine with sodium bicarbonate added is used (2 ml of 8.4% NaHCO3 in 20 ml 1% lidocaine) will cause less burning during injection.

Paraumbilical Approach

This is the author's preferred choice and therefore described in detail. In this case, skin incision is placed approximately 2 cm on either side of the umbilicus (Fig. 6.1). The abdominal wall anatomy in relation to catheter placement is depicted in Fig. 6.3A. Dissection is carried down sharply and/or with electrocautery, maintaining exact hemostasis. The anterior rectus muscle fascia is divided longitudinally. The rectus muscle layers are split bluntly. Intrarectus muscle vessels are electrocauterized or clipped. The posterior fascia is exposed. Exposure is improved by using a Weitlander retractor (Fig. 6.3A,B).

Using fine scissors, the peritoneum is opened for 3 mm (Fig. 6.3A-B). A 2-0 Prolene® purse string suture, including the posterior rectus fossa and the peritoneum, is placed in a manner that the tie will be located behind and cephalad of the cuff.

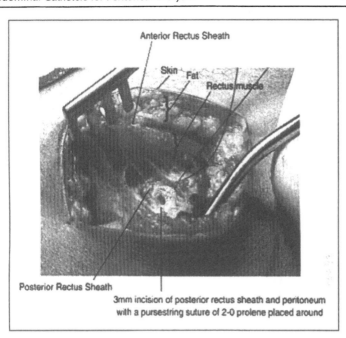

Fig. 6.3A. Incision, dissection of abdominal wall anatomy with posterior rectus sheath exposed and exposure of wound for PD catheter placement. The rectus muscle is split bluntly. Hemostasis obtained with electrocautery. A 3 mm hole is made through the posterior rectus fascia and the peritoneum. A pursestring suture (i.e., 2-0 Prolene®) is placed. The catheter is ready for insertion.

The catheter is placed using the stiff guidewire (Fig. 6.4A) inserted into the catheter to make the curled position straight (Fig. 6.4B-C) facilitating insertion through the small peritoneal opening (Figs. 6.3 and 6.5). Also, the semi-sharp tip of the stiff guidewire must not pass outside the straightened pigtail to avoid visceral injuries when inserted (Fig. 6.4D). Once inside the abdomen, the catheter with the guide slides down along the anterior abdominal wall. No force must be used. Should resistance occur, redirection of the catheter is attempted until the pelvic area is reached. At this point the stiff guidewire is slowly removed with one hand holding the PD catheter in place at the exit site. Sliding of the guide is greatly facilitated by flushing the PD catheter with saline prior to its insertion. Once fully inserted the pursestring suture is tied snugly around the catheter (Fig. 6.5). The stitch is tied behind in the cephalad direction and attached to the posterior aspect of the cuff. A second pursestring suture is not needed. The catheter is held in the cephalad direction to maintain the pelvic position of the curled portion. Next the catheter is pulled through the anterior rectus fascia (Fig. 6.6), using a sharp stylet tunneling device (Fig. 6.7A-B). A mosquito hemostat may be necessary to widen the anterior rectus fascia (Fig. 6.6) to allow passage of the second dacron cuff to be placed at the skin exit site. This step will definitely secure the vertical direction of the catheter making malposition of the catheter unlikely.

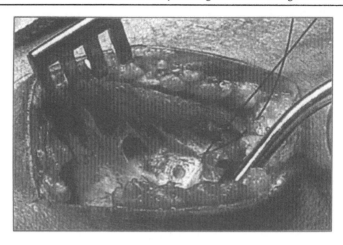

Fig. 6.3B. A pursestring, i.e., 2.0 Prolene® suture (including the posterior rectus fascia and the peritoneum) is placed around the small abdominal opening.

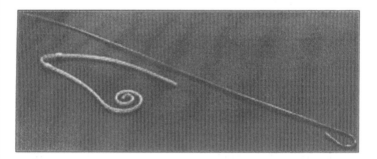

Fig. 6.4A. The stiff guidewire is used to insert the pig tailed PD catheter into the pelvic location.

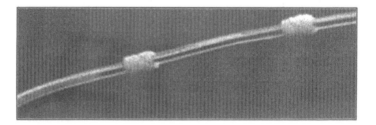

Fig. 6.4B. Swan neck curve eliminated by guidewire.

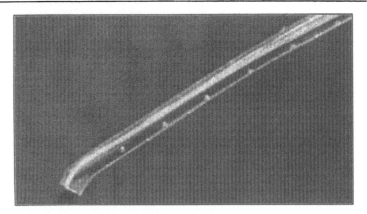

Fig. 6.4C. Ideal position of guidewire

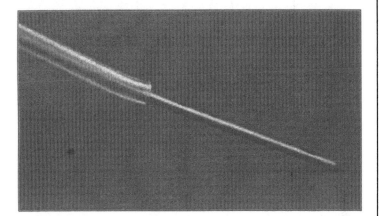

Fig. 6.4D. Too far!

The catheter is next pulled out to the exit site (Fig. 6.8A-B). This step places the catheter in a 180 degree fairly sharp curve (Fig. 6.8B). If a swan neck type catheter is used this curve is predetermined by catheter design (Fig. 6.2B). If the sharp stylet tunneler device is used, both these steps could be combined in one move using the sharp curved tunneler (Fig. 6.7B). The second dacron cuff ideally is placed 1-1.5 cm from the exit site (Fig. 6.9A,B). With the stylet type device creating a snug fit between catheter and skin at the exit site, cuff migration is unlikely. In cases of immediate exteriorization the exit site wound, skin fits snugly around the catheter (Fig. 6.8C).

In cases of the so-called Moncrief procedure the catheter is re-inserted and this larger (~1 cm) exit wound is closed with subcuticular 5-0 PDS. The Moncrief catheter placement, where the external portion of the catheter is placed subcutaneously for 4-6 weeks allowing skin ingrowth of dacron cuffs prior to externalization, is outlined in Fig. 6.9A-B.

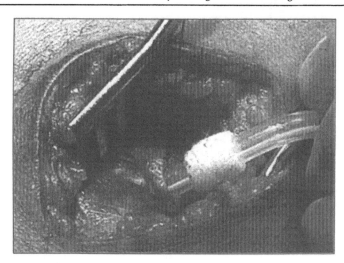

Fig. 6.5. The double cuffed Tenckhoff catheter has been inserted into the peritoneal cavity. The guidewire has been retracted and the pursestring suture is tied. The first dacron cuff is placed next to the posterior rectus fascia; the author also places a superficial stitch through the posterior aspect of the dacron cuff (with the swan neck catheter if present properly aligned, in this case, to the left). This maintains the cephalad direction of the catheter at all times, which keeps the pigtail end of the catheter in the pelvic area.

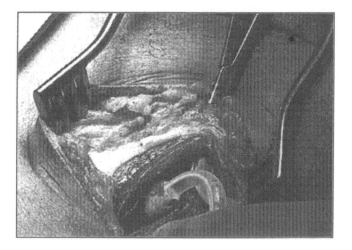

Fig. 6.6. The dacron cuff rests on the posterior rectus fascia. To ensure the downward direction of the intraperitoneal catheter the first few cm run in or behind the rectus muscle to exit through and above the proximal end of the anterior rectus fascia incision. This step is accomplished with a mosquito hemostat or using the sharp tunneller (Fig. 6.7A-B). The second dacron cuff (not shown) is placed approximately 1 cm from the skin exit site.

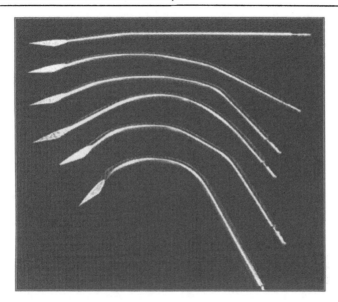

Fig. 6.7A. These custom made curved subcutaneous tunnelers (from JP drains) are used to pull the PD catheter in the desired direction to the skin exit site. It creates a tight fit between catheter and skin at the exit site. Since the catheter is pulled from inside out, tract contamination is eliminated. The catheter fits snugly at the blunt end.

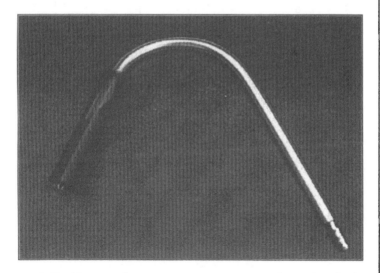

Fig. 6.7B. The OR personnel must exercise great caution to avoid injury from the sharp end.

6

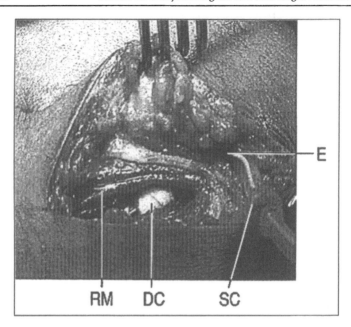

Fig. 6.8A. The dacron cuff (DC) rests behind the rectus muscle (RM). The catheter exits through the anterior rectus sheath at the beginning of the swan neck curve (E). The second cuff (SC) will be placed approximately 1 cm from the skin incision.

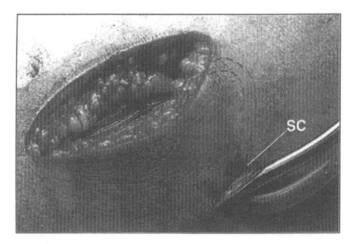

Fig. 6.8B. In cases of the so called Moncrief method, the entire Tenckhoff catheter is placed (buried) in the subcutaneous space. After 4-5 weeks the catheter is externalized at the site 1-2 cm distal to the second dacron cuff (SC). The placement of the catheter is facilitated using different curved, sharp subcutaneous tunnelers, shown in the figure on 6.7A-B.

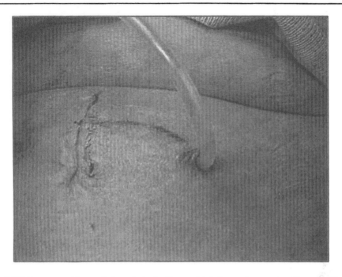

Fig. 6.8C. In cases of immediate exteriorization, which is most common, the exit site is snug around the catheter.

Closing the Wound

The muscle layer may be approximated loosely with running 2-0 PDS. The anterior fascia is also closed in a running fashion with a 2-0 PDS or Prolene® suture. Depending on the amount of subcutaneous tissue, one or two layers are closed with running 4-0 PDS or vicryl on an SH needle. Skin is closed with 4-0 PDS or vicryl subcuticular and covered with 1/2" steri-strips, gauze and Tegaderm®.

Catheter Connecting Devices

Depending on the type of peritoneal dialysis system used by the dialysis center, different adapters will be applied to the catheter. The details of the two infusion sets available (Baxter and Fresenius) are described in Chapter 9. In the operating room unless otherwise agreed upon, the surgeon will place the plastic end cap supplied in the peritoneal catheter kit. When the patient begins peritoneal dialysis training, they will be fitted with a transfer set, connecting the peritoneal dialysis catheter to the dialysis fill and drain system. The transfer set is connected to the indwelling catheter by means of an adapter. The adapter can be plastic or metal (titanium) but should be of a two piece construction to completely seal and cover the end of the catheter. Centers utilizing the Baxter dialysis systems generally require the placement of the titanium adapter, covered by a povidone iodine sealing cap, at the time of catheter placement. Centers using the Fresenius dialysis system may require placement of the Fresenius plastic adapter or may convert the plastic end cap supplied with the peritoneal catheter (Chapter 9).

Removal of Tenckhoff Catheter

A Tenckhoff catheter will require removal after a successful kidney transplant, switch to hemodialysis, infected catheter, recurrent peritonitis or malfunction.

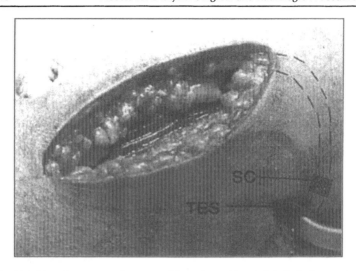

Fig. 6.9A. The subcutaneous placement of the catheter (Moncrief method) also requires a longer temporary exit site incision (TES).

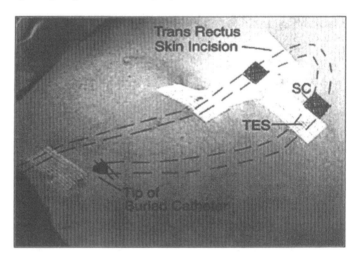

Fig. 6.9B. Outline of PD catheter tract in the Moncrief method. Four to five weeks after placement the distal catheter will be exteriorized through a small incision at TES.

Generally, removal of a PD catheter is performed under local anesthesia. The old incision is reentered to find the catheter. The catheter is clamped with a hemostat and divided distal to the clamp. If there is only one cuff at the peritoneum level, the catheter at skin level can be pulled out. The catheter towards the abdomen has to be sharply dissected to the cuff. As the cuff is freed the peritoneal cavity will be reentered, often evident by peritoneal fluid showing. The author at this point places a

2-0 Prolene® suture across this opening to secure the lumen as the rest of the catheter is excised and removed. Using the previously placed suture, the defect can now be closed with a figure-of-eight. When no infection is present, the subcutaneous tissue and skin is closed with running suture, i.e., 4-0 PDS on an SH needle.

Management of Specific Problems

Catheter related complications (leaks, peritonitis, obstruction, hernias) may occur in 70% of cases.

Leaks

Leaks can occur at any time at the catheter exit site or through the wound. An early leak represents poor surgical technique, i.e., loosened knot or inexact suture placement. Exact watertight closures are not associated with leaks and will allow the catheter to be used early. An early small leak is managed by not using the catheter for 2-3 weeks. In fact, most PD dialysis centers wait 3-4 weeks before starting fluid exchanges, allowing cuff ingrowth and healing to take place. Large leaks weeks after placement represent a major technical defect and are unlikely to close by waiting. Repairng a defect may be difficult and also tends to induce other problems such as infection. Replacing the catheter at a new site may be the most appropriate option in some instances. Removing the catheter to allow healing and recovery from subclinical infection (i.e., *S. epidermidis*) is always an option. The surgical decision involves much clinical judgment, factoring in patient's overall medical risk, time after catheter placement, other dialysis options available and surgeon's experience.

Obstruction

Obstruction is usually caused by dislodgment and omentum migrating into the catheter; this complication is quite uncommon with the current curled catheters, properly placed in the pelvic region (Fig. 6.10). Temporary relief is sometimes obtained by forcefully injecting saline, blowing away omental tissue stuck in the catheter. Permanent corrective intervention requires a mini-laparotomy or laparoscopy to remove omental tissue off the catheter and place a stitch to redirect the catheter to the pelvic area. Partial omentectomy will permanently prevent recurrent obstruction.

A dislodged catheter needs to be returned to the pelvic area, either by a small, below umbilicus midline incision or through a laparoscopic procedure. Dislocation is usually detected by concomitant obstruction, with the catheter flipping up in the right or left flank catching omentum. During the initial catheter placement, directing the first portion of the catheter cephalad as it comes out from the abdomen prevents later dislodgment. Occasionally, a stiff guidewire can be used to redirect the catheter to the pelvis area under fluoroscopic guidance. However, omentum often firmly stuck to the catheter prevents this maneuver (Fig. 6.11). Laparoscopic or open reposition, including removal of omentum from the catheter and placing a stitch around the catheter to the anterior abdominal wall is the definitive procedure.

Catheter tract infection with abscess formation has traditionally been an indication for removal. In the absence of peritonitis the authors have successfully salvaged such catheters by a two-step procedure. First the abscess is widely drained, the wound packed with wet antibiotic gauze; the skin level dacron cuff is shaved off the catheter. When the wound is clean and granulating, the catheter is rerouted to a new exit site and the clean cavity allowed to granulate from the bottom.

6

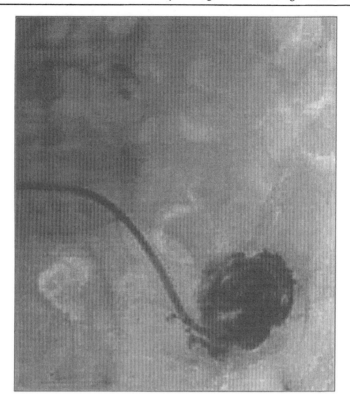

Fig. 6.10. PD catheter obstructed by overlying omentum, with contrast injected in the catheter.

Peritonitis is caused by a break in sterility, and is characterized by pain, fever and cloudy or bloody drainage fluid. After bacterial and fungal cultures are taken, treatment with IV antibiotics is begun. The author also prefers to continue exchanges with antibiotics added to the exchange fluid. Peritonitis resistance to treatment (for > 72 h) or recurrent episodes of peritonitis, i.e., relapses after discontinuation of antibiotics are indications for catheter removal. Fungal peritonitis is very unlikely to be permanently eradicated.

Abdominal Hernia

Because of increased intraabdominal pressure, a previously asymptomatic abdominal hernia often manifests itself. The most common hernias are inguinal and umbilical, with various forms of clinical manifestations. Surgical repair is indicated. After surgical intervention, dialysis is delivered through an alternative method, usually with a temporary cuffed tunneled central vein catheters. After 4-6 weeks, peritoneal dialysis is resumed.

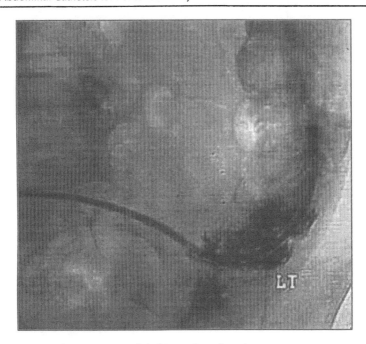

Fig. 6.11. Guidewire attempt at dislodging catheter from the omentum.

Selected References

1. Moncrief JW, Popovich RP, Broadrick LJ et al. The Moncrief-Popovich catheter. A new peritoneal access technique for patients on peritoneal dialysis. ASAIO J 39:62-5.

CHAPTER 7

Image-Guided Endovascular Evaluation and Intervention for Hemodialysis Vascular Access

W. Perry Arnold

Introduction

This chapter is meant to provide a brief overview of hemodialysis vascular access (HVA) interventions using the NKF Dialysis Outcomes Quality Initiative (DOQI) guidelines as a working matrix for the presentation of a problem-oriented approach to "best care." It is the result of influential literature, discussion with others, trial, error, successes, failures, observations, meticulous data collection and analysis, patient follow-up, and complication analysis from my practice as an interventional radiology physician engaged in HVA over the past 12 years. The last 4 years have been dedicated exclusively to HVA in a freestanding access center in the Baltimore, Maryland metropolitan area. My experience in the field consists of more than 7,000 HVA procedures ranging from simple diagnostic angiography to promotion of maturation of undeveloping AVF by serial dilatation procedures. Included in this experience are 3,000 access thrombosis recanalizations, 2,000 angio/percutaneous transluminal angioplasties (PTA) of dysfunctional but not thrombosed accesses, and 1,600 tunneled cuffed catheter accesses. I have performed over 1,000 procedures on native arterio-venous fistulae (AVFs). Success for thrombosed accesses is 95% for grafts and 88% for AVFs, both with a significant complication rate of less than 2%. Success with both AVFs and grafts that underwent angio/PTA for dysfunction was 99% with less than 1% significant complications. Tunneled catheter procedures were successful in all but four cases with fewer than 1% complications. Thus the information shared in this chapter is presented under the influence of guidelines tempered by the experience of a single physician practicing dedicated hemodialysis vascular access care as part of team of health care professionals.

Background

Vascular access for hemodialysis is a critical problem for all patients, dialysis centers and physicians providing care for patients with end stage renal disease (ESRD). Missed dialysis is not only responsible for much of the morbidity encountered in multisystem failure,[1] but it is also implicated in accelerated mortality.[2] The average life of synthetic bridge grafts (SBGs), usually made of polytetrafluoroethylene (PTFE) and which comprise approximately 80% of all accesses in the United States is about 20 months per site.[3] Thus for most ESRD patients, the six upper extremity access sites (one forearm and two sites above the elbow in each arm) will most likely be exhausted within 10 years. For the 58% of ESRD patients between the ages of 40 and 70 preservation of access function and access site usefulness is or will become a critical issue.

Access for Dialysis: Surgical and Radiologic Procedures, 2nd ed., edited by Ingemar J.A. Davidson. ©2002 Landes Bioscience.

Furthermore, the incidence of ESRD has been steadily rising. Over the period from 1985 to present, the number of patients with ESRD increased from 84,797[4] to more than 300,000, the majority of which are maintained on hemodialysis.[5] Nationally the ESRD population continues to grow by approximately 7-12% per year. Current incidence of heart disease, diabetes, and hypertension indicates that the incidence of ESRD will become even greater. The financial implications to the health care system are staggering, having already reached the $1 billion level for vascular access management alone. The development of image-guided percutaneous interventional techniques has afforded an alternative to surgical thrombectomy and revision in prolonging access function.

Over the past decade, image-guided endovascular diagnostic and interventional procedures have evolved as major techniques for maintenance of HVA in spite of some controversy and "turf battles".[6] These endovascular interventions, of which PTA is the foundation, have slowly replaced surgical intervention as the "gold standard" for maintaining and restoring patency to dysfunctional and failed synthetic and native accesses because of both outcomes success and preservation of "venous real estate." The progression of technology and experience through thrombolytic enzyme techniques, maceration techniques, mechanical device development, and thrombo-aspiration techniques has resulted in improved care for this group of patients. As the technology for care has developed, the challenge of designing the "ideal access" has continuously faced the surgical community. During the last two decades in the United States, native AVFs, the common access in use throughout most of the rest of the world, were largely replaced by synthetic bridge grafts made of PTFE (polytetrafloroethylene). The advantages and disadvantages of each have been the subject of much discussion, but the general international consensus is that the AVF is the "gold standard" once maturity has developed.[7] Tunneled large-bore, double-lumen catheters have also evolved as acceptable accesses for both short and long term use. None of these accesses are ideal, however, so the search for an access that works better, is cosmetically acceptable, and lasts longer with less pain and dysfunction continues.

In the meantime, we must deal with the accesses available in preserving the best possible quality of life for those ESRD patients maintained by hemodialysis for their renal replacement therapy permanently or while awaiting renal transplantation. The first step in establishing standard guidelines for the care of these patients was the monumental task undertaken by the National Kidney Foundation, through interdisciplinary teams reviewing available literature, which resulted in the publication of the NKF Dialysis Outcomes Quality Initiative (DOQI) guidelines in 1997. These guidelines currently referred to as K-DOQI have since been updated and are available to all via the Internet at the NKF web site (www.kidney.org/professionals/doqi/guidelines/doqiupva_intro.html). It is imperative that everyone involved in the care of hemodialysis patients and use or preservation of vascular access function is intimately familiar with these guidelines as they apply to hemodialysis vascular access.

The DOQI Guidelines Applied to Image-Guided Endovascular Procedures

In the first section of the guidelines, patient evaluation prior to access placement is addressed, including guidelines specific to surgical evaluation and placement of new accesses. The emphasis is on placement of a native AVF and the pre-procedural evaluation to insure adequate arteries and veins. This may include venography, arteriography, MRI evaluation, ultrasonic mapping, and always a detailed physical examination. If the native vasculature is deemed inadequate to support an AVF, a synthetic bridge graft (SBG) is the next choice for a permanent access. If required, tunneled catheters should be inserted any time there is a projection of their need for more than 3 weeks. Tunneled catheters may be utilized while a permanent access matures. Tunneled catheters should be placed under direct ultrasonic localization and fluoroscopic guidance in order to decrease the potential for significant complications. Complications have been reported including:

- puncture of the carotid or subclavian arteries (with hematoma formation or hemothorax),
- rupture of major venous channels,
- malposition of the tip of the catheter into an inadequate vessel such as the azygous vein or left innominate vein, and
- inadequate positioning of the ports so that dialysis is unsuccessful because of apposition of the catheter tip to the wall of the superior vena cava or lateral wall of the right atrium.

The position of any hemodialysis catheter placement in descending order of preference should be: right internal jugular vein, right external jugular vein, left internal jugular vein, left external jugular vein, and as a last resort in the upper extremities, the subclavian vein. Subclavian dialysis catheterization has a very high incidence of complications by stenosis and/or occlusion and should be avoided until the extremity involved has been totally abandoned and no further access in that extremity is contemplated. In the case of a native AVF, maturation may require up to 4 months. Sites of creation are preferred from distal to proximal: radio-cephalic at the wrist, brachio-cephalic above the elbow, brachio-basilic transposition above the elbow, and femoro-saphenous in the thigh. If slow maturation is noted, endovascular interventional procedures may assist with improved flow and maturation. Graft healing is usually stable by 4 weeks; however, our recent studies have shown a high incidence of thrombosis in grafts used for dialysis in less than 4 weeks. Early thrombosis (less than 3 months) of an SBG places the patient at risk to become a "frequent failure" patient. Communication between the surgeon, interventional radiologist and nephrologist is crucial in coordinating the onset of dialysis utilizing an access that best benefits the patient.

The second section of the DOQI guidelines deals with the routine monitoring, surveillance, and maintenance of an access, and is primarily directed to those who see the patient on a regular basis under the guidance of the patient's nephrologist. Monitoring of the access by physical examination at each dialysis visit or at least a minimum of a weekly exam is performed as a part of general dialysis care. Surveillance of an access for dysfunction associated with stenosis requires personnel- and time-intensive evaluation and thus should be performed monthly or quarterly. It can be as simple as evaluation of static and dynamic venous pressures (monthly), or by more sophisticated but effective intra-access flow by direct Doppler ultrasound or dilution methods such as the Transonic HDO1 method (quarterly). The possibility and amount of compensation for these surveillance methods is still under consideration by payers. Other surveillance studies involve dilution techniques and urea concentrations to evaluate dialysis efficiency by urea reduction ratio (URR) or overall dialysis kinetics (KT/V). These findings usually are less specific and accurate than either pressure or direct intravascular flow measurements. All surveillance techniques tend to be less accurate and applicable in native AVFs than in SBGs.

Physical findings, the basis of access monitoring, are of great importance and should be employed every time the access is cannulated for dialysis. Physical examination should consist of evaluation by observation, palpation, and auscultation. Observation of arm swelling and superficial venous prominence probably indicates central venous stenosis or occlusion. Heat, redness, and pain may well indicate infection, the absolute contraindication to endovascular instrumentation. Prolonged bleeding after needle withdrawal following dialysis probably indicates a significant increase in intra-access pressure and the presence of a significant venous stenosis that could result in thrombosis. Palpation for the presence of a continuous thrill, most pronounced at the arterial or inflow segment of the access, is imperative. If a thrill is not present the flow is either inadequate or is laminar in nature. A systolic thrill indicates increased intra-access pressure. Complete loss of the thrill in an access with only strong pulsations present indicates outflow stenosis. In our practice, a sign that has proven to be 80% reliable in predicting the need for a PTA is the firmness of the access and quality of the thrill when the access is evaluated with and without venous outflow compression/occlusion. A minimal or imperceptible change in pressure with compression indicates an anatomically and hemodynamically significant stenosis. Venous outflow compression is a very reliable sign of stenosis and the need for a venous PTA. Auscultation of an access is used to determine the pitch and continuity of a bruit. A normal access is associated with a continuous low-pitched bruit with superimposed systolic accentuation. Anytime the background sound disappears and the access becomes silent even for a fraction of a second, there is cessation of flow indicating a significant stenosis causing a "systolic bruit." A systolic bruit can almost always be correlated with both a significant venous stenosis and with a lack of pressure changes to venous outflow compression. If the pitch of the bruit changes along the course of the access, internal irregularity is present, necessitating angiographic attention during intervention.

The physical basis for these findings involves unimpeded flow through a tubular conduit. As stenoses become significant, pressure increases within the graft and venous outflow compression pressure changes may be imperceptible to palpation. A normally functioning access will have wide pressure changes when comparing the firmness of the access with and without compression of the venous outflow. Primary AVFs should be monitored in the same manner as SBG accesses; however, their surveillance by venous pressures tends to be much less accurate. It is preferable in native AV fistulas to perform direct flow measurements or Doppler analysis for confirmation of adequate flow. Abnormal monitoring or surveillance changes coupled with at least one physical finding are an indication of the need for angiographic evaluation and intervention.

Routine access monitoring and surveillance can decrease the thrombosis rate by 20-40%. A comprehensive physical exam is 80% accurate for dysfunction, and can be performed in the outpatient setting of the dialysis unit. Static pressure is easy to perform during the dialysis treatment, as is dynamic flow monitoring (Transonics/Doppler Flow), which can give actual flow data. Pressure ratios (Pa:Ps) are effective for trends and as absolute values (0.40 is upper normal). The ideal situation is that access monitoring and surveillance techniques in the presence of abnormal data should provide referral for timely outpatient evaluation to an interventional service that is convenient for the patient and dialysis schedule, can improve primary access patency and access life, and is designed to decrease procedure time and cost. Prospective PTA in our experience gives primary patency of

9-10 months, and is performed away from the dialysis unit in a pleasant setting as a separate procedure.

The third and major focus of the K-DOQI guidelines is management of vascular access complications; this facet of care is the primary focus of this chapter. In determining an appropriate time for intervention of a permanent dialysis access, complications must be considered, including:

- inadequate flow to support adequate dialysis,
- thrombosis,
- infection,
- access degeneration with associated breakdown of the wall of the access or the overlying skin, and
- the development of ischemia in the distal portion of the involved extremity.

The practitioners must bear in mind that the time of intervention can only be justified (and compensated by third party payers) in the presence of a detectable abnormality with associated clinical signs and symptoms ("medically necessary"). For instance, an ultrasonically detected 50% venous outflow stenosis in the presence of normal URR, normal KT/V, normal pressure, and no cannulation or bleeding difficulties would not in itself be an indication for prospective PTA. This same lesion in the presence of elevated static venous pressure, deteriorating URR or KT/V, prolonged bleeding, or progressive enlargement of pseudoaneurysms should be appropriately treated in the most efficient and vessel-sparing manner possible in a specific locale.

It is important to become familiar with the history and development of access declotting techniques. Recommended reading specifically is Richard Gray's review article from the *JVIR*,[6] which gives some comparison figures of surgical patency with percutaneous patency and notes that surgery is held to a higher standard because of surgical consumption of "venous real estate." Other useful literature is the section on hemodialysis vascular access from recent SCVIR annual meeting workshops. The spirit of the workshops is used throughout this chapter with permission of the SCVIR; technique summaries comprise Appendix A. Both surgery and percutaneous endovascular techniques for re-canalization are considered satisfactory depending on the expertise and availability of local service. DOQI states that percutaneous declotting with PTA should achieve a 40% unassisted patency at 3 months; with surgery it is held to 50% unassisted patency at 6 months and 40% functionality at 1 year. DOQI guidelines however state that it is essential that the following six parameters be met:

- Treatment is performed in a timely fashion to prevent missed dialysis as well as minimize temporary catheter placement.
- It is essential that the access be evaluated by angiography for residual stenosis after any re-canalization procedure, whether surgical or endovascular.
- PTA procedures are expected in any stenosis that is 50% or greater in a thrombosed graft. (We find venous stenoses present more than 95% of the time in our patient population, as well as arterial inflow stenoses as measured by balloon deformity in 48% of thrombosed grafts. In addition intragraft stenoses associated with either saccular pseudoaneurysm or regions of neointimal hyperplasia secondary to numerous puncture sites require PTA. Our experience shows prolonged patency of 50-60% unassisted primary patency at 3 months and 45% unassisted patency at 6 months. This has been achieved by the combination of arterial and venous PTA, as well as

PTA of the entire length of the dialysis access graft in the process of totally eliminating thrombus from the graft without embolization into either the arterial or central venous circulation).

- Because of the low complication rate, the procedure should be performed as an outpatient procedure under local anesthesia. It is sometimes necessary to administer IV conscious sedation to achieve full venous balloon angioplasty. Equally or more effective than conscious sedation is the practice of infiltrating local anesthetic around the known firm stenosis. Patient recovery time is decreased by the latter technique.

- Meticulous records regarding prospective monitoring and surveillance of access function are necessary to promote access maintenance prior to thrombosis.

- Monitoring of the outcome results of dialysis access patency following access-related procedures should be instituted in order to evaluate success of all access procedures. The achievement of greater than 40% unassisted patency at 3 months for declotting and PTA is to be equated against 50% 6-month and 40% 1-year patency for surgery after thrombectomy and revision as the aim of minimal achievement for any access service.

Venous Stenosis: The Primary Offender in Access Dysfunction

By far the majority of dysfunctional or failed permanent accesses are primarily related to venous stenoses. The pathophysiology of these venous stenoses is based on the premise that turbulent non-laminar flow promotes vessel vibration in a thin-walled vein with different wall compliance than the feeding artery (AVF) or graft (SBG). The turbulence promotes thickening of the vein wall by intimal and smooth muscle hyperplasia, thus compromising the lumen caliber.[8] This is most commonly seen near the venous anastomosis (artery-to-vein anastomosis in AVF and graft-to-vein anastomosis in SBG), but can also be associated with sharp turns, the presence of intravascular catheters or wires, or accompanying other flow impeding anatomic defects such as hypertrophic valves, stenoses associated with previous catheter placement, trauma, and needle puncture sites with infiltration of a native vein (Fig. 7.1). This finding is seen at or within 5 cm of the venous anastomosis in the vast majority (>90%) of patients with dysfunctional or failed access grafts, and in more central draining veins or specifically in the central vasculature from the superior vena cava to the apex of the first rib in a much smaller percentage of cases usually associated with previous catheter placement. In native AVF the common site of stenosis is the juxta-anastomotic segment (first 5 cm of vein above the anastomosis) or at the "swing site" (the site where a transposed vessel returns to its normal anatomic position). Other common sites are associated with fistula infiltration sites and repeated needle puncture sites. Grafts most often show stenosis at or within 5 cm of the venous anastomosis, at the arterial anastomosis associated with the arterial plug or in draining veins at hypertrophic valves and vessel divisions. Central cephalic vein stenoses are often seen in accesses in which the cephalic vein is the primary drainage. These stenoses are usually associated with sharp angulation and thickened valves in the proximal 5 cm of the vein. They are sometimes resistant but do respond to PTA.

The Ideal Vascular Access

The "ideal" vascular access, a conduit for high volume blood flow between an artery and a vein, is used to facilitate hemodialysis. It should:

7

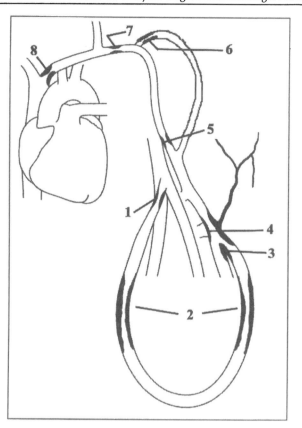

Fig. 7.1. Sites of stenosis associated with graft access circuits
1.   Arterial anastomosis
2.   Repeated puncture sites of the access
3.   Venous anastomosis
4.   Central extension of vein from anastomosis
5.   Mid brachial vein (usually associated with thick valves or venous division
6.   Central cephalic vein
7.   Subclavian vein (catheter related)
8.   Innominate vein (catheter related)
Note that a graft is illustrated, but the same principles apply to AVF.

•   be effectively placed as an outpatient procedure,
•   be cosmetically acceptable to the patient,
•   be easily and consistently cannulated with minimal patient discomfort,
•   provide consistently adequate dialysis flow,
•   be able to be reliably and routinely monitored, and
•   last for the life of the patient.

Needless to say, such an access does not exist today. The native AVF most nearly approaches the above criteria, but maturation of AVFs continues to be problematic for the dialysis care team, in some areas approaching 40% "failure to mature" in the first 4 months after creation. After creation, access complications do occur, both in native AVFs and AV grafts. AVF complications include:

- poor maturation
- stenosis
- repeated infiltration
- steal syndrome or nerve damage
- pseudoaneurysms
- thrombosis and
- rarely, infection

AV graft complications include:

- venous stenosis
- pseudoaneurysms
- thrombosis
- difficult cannulation
- infection, and
- steal syndrome or nerve damage

Any permanent access can be associated with extremity swelling secondary to central venous stenosis. In my practice I have encountered only five significant central lesions that were not associated with previous catheter access.

Collaborative Care of Vascular Access—The Team Approach

The team approach to dialysis access creation and utilization can lead to improved patency and longer access life.[9] Maintenance of dialysis access is a complex task requiring dedication and focus by all caregivers. Communication between team members is the cornerstone of the quest for "best care."

Surgeons create and revise permanent access, placing AV fistulae if possible, and PTFE grafts when necessary. Nephrologists provide patient stability through adequate dialysis prescription, monitoring, and surveillance. Interventionalists maintain patency of accesses by image-guided endovascular procedures, including prospective PTA, percutaneous thrombectomy with PTA and tunneled catheter insertion when necessary. PTA is the framework around which all endovascular dialysis access interventions are performed. (We should all be grateful to the space age technology that provided the development of the plastics now used in high-pressure balloons.) Interventionalists performing endovascular procedures on dialysis vascular accesses will often become a leading source of patient referral to their surgical colleagues. This will be associated with interposition access revisions, venous extension revision, and creation of new permanent accesses.

Since the life span of a dialysis access graft averages approximately 20 months, endovascular procedures treating dysfunctional and failed access can significantly prolong median access life patency. After 27 months of observation and follow-up, we observed the median patency of access life was extended by a minimum of 14 months when compared with those treated prior to 1997 in a hospital setting using surgical methods. Nurses play a pivotal role in vascular access care through assertive preservation of existing accesses; patient and staff education; interaction with nephrologists, radiologists and surgeons; promoting expert cannulation and self-cannulation; facilitation of access monitoring and minimizing venous abuse in "virgin"

Table 7.1. Best care algorithum for vascular access.

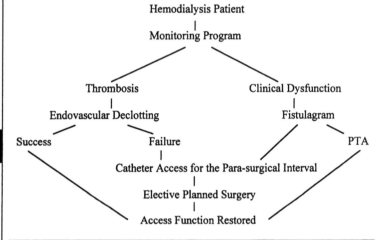

Algorithm for "Best Care" for Hemodialysis Patients

Pathways for Optimizing Vascular Access Care

Hemodialysis Patient
|
Monitoring Program

Thrombosis   Clinical Dysfunction
| |
Endovascular Declotting   Fistulagram

Success   Failure   PTA
|
Catheter Access for the Para-surgical Interval
|
Elective Planned Surgery
|
Access Function Restored

limbs. Finally dialysis technicians, social workers and dieticians work directly with the patients to achieve continuing clinical and psychological stability.

Managing Hemodialysis Vascular Access Dysfunction and Failure

The following sections provide an in-depth look at the usefulness of image-guided endovascular interventions in the management of the complications associated with the three types of hemodialysis vascular access: catheters, native AVFs, and SBGs.

Catheters

In our practice, because of the efficiency of delivering a declotting procedure to a patient with graft thrombosis, temporary catheters have almost totally been deleted from our patient population. Philosophically, if we cannot re-canalize a permanent access by endovascular means, the patient must have restoration of permanent access function by either surgical revision or creation of a new access. For that reason we use only tunneled catheters of which the preferred site of insertion is the right internal jugular vein low in the neck with a subcutaneous tunnel exiting the skin in the right infraclavicular region. Currently, tunneled, cuffed catheters exist in two basic tip designs: Split tip (Ash split-cath) and dual lumen single catheter (Hickman type) (Fig. 7.2). They are considerably larger than temporary catheters (14.5 F vs 12 F; see Fig. 7.3) to provide greater dialysis flow.

Tunneled catheters should ideally be placed while using ultrasonic guidance and fluoroscopic monitoring (DOQI guideline 5) to insure minimal complication rates

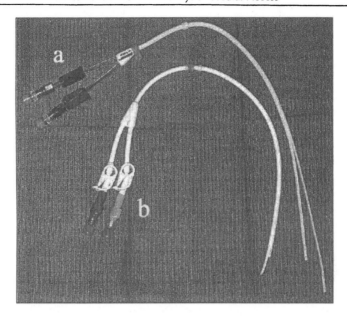

Fig. 7.2. Two types of tunneled dialysis catheters: A) Double tip or split catheter, as typified by the Ash split catheter (polyurethane) and B) the single catheter double lumen catheter shown here as a Hickman catheter (silicone rubber).

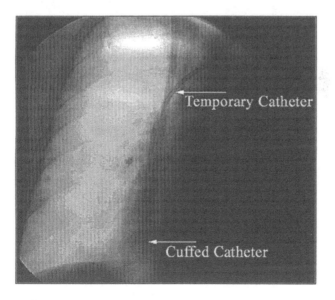

Fig. 7.3. Catheter size has a direct effect on flow. This explains the greater dialysis efficiency with a cuffed catheter 14.5F, and a temporary catheter 12F.

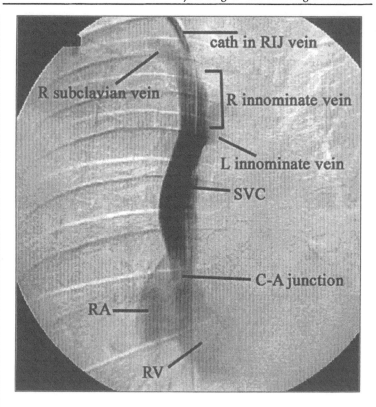

Fig. 7.4. Anatomy of the superior vena cava and right atrium and ventricle.

and optimal catheter tip length and positioning. Knowledge of central venous anatomy is imperative when placing central catheters (Fig. 7.4). I find the best function is achieved when (Fig. 7.5):

- the catheter is low in the neck in the internal or external jugular vein, with the tips extending inferior to the cavo-atrial junction in the right atrium,
- the exit site is just lateral to the mid-clavicle, and
- the tissue in-growth cuff is 1-2 cm deep to the skin.

These criteria necessitate measurement of the distance to the right atrium to correctly choose a catheter of appropriate length. Also breasts should be retracted caudally when choosing the catheter exit site to prevent displacement of the catheter by gravity. Split catheters are not position dependent, but single dual-lumen catheters should have the arterial port facing centrally toward the major pool or right atrial blood. Remember that the catheter is placed when the patient is recumbent and will foreshorten when the patient sits or stands erect.

Catheter dysfunction is often related to the catheter tip design employing multiple side holes in the catheter tip to improve flow during dialysis, but allowing thrombus to form between dialysis treatments because of hemostasis within the catheter tip (Fig. 7.6). The routine practice of explosive catheter irrigation before

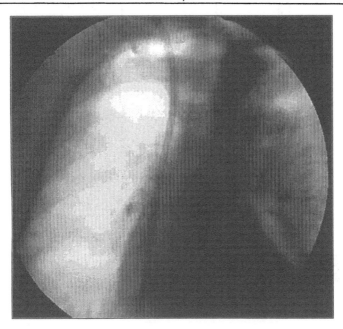

Fig. 7.5. Ideal position for a catheter. Right IJ vein access. Extends with tip inferior to cavo atrial junction.

7

connection to the dialysis machine will greatly improve long-term catheter function. Nursing manuals, however, prohibit most dialysis nurses from performing this simple preventive maneuver. Access physicians should give specific orders to have this maneuver performed. The other common cause of catheter dysfunction is the development of a fibrin "biofilm" sheath that often acts as a valve to prevent aspiration through the arterial port (Fig. 7.7). In 100 consecutive catheter exchanges with angiography to evaluate for the presence of a sheath, we found a significant sheath in 42%. The recent availability of t-PA in 2 mg doses has provided dialysis centers a means to treat access catheters with thrombolytic enzymes as part of delivery of dialysis. Whether or not payers will fund catheter thrombolysis remains to be seen.

When catheter clearing fails, we have chosen to exchange catheters demonstrating poor function. Even though thrombolytic enzymes can effectively be utilized to clear a tunneled catheter,[10,12] we have often observed only a short-term improvement usually resulting in a rapid return for additional treatment. Fibrin sheath stripping has not been specifically effective,[11,12] even though the technique itself can be relatively easily mastered. Placing a through-and-through guide wire down the catheter to be stripped and out a femoral venotomy sheath can greatly facilitate the ease of snaring and stripping the catheter by placing the snare around the guide wire. Because of the lack of efficiency and relatively high cost in both time and material, we have found it more effective to remove a non-functioning catheter, destroy the fibrin sheath, if present, by PTA,[13] and replace the catheter with one of more optimal size and position (Fig. 7.8). Our current preferred catheters are 14.5 polyurethane

7

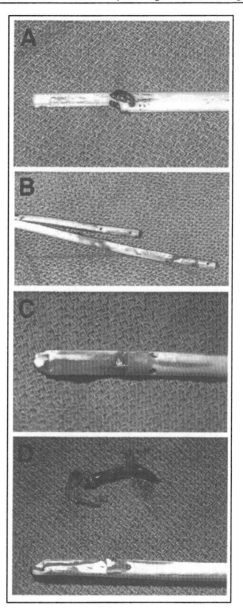

Fig. 7.6. A) Dual lumen catheter showing soft thrombus occluding arterial port from side holes to tip. B) Split tip catheter showing thrombus plug occluding from most proximal side hole to tip. There is also some thin fibrin "biofilm" on the catheter. C) Fibrin sheath causing poor flow (arterial) seen on face. D) Arterial lumen plug of thrombus and fibrin biofilm removed by "explosive irrigation".

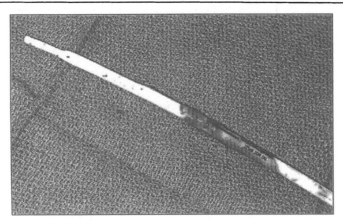

Fig. 7.7A. Thin biofilm on a split tip catheter.

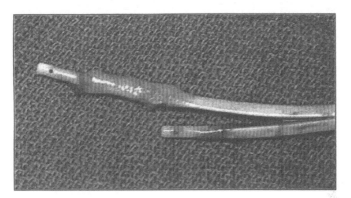

Fig. 7.7B. Thick biofilm on a split tip catheter.

catheters of the split type (example: Medcomp Ash Split-catheter) or the dual-lumen catheter (example: the Bard Optiflow) because of their high flow volumes. These catheters are relatively large bore, are inserted through a single needle puncture site, and tend to have longer patency with fewer catheter infections than silicone rubber tubes in our experience. It must be noted that silicone rubber catheters are sensitive to degradation by iodine and polyurethane catheters are sensitive to degradation by alcohol and glycol (Fig. 7.9). Thus iodine solution is the preferred antiseptic for polyurethane catheters and antibiotic ointment is preferred for silicone rubber catheters. Many other polyurethane catheters are now appearing and should also give flow in the range of 400 cc per minute.

Even though catheters now have a lumen size that a small balloon catheter or brush type catheter could be utilized for clearing the lumen of thrombus, the time and cost efficacy of such maneuvers limit their usefulness both in the dialysis unit and the interventional suite. From both a functional and cost perspective, catheter exchange has proven most effective in my practice. If there is no evidence of tract or

Fig. 7.8. A) Contrast injection shows fibrin sheath after the split tip catheter has been removed. It must be destroyed to achieve adequate function of a new catheter. B) Contrast injection shows fibrin sheath remaining after removal of a dual lumen catheter. C) 8 mm PTA of the entire sheath up to the venotomy site destroys sheath. D) Post fibrin sheath PTA angio shows normal anatomy.

7

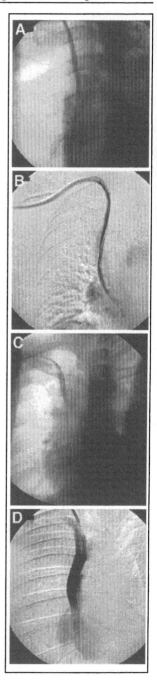

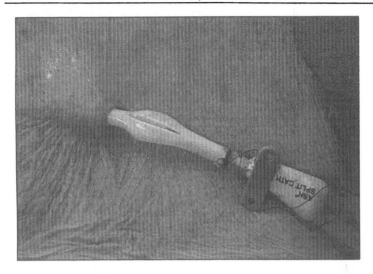

Fig. 7.9. Chronic antibiotic ointment use at the exit site of a polyurethane catheter has caused wall weakness and "catheter aneurysm".

exit site infection, exchange of a catheter over a guide wire with appropriate angiographic evaluation for a fibrin sheath is acceptable. If a sheath is visualized, I use an 8 mm PTA balloon to destroy it. Other methods such as twirling a pigtail catheter have not been particularly effective for me.

The placement site of tunneled catheters is extremely important in preventing central venous stenoses. The natural history of catheter access, however, is that central venous stenosis can well occur within one year (Fig. 7.10). Subclavian stenoses associated with dialysis access catheters of any type have been noted in my practice to occur in as little as 2 weeks (Fig. 7.11). This stenosis develops primarily due to trauma to the venous wall by the crushing action of the clavicle in apposition with the first rib during motion. Based on my experience, I strongly feel that the placement of a subclavian catheter in a dialysis patient whose extremity may eventually be used for a permanent access (or at the same time a new access is placed in the ipsilateral extremity) is a manifestation of lack of knowledge, inadequate technique or equipment, or patient neglect, and constitutes malpractice. The subclavian approach should be used only if there is no plan for further permanent access placement in that extremity and only as a last resort (Fig. 7.12).

I again stress that it is imperative to place tunneled catheters under ultrasonic localization and fluoroscopic guidance, not only to select the proper vein and avoid complications associated with the puncture, but also to avoid malposition secondary to inadequate catheter length or position of the catheter tip ports into aberrant veins or in apposition with the right atrial wall (Fig. 7.13).

I now almost always exchange dysfunctional catheters through a new tunnel because of six procedure-related infections (sepsis) that occurred after exchange over a guide wire through the same tunnel where the exit site and tunnel showed no clinical suggestion of infection. Even though this number is well below the accepted

Fig. 7.10A-C. Natural history of catheter related stenosis. A) External jugular catheter on the same side as a new forearm graft present for 3-4 weeks. B) Veins are still patent. C) Occlusion present 1 year post placement.

7

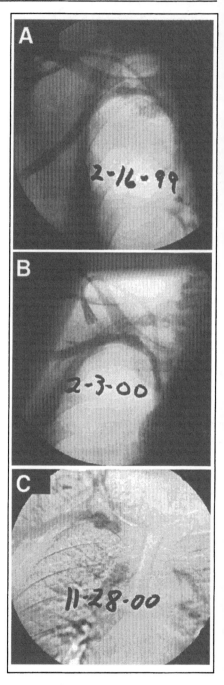

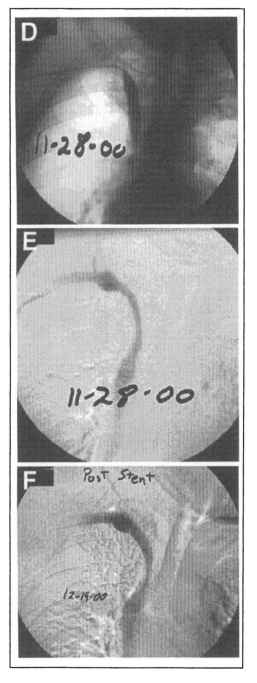

Fig. 7.10D-F. D) Central occlusion treated with 12 mm PTA. E) Angio post 12 mm PTA. F) Recurrent occlusion in 1 month required stent for prolonged patency.

Fig. 7.11A-C. A) Central venous occlusion related to 2 weeks presence
of a left subclavian catheter placed
at initial creation of a left forearm
graft. B,C) Central venous 10 mm
PTA.

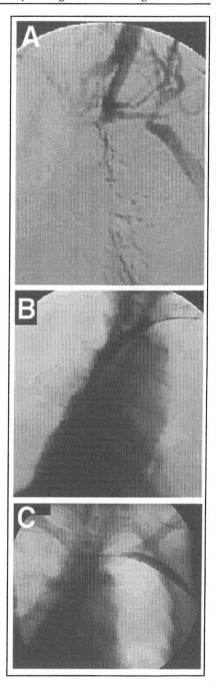

7

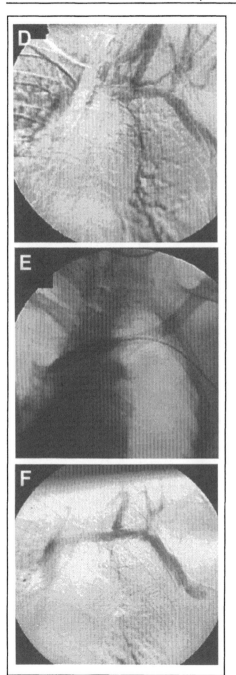

Fig. 7.11D-F. D) Persistent central occlusion after PTA. E) Stent placement and 12 mm PTA. F) Unrestricted flow post stenting.

7

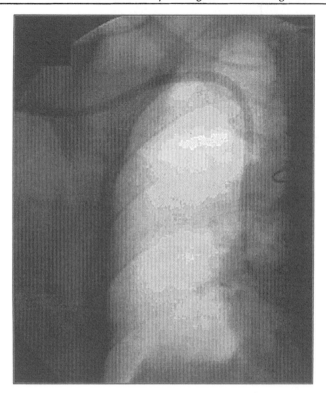

Fig. 7.12. Subclavian catheter should only be placed after extremity is abandoned. Note relationship of catheter to 1st rib and clavicle accentuating the probability of vein wall injury. Also note inadequate catheter length.

rate of 1-2%, I find it unacceptable to place patients at even this small risk for such a serious complication. The technique is simple:

- Expose the old catheter central to the cuff through a small incision and elevate and divide the catheter, being sure to control and clamp the central fragment.
- Remove the external portion after loosening the cuff by blunt dissection.
- Irrigate the old tunnel and re-apply antiseptic solution to the field.
- Place a guide wire through the central portion of the catheter into the inferior vena cava, being careful not to lose control of the catheter and also to avoid air embolism.
- Retract the catheter tip to the venotomy site and inject contrast material to search for a fibrin sheath. If present, destroy the sheath (I use an 8 mm PTA balloon), create a new subcutaneous tunnel, and place the new catheter into position over the guide wire.
- Suture in place and dress as usual. (Again refer to Fig. 7.8)

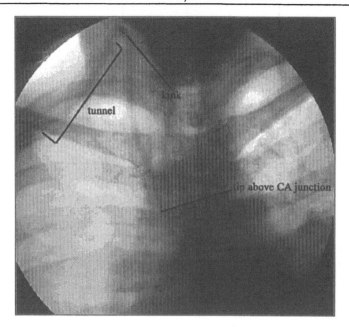

Fig. 7.13. High jugular punctures are often associated with catheter kinks, inadequate tip position above the C-A junction and overall poor function.

The treatment of infection in tunneled catheters must be coordinated and based on actual catheter involvement. If blood cultures are positive and there is evidence of systemic sepsis without evidence of either exit site or tunnel infection, dialysis access can be maintained by removing the old catheter and replacement with a new catheter through a new tunnel following the initiation of antibiotic therapy (at least two IV doses). This exchange is preferably done without the use of a guide wire during catheter removal. Re-entry can then be done using an angiographic catheter and guide wire combination when re-accessing the tract from which the old catheter was removed. This procedure is performed only if the patient has been adequately covered with antibiotics for two dialysis treatments. The new catheter is then inserted as above in the technique for exchange through a new tunnel. Incidentally, if the tissue in-growth cuff of a tunneled catheter separates during removal, it must be removed because it is the junction of the exit site with the fibrin sheath of the tunnel and probably contains significant bacterial inclusions. If left in place, it can cause an abscess with a foreign body as the nidus (Fig. 7.14).

Local exit site infections can usually be adequately treated by application of an antiseptic ointment or solution along with systemic antibiotics. Remember that polyurethane catheters are sensitive to alcohol and glycol, and silicone rubber catheters are sensitive to iodine. If the patient is unresponsive to local antibiotic therapy over a short time, creation of a new tunnel is an effective way to treat exit-site infections. Again the technique described above for exchange through a new tunnel is utilized. Tunnel infection, the equivalent of an abscess, must be aggressively dealt

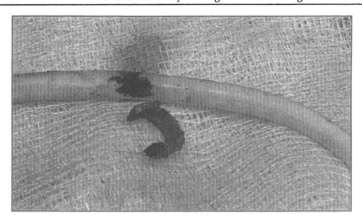

Fig. 7.14. Tissue cuff separation at catheter removal requires foreign body retrieval from the tunnel.

with through IV antibiotic therapy, catheter removal, and placement of a new catheter in a different site.

Native AV Fistulae

Although traditional treatment has been surgical revision or abandonment of an immature or thrombosed fistula to place a graft, I have used endovascular procedures to treat dialysis patients who have native arteriovenous fistulae (AVF) accesses for the past ten years. My opinion based on experience and outcomes analysis in treating immature AVFs, as well as recent literature[14-16] is that abandonment without a trial of endovascular therapy is a disservice to the patient. These procedures can promote maturation, improve flow, and achieve long-term usefulness of the access. Failure of AVFs to mature, defined as inability to support adequate dialysis at four months after creation, is commonly related to inflow stenosis at or within 5 cm of the anastomosis, the juxta-anastomotic (J-A) segment. These stenoses cause low inflow volume and thus low pressure in the fistula. Arterial flow and elevated pressure are necessary to promote maturity of the fistula; so in this setting the system does not develop. These lesions respond well to PTA that restores increased pressure and volume to the fistula to promote maturation. Long stenoses and small-sized main veins respond well to serial PTA at 2-3 week intervals (Fig. 7.15).

Sometimes side veins divert flow and pressure from the major draining vein of the AVF. Although this is usually related to a stenosis further up the fistula, some diverting side veins persist even after adequate PTA. These side branches can be occluded by ligation[14] or intravascular coil occlusion,[16] returning all blood coming into the AVF to the main vein and resulting in ultimate development into a useful access (Figs. 7.16 and 7.17). In my practice if a fistula does not show signs of progressive maturation by 2 months, it is studied angiographically to look for lesions that can be treated to promote maturity. Some developing fistulae are of inadequate size for use, but respond to serial dilatation by increasingly larger balloons at 2- to 3-week intervals until adequate size is achieved.[16,18] I consider minimal adequate size to be 6 mm, and ideal size for dialysis 7-8 mm.

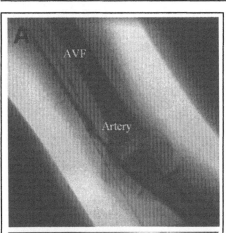

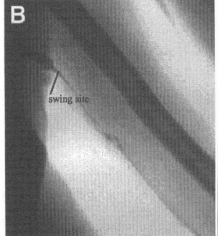

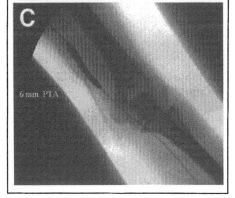

Fig. 7.15A-C. PTA for immature brachio-basilic transposition AVF. A,B) Angio shows 6 mm brachial artery with a 17 cm long segment 2 mm caliber firm scarred vein. C) 6 mm PTA of entire long segment of small caliber vein.

7

Fig. 7.15D-F. D) Central veins are normal above the "swing site". E) Angio post 6 mm PTA at "swing site" shows persistent stenosis. F) Swing site dilated to 8 mm.

7

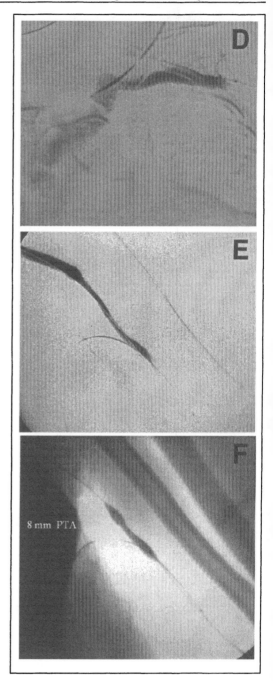

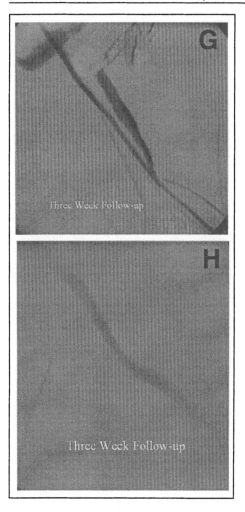

Fig. 7.15G,H. Three week follow up shows a normal AVF-now in use for 6 months.

Problems occur even after maturity has progressed and cannulation begun. If the vein has been harvested from a deeper location and superficialized after ligation of numerous side branches, multiple stenoses may be encountered. Common sites for stenoses in this setting are at the juxta-anastomotic segment, the "swing site"[17] where the vein has been mobilized from its normal position, and at the site of ligation of venous side branches (Fig. 7.18). Even though the etiology of these stenoses is unsure, it may well be associated with adventitial stripping of the vessel during preparation for the arteriovenous anastomosis or granulomata associated with surgical ligatures and electrocautery. The stenoses might even be accentuated by the placement of vascular clamps, however gentle, on the vein at the time of the anastomosis. Infiltration sites and hypertrophic valves will also result in native vein stenoses. These are usually within the usable portion of the fistula or the draining veins at a longer

Fig. 7.16A-C. PTA and side
vein occlusion to promote
maturation. A,B) Angio of immature AVF—note beaded
juxta-anastomotic (J-A) segment, tiny primary vein and
large diverting side vein. C)
6 mm PTA of J-A segment and
entire draining primary vein.

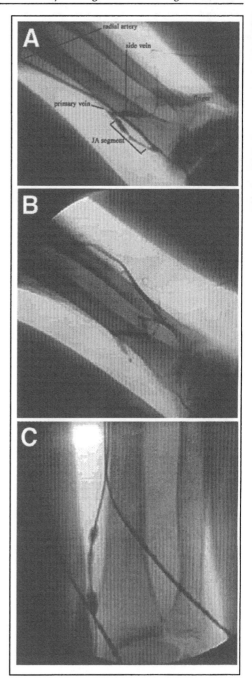

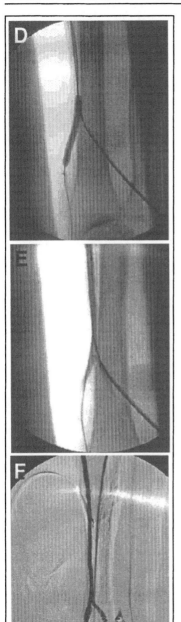

Fig. 7.16D-F. D) 6 mm PTA of J-A segment and entire draining primary vein. E) Flow past 6 mm PTA with side vein manually occluded. F) Fluoro following coil occlusion of diverting side vein. Note that thrombosis has not yet been completed. Follow up— this fistula is now 8 mm caliber and has had uninterrupted use for 18 months.

7

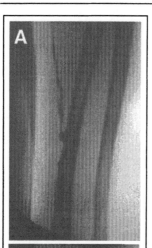

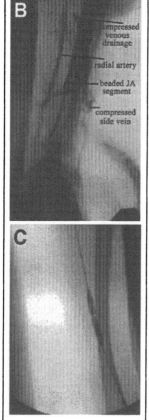

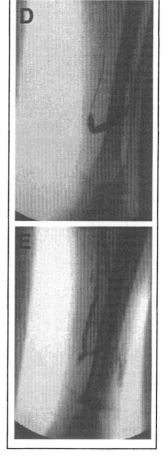

Fig. 7.17A-C. AVF failure to mature: PTA and side vein occlusion. A,B) Angio of immature AVF with clinically large side vein compressed. C,D) Entire primary vein was dilated to 6 mm. E) Persistent side vein diversion of flow.

Fig. 7.17F,G. Two week post PTA/coil follow up shows a well matured AVF ready for use. It has now been used for 6 months.

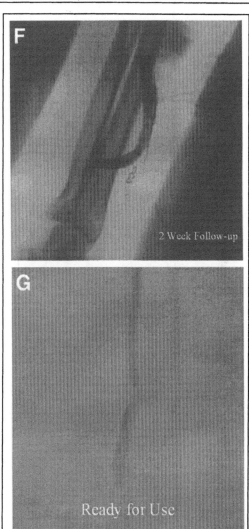

7

distance from the arteriovenous anastomosis and may well be associated with use before full maturation has occurred. Ineptness of needle cannulation for dialysis with resultant infiltration and hematoma formation leads to vasospasm, perivascular scarring and stenosis. Hypertrophic valves and venous divisions also are sites of stenosis that respond to PTA (Fig. 7.19).

Any of these stenoses may result in the development of flow-diverting side branches or "stealing veins." Stealing veins from AVFs divert flow from the main draining vein and should be ligated or occluded with coils to improve flow through

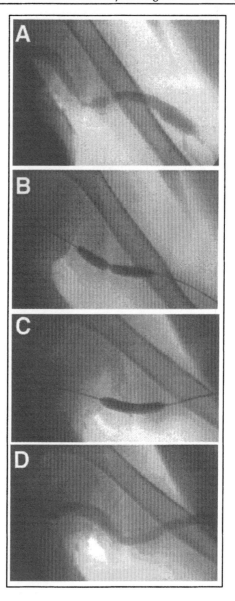

Fig. 7.18. Mature brachio-basilic transposition AVF dysfunction related to a "swing site" stenosis. A) High grade stenosis at the "swing site" of a brachio-basilic transposition AVF causing decreasing URR after 4 months of good function. B) Stenosis is often resistant and may require local infiltration and high pressure balloon PTA. C) Successful 8 mm PTA with a Blue Max catheter. D) Final angio shows an excellent result.

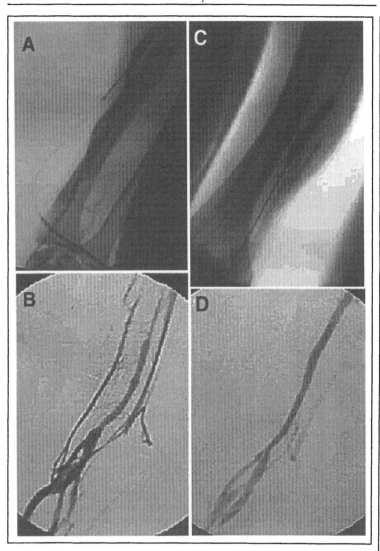

Fig. 7.19. AVF dysfunction related to multiple draining vein stenoses. A) Radio-cephalic AVF with high pressure and low Kt/V. B) Angio shows stenoses in all draining veins. C) PTA of a vein chosen to be the primary drainage. D) Straight line venous drainage post PTA. This fistula is now 2 years post procedure with continuous use.

Fig. 7.20. Central cephalic stenoses are frequent, but respond to PTA. A) Angio showing a central cephalic vein stenosis. The apparent stenosis of the left innominate vein was not anatomic and allowed easy passage of an 8 mm balloon. Note that there are no collateral vessels. B) 8 mm PTA of central cephalic stenosis for 2 minutes at full effacement. C) Final angio shows restored flow.

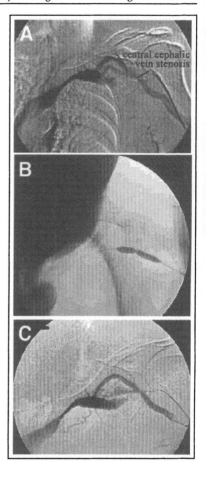

7

the fistula if they persist after adequate PTA,. The first defense against stealing veins is effective venous PTA to promote primary flow. This may necessitate dilatation to 10 mm. If the vessels continue to steal from the primary venous conduit following adequate PTA, occlusion can lead to complete maturation or improved function with optimal flow.[14,16,18] This can be effectively accomplished in smaller vessels (up to 4-5 mm) by employment of embolization coils (Cook Inc). Care must be exercised in placing these coils to appropriately size the vessels to prevent an embolization coil escaping to the lung and yet to achieve permanent occlusion. Vessels of greater than 6 mm should probably be surgically ligated. This can be done as a surgical referral or in the interventional suite after definitive localization with a catheter and guide wire, allowing for very focused cut-down and double ligation of a stealing vein.

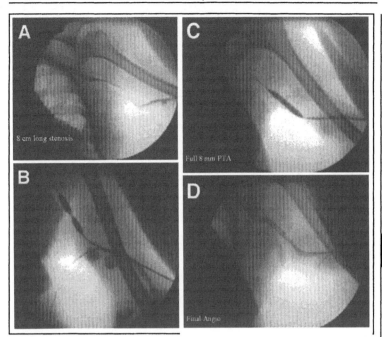

Fig. 7.21. Angio/PTA long segment venous stenosis. A) Angio showing long segment stenosis extending 8 mm centrally from the venous anastomosis of an upper arm graft. B) PTA showing resistant area after dilatation of the rest of the stenosis. C) Full 8 mm PTA. D) Final angio shows no residual.

Central stenoses are noted no more frequently in patients with native fistulae than in patients with grafts; however, often fistulae (as well as grafts) in which the cephalic vein is the major draining vein tend to have an area of relative narrowing near the junction of the cephalic vein to the subclavian vein (Fig. 7.20). This stenosis is usually beaded in appearance, 4 to 5 cm long with a variable number and caliber of stenoses, probably related to hypertrophic venous valves. All of these stenoses generally respond well to PTA but resistant stenoses may require aggressive endovascular techniques for resolution.[19] Especially in central lesions, operative interventions often require difficult exposure making endovascular methods the procedures of choice. Long draining vein stenoses (> 6 cm seen in the draining veins of both AVF and SBG) were traditionally cause for access abandonment, but they can be effectively treated by endovascular methods[20] (Fig. 7.21). One should note that the venous tree is very resilient often giving excellent patency in spite of a terrible looking initial result (Fig. 7.22).

Thrombosis occurs in only about 10-15% of dysfunctional AVF accesses and responds well to endovascular recanalization by aspiration thrombectomy[18,21-23] within 72 hours of the clotting episode. After 3 days, thrombectomy is more difficult because of the adherence of the clot to the wall of the vein. I have performed successful

Fig. 7.22A-C. Long segment venous stenosis above a graft. A) Angio of 12 cm long stenosis of draining vein causing thrombosis. B) 8 mm PTA of entire length of stenosis. C) Improved caliber and flow. Irregularity and minor extravasation in vein just above anastomosis.

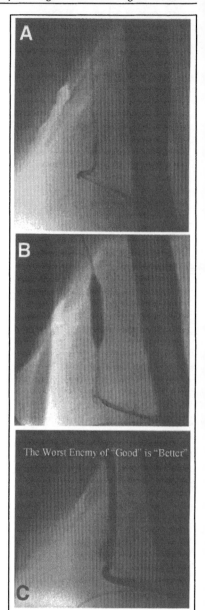

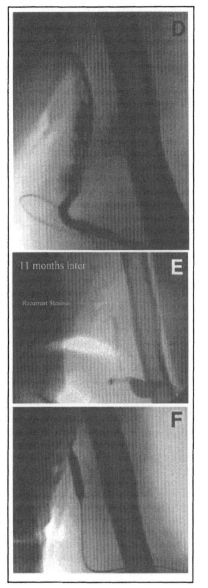

Fig. 7.22D-F. D) Appearance following second 8 mm inflation. Flow was occluded for 3 minutes with balloon tamponade to preserve flow and the partially inflated balloon used to reduce spasm. Patient went to dialysis the next day with excellent result and no hematoma. E) Short recurrent stenosis 11 months later. F) 8 mm PTA.

7

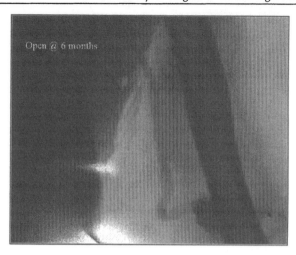

Fig. 7.22G. Final angio in follow up, often at 6 months.

fistula declotting up to three weeks following thrombosis. In some AVFs, thrombosis may result from a small thrombus plug in a tight stenosis; it responds well to PTA and requires little aspiration (Fig. 7.23). However, massive thrombosis, most often seen in an upper arm AVF extending to a central stenosis (cephalic vein), can result in difficult clot removal that is time consuming and tedious in order to avoid massive pulmonary emboli (Fig. 7.24). Thrombus in AVF should be removed from venous to arterial with flow restoration at the anastomosis as the last step. The advent of t-PA has added another tool to the recanalization process of AVF (Shoen), especially when the clot load is massive, enabling improved aspiration of the softened thrombus. Following complete aspiration, the fistula should undergo PTA of all stenoses.

Some thrombosed accesses have chronic mural thrombus in aneurysms that occur along the course of the fistula. This tissue is often difficult to remove and may necessitate the employment of a mechanical device such as the arrow trerotola percutaneous thrombectomy device (ATPTD). When creating a slurry of this firm tissue, care must be used to control flow through the access to prevent large numbers of microemboli from entering the systemic venous circulation and pulmonary bed. This can be easily accomplished by the employment of a large caliber (9 F) sheath for concomitant aspiration when activating the device. Occluding balloons can be placed through separate sheaths in order to control both inflow and outflow in a native fistula containing a huge clot load, thus preventing massive embolization while thrombus removal is completed. In some instances the mural thrombus is organized, will not fragment, and becomes loose after declotting to act as a "flap valve" which continues to compromise flow through the access. Open removal of this tissue with surgical reduction of the aneurysm size can return prolonged use to the fistula (Fig. 7.25).

The fact that more native fistulae are currently being placed will have an impact on the overall practice of vascular access in dealing with dysfunctional fistulae. We

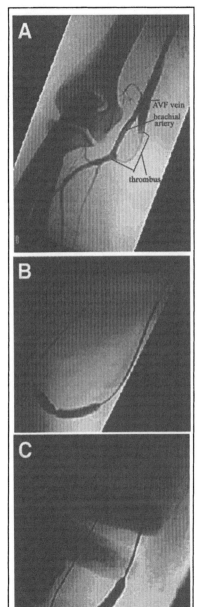

Fig. 7.23. Short segment AVF recanalization. A) Angio after passage of a catheter through the clot showing a short segment thrombosis. B) Balloon deformity showing an anastomotic stenosis as the underlying etiology of thrombosis. C) Full 6 mm PTA of lesion and thrombus.

Fig. 7.23D,E. Full angio of AVF post procedure. The small volume of clot does not require aspiration or lysis.

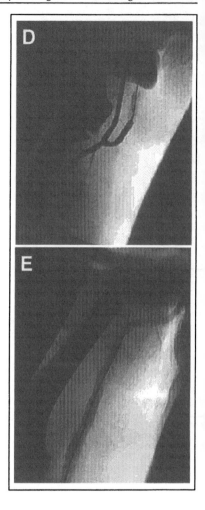

would all be well served to review the experience of European interventionalists,[21-25] specifically Poulain, Haage, Vorwerk, and Luc Turmel-Rodrigues, in creating protocols for treatment of native fistulae. Even though immature fistulae are still largely treated surgically in Europe, there is now interest in currently applied U.S. techniques for promotion of maturity. However, even in Europe where the majority of accesses are native fistulae, synthetic grafts are responsible for a high percentage of access failures.

Synthetic Bridge Grafts

I have experienced great success using endovascular procedures to maintain vascular access graft patency. Our practice has endeavored to present a cost-effective method for resolving graft dysfunction and restoring usefulness of thrombosed

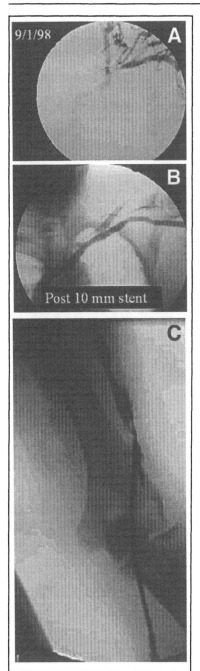

Fig. 7.24. Long segment thrombosis AVF. A) Central vein occlusion secondary to previous left subclavian catheter. B) Patent after 10 mm stent insertion. C) Long segment thrombosis (up to and including the stent) eight months after initial stenting-thrombosis was 1 week old.

7

Fig. 7.24D-F. D) Long segment thrombosis (up to and including the stent)
eight months after initial stenting-thrombosis was 1 week old. E) AVF contained
two aneurysms and multiple stenoses
which were dilated post aspiration
thrombectomy. F) Arterial inflow post
aspiration thrombectomy and 6 mm
PTA.

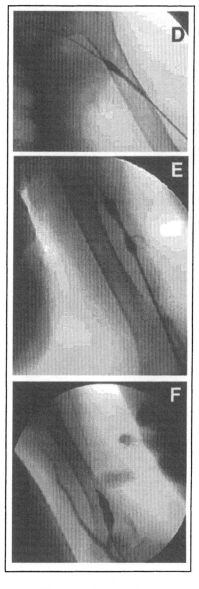

accesses, identify the subgroup of patients with "frequent failure" of dialysis access
grafts, and to present the algorithm for "best care" for hemodialysis patients in our
area. A flow diagram of care was presented earlier in this chapter. Prevention of
thrombosis by prospective PTA is desirable from both an economic and practical
perspective.[9]

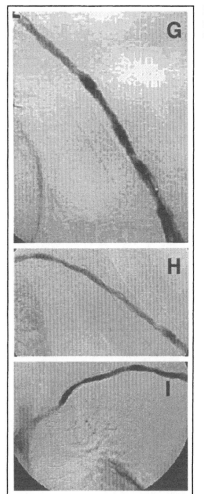

Fig. 7.24G-I. Some vessel irregularity related to adherent clot after PTA and aspiration.

The Dysfunctional Access Graft

In dealing with dialysis access synthetic grafts, one deals with both dysfunctional and failed (thrombosed) accesses. When dealing with the treatment of stenosis without thrombosis, it is necessary to have clinical or physiologic abnormalities associated with anatomic stenosis of greater than 50% to justify as "medically necessary" any procedure to be compensated by the payer. Many risk factors can indicate a need for PTA if the stenosis is 50% or greater and associated with clinical findings. These clinical signs include:

- previous thrombosis of the access,
- elevated venous pressure within the access graft, and
- abnormal dialysis delivery or recirculation measurement.

Fig. 7.24J-L. J) Thrombectomy device to fragment adherent thrombus. K) Vessel wall looks terrible at the end of procedure-use access for dialysis. L) Two weeks later angio shows smooth wall with good flow.

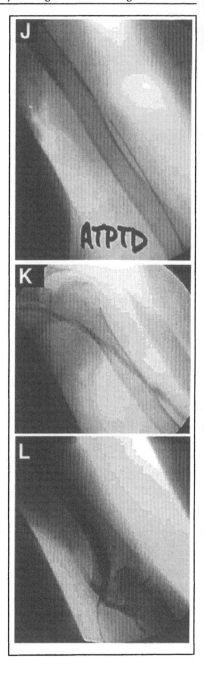

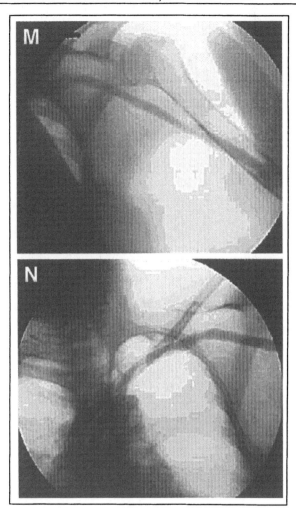

Fig. 7.24M,N. Two weeks later angio shows smooth wall with good flow.

Physical findings including:
- the loss of a thrill
- a systolic bruit
- prolonged bleeding after dialysis needle removal
- "black blood" or
- a lack of pressure differential to venous outflow compression

can also justify intervention and are findings usually alleviated by prospective interventions. Thus, these objective clinical findings indicate a need for angiographic evaluation and PTA to be performed to promote improved flow and prevent thrombosis (Fig. 7.26).

Fig. 7.25. Pseudoaneurysm with mural thrombus can compress an access to cause thrombosis. A) Saccular aneurysm containing mural thrombus compresses inflow. B) Mural thrombus is loose after aspiration thrombectomy. It was hard and would not break up with an ATPTD or external massage. C) Aneurysm was opened laterally about 1 cm and the occluding "flapper valve" thrombus removed and aneurysm closed.

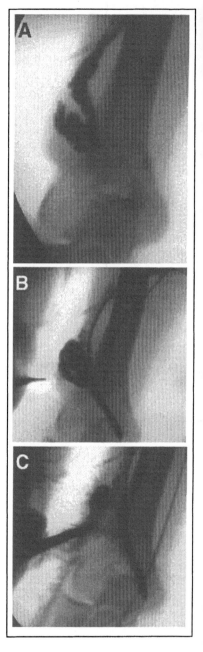

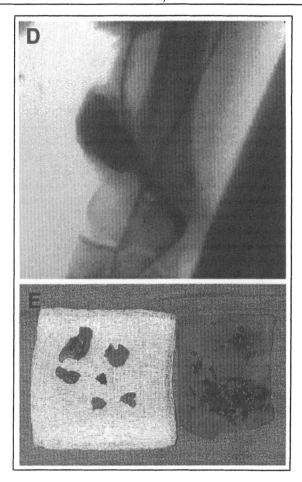

Fig. 7.25D,E. D) Complete clearing of aneurysm. E) Material removed from the graft: 1) hard organized mural thrombus; 2) soft fresh aspirated thrombus.

Even though PTA in non-thrombosed grafts with stenosis is approximately 98% successful, a small cohort of patients will experience difficulties during or following the PTA, resulting in a need for a surgical procedure or abandonment of the access. After correction of stenoses in dysfunctional grafts, surgical revision is held to a slightly higher standard than PTA because of the fact that surgery uses up venous real estate. DOQI guidelines recommend 50% unassisted patency for PTA at 3 months (our analysis demonstrates 83% primary patency at 3 months and 64% patency at 6 months). No more than 30% residual stenosis is acceptable following a PTA procedure. In contradistinction, surgical revision is held to an unassisted patency at 1 year of 50%. In an opinion set forth in the DOQI guidelines, it has been

Fig. 7.26. Hypertrophic valve at the venous anastomosis causing stenosis and thrombosis. A) Angio of venous limb after aspiration thrombectomy shows valve hypertrophy. B) 10 mm PTA. C) Restored flow. Note the small rounded remnant of the valve.

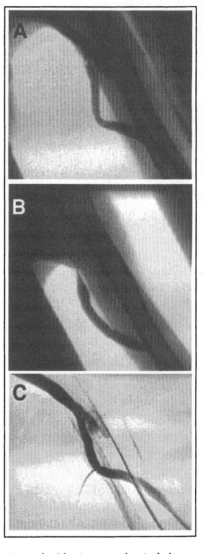

7

stated that two PTA procedures within 3 months (also interpreted to include two declotting procedures within 3 months) should be an indication for referral for surgical revision. However, our experience shows that if difficulties occur within the first 3 months after creation of a dialysis access graft, three or more procedures may be necessary to achieve prolonged patency until the healing process has stabilized at about 6 months. Patency after 6 months is then often similar to that of new mature access grafts.

Recanalization of Thrombosed Access

Many techniques have been utilized for percutaneous endovascular thrombectomy of AV grafts. The premier review article about dialysis access technique and outcomes is by Richard Gray.[6] Original articles detailing infusion techniques include Valji et al[26] presenting pulse spray thrombolysis with thrombolytic enzymes, Gerald Beathard[27] showing the utility of thrombolysis by hydrodynamic mechanical means using pulsed spray heparinized saline, and the "lyse and wait technique" as popularized by Jacob Cynamon.[28] Aspiration techniques as described by both Poulain[24] and Luc Turmel-Rodrigues,[25] and aspiration and extrusion thrombectomy with PTA (AET/PTA) that I have developed should also be reviewed[29] in Appendix A and the figures (Fig. 7.27). There is also an excellent mechanical device review by Sharafuddin.[30,31] The employment of mechanical devices can be a helpful adjunct for difficult cases, but cost can be a significant factor in the overall efficacy of mechanical device use, especially in a freestanding access center. To me it seems unnecessary to routinely employ an expensive device when less expensive techniques give equal or better results.

A brief technique review is presented in Appendix A to this chapter courtesy of the Society of Cardiovascular and Interventional Radiology Workshop Manual from the 2000 annual meeting. All of these methods successfully remove thrombus, but it is the proper use of PTA in all thrombectomy procedures that results in prolonged patency. Venous PTA balloon size certainly has an influence on the outcomes because adequate wall contact in the graft is necessary to remove all thrombus and achieve resolution of all stenoses. I stress the importance of venous PTA balloon size in promoting prolonged patency of recanalized thrombosed access grafts as accentuated in our retrospective review of a series of 69 patients who had graft thrombosis treated on at least two occasions with different size balloons. We treated and retrospectively compared patency of thrombosed grafts with venous stenosis at the same site, in the same patients, treated by the same technique using a 6 mm PTA balloon on one occasion and an 8 mm PTA balloon on another occasion with no intervening treatment other than routine dialysis. Median primary patency when treated with the 6 mm balloon was 2.5 months compared to median patency with the 8 mm balloon of 4.5 months. In this series of 69 patients, there was no difference in the complications.[32]

Although it is ideal to treat thrombosis of a vascular access within a short time in order to deal with fresh thrombus rather than organized thrombus, our practice has had success in recanalizing access grafts up to 3 months after initial thrombosis. Our minimum time for treatment of graft thrombosis is 2 weeks following the initial creation of the access. If perigraft fluid is present, I order that the access be rested for 2-4 weeks before resumption of use. It must be noted that there is one absolute contraindication to access instrumentation for declotting. Under no circumstances should infected grafts be subjected to treatment for recanalization. The presence of purulent fluid in the perigraft space or within the graft itself should be dealt with surgically, requiring immediate gram stain and culture, broad-spectrum antibiotic therapy, and in many cases urgent surgical removal. Relative contraindications to instrumentation of a thrombosed access include:

- contrast allergy
- severe COPD
- pulmonary edema and
- left-to-right intra-cardiac shunt

Fig. 7.27. Aspiration and extrusion thrombectomy with PTA—an efficient means of graft recanalization. A) Insert a 6 or 7 F aspiration sheath in venous limb 10-12 cm from and toward the venous anastomosis. B,C) After administering 3000 U of heparin, perform extremity venogram.

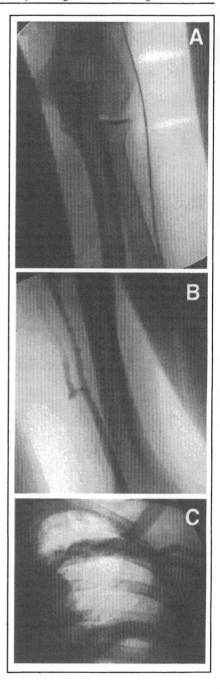

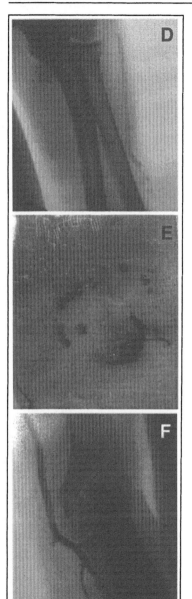

Fig. 7.27D-F. D) Remove dilator, attach aspiration syringe and slowly withdraw sheath about 75% and apply external massage to remove venous limb thrombus-repeat at least twice. E) Thrombotic products from initial aspiration. F) After crossed catheter insertion of a 6 mm PTA balloon advanced into the artery, perform an inflow arterial angiogram—this confirms graft thrombosis and shows the size and runoff of the feeding artery.

Fig. 7.27G-I. G) Partially inflate the arterial balloon (good wall contact) and withdraw it into the arterial anastomosis to displace the arterial plug and show the inflow stenosis by balloon deformity. H) Full arterial anastomotic 6 mm PTA. When compared to previous image it is >50%. I) Use the PTA balloon as a Fogarty catheter with complete inflation about every 2 cm as the catheter is withdrawn to the venous puncture site to displace all thrombus to the venous limb. Repeat several times. Then perform a second aspiration from the venous sheath. Confirm central limb clearing by a gentle retrograde angio.

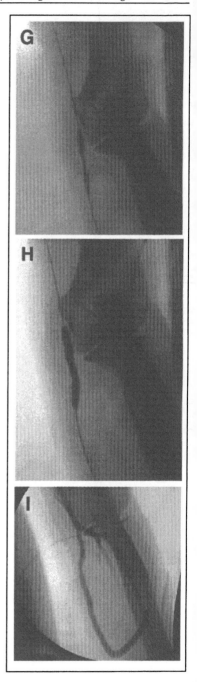

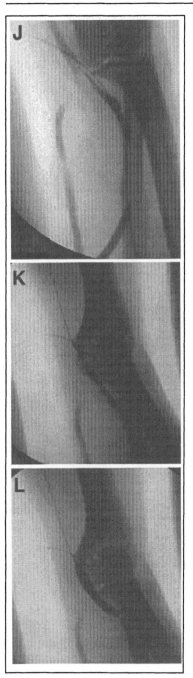

Fig. 7.27J-L. J) After second venous aspiration, check the venous outflow for stenosis. It is usually in the vein extending from or near (within 5 cm) of the graft/vein anastomosis. K) Insert an 8 mm PTA balloon for treatment of the stenosis. L) Full venous 8 mm PTA for 1 minute. This creates a closed system between the balloons.

M   PTA Balloon Size Results
69 Patients

• Size	#	1m(%)	3m(%)	6m(%)	9m(%)	Med.
• 6mm	86	57(66)	37(43)	25(29)	11(13)	2.5m
• 8mm	112	96(86)	75(67)	46(41)	23(21)	4.5m

N

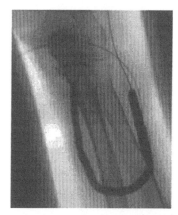

O

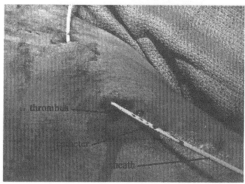

Fig. 7.27M-O. M) Influence of balloon size on primary patency—8 mm is nearly double that of 6 mm without increased complication rate in our experience. N) Remove the sheath from the graft onto the shaft of the 8 mm catheter and incrementally move the 8 mm balloon to its puncture site to extrude trapped and mural thrombus when the balloons kiss. Repeat until no more thrombus is extruded. O) Extruded thrombus around the catheter shaft. If the aspiration has been effective there may be little extruded thrombus.

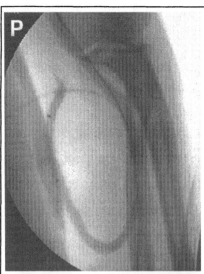

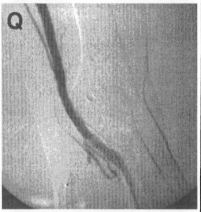

Fig. 7.27P-R P) After the sheath is replaced, deflate the balloons, move the venous balloon into the vein out of site, advance the arterial catheter to within 5 cm of the inflow anastomosis and perform an angio of the graft to confirm complete clearing of the thrombus. Q) Remove venous balloon and perform angio through the sheath to show resolution of the venous stenosis. If satisfied with the result, remove catheters and achieve hemostasis by compression or suture. R) Thrombus removed by aspiration and extrusion thrombectomy.

7

Patients with contrast allergy can be treated after appropriate pre-medication with antihistamines (Benadryl), steroids (prednisone), or free radical blockers (Tagamet). Patients in cardiac failure who cannot lie supine for a procedure (even insertion of a temporary catheter) can be treated sitting erect with the arm on the angiographic table while receiving oxygen support. Patients with severe COPD can receive supplemental oxygen therapy while employing a technique that minimizes the potential for pulmonary emboli (AET/PTA). Patients with right-to-left shunts require extreme care to occlude flow during complete removal of all thrombus prior to restoring flow. These patients are also at risk when exchanging catheters.

Even though one author has reported positive cultures from thrombotic material removed from a number of grafts,[33] we do not routinely use prophylactic antibiotics. In the more than 3,000 declotting procedures by different techniques with which I have been intimately involved, the infection rate has been well below 0.5% with no deaths attributable to sepsis. Anecdotally, I am aware of one death attributed to pre-procedure IV antibiotic administration. If meticulous sterile technique it utilized during performance of recanalization by actual clot removal rather than endovascular dissolution or fragmentation with displacement into the venous circulation, the infection rate related to the procedure is extremely low. Septic dialysis grafts should be clinically evident and referred for hospitalization. They usually respond well to treatment by a combination of intravenous antibiotic therapy and removal of the infected graft. The rare infected primary AV fistula, however, can be left in place and treated with 6 weeks of intravenous antibiotic therapy as if bacterial endocarditis were present.

Percutaneous declotting should be thought of as a two-component procedure, the first being clot removal and the second being treatment of the underlying stenosis(es) which caused the access to clot. The latter is almost always present and should be actively pursued for confirmation by visualization of the anastomosis in profile. My experience, having used nearly all methods of clot removal over the past 10 years, suggests that treatment of the underlying causative factors for thrombosis have more effect on patency than how the clot is removed, as long as all clot is totally cleared from the graft. The arterial plug, a firm, rubbery piece of clot present at the arterial anastomosis in nearly all clotted dialysis grafts deserves special attention. It is almost never seen in the absence of thrombosis, even in grafts in which there is flow in spite of 75% venous anastomotic stenosis; thus the mere presence of the plug compromising inflow at the artery-graft anastomosis must have an influence on thrombosis and must be removed. This plug responds poorly to enzymatic thrombolysis, probably due to its physical structure of multiple layers of fibrin and red cells. Therefore, most interventionalists have abandoned infusion techniques in favor of mechanical maceration or removal in declotting procedures.

Devices used to treat the arterial plug include:
- the regular 4-5 F Fogarty
- over-the-wire Fogarty catheter
- occlusion balloon
- PTA balloon
- mechanical catheter manipulation with thrombo-aspiration and
- one mechanical thrombolytic device, the ATPTD

In cases of difficult plugs where residual adherent clot is present, the Fogarty adherent clot catheter can be a valuable tool. Some believe that angioplasty of the arterial plug may leave residual material or injury behind that may serve as a nidus for further intimal hyperplasia and/or re-thrombosis; however in my experience of 3,000 declots, I have seen only five cases in which this might have occurred. At a second thrombectomy, it seemed more likely that a small fragment of adherent plug was not appreciated angiographically at the first procedure and resulted in re-thrombosis in less than 2 weeks. All five of these cases achieved long term patency following the second procedure.

I routinely dilate the arterial anastomosis with a 6 mm balloon when performing a percutaneous thrombectomy. The PTA balloon not only gives me objective evidence of a functional inflow stenosis, but it also eliminates unsuspected narrowing while removing the arterial plug. As noted above, true stenoses (i.e., those not representing clot or tapered graft) of the arterial anastomosis are rare, but adherent and resistant arterial plugs causing stenoses occur much more frequently than appreciated in the past, with a current recognized incidence of "resistant plugs" at 33.5%;[34] our incidence in 1,624 declot procedures is 48% confirmed by images showing balloon deformity greater than 50%, thus warranting PTA.

In dialysis grafts, thrombosis is most often associated with decreased flow through the access. Myriad factors contribute to thrombosis and can be divided into anatomic and physiologic categories. Anatomic categories deal directly with arterial, intragraft, or venous stenoses. These defects can be secondary to surgical technique, intimal hyperplasia, and development of saccular pseudoaneurysms that compress the access. External compression when a patient sleeps on the arm or carries objects that rest on the access (grandchildren) can cause thrombosis, but the most notorious external compression that causes thrombosis is the use of hemostatic clamps following dialysis. Hemoconcentration following dialysis, particularly when associated with a hypovolemic, hypotensive episode can effectively decrease the flow through a graft to the extent that it can achieve thrombosis without anatomic defects. A hypercoagulable state such as one sees in connective tissue diseases, paraneoplastic syndrome, and compromised flow secondary to generalized atherosclerotic vascular disease can also contribute to thrombosis.

Theoretically an access with borderline flow (the same or nearly the same as the dialysis flow rate) could have almost complete stasis of blood between the cannulation needles leading to the development of a soft plug of clot that will travel to the nearest stenosis when the needles are removed. We should be able to confirm or negate this theory once small ultrasonic Doppler probes now under development are available. This occlusion will lead to clot propagation over a matter of minutes to hours to result in a thrombosed access graft by the next day or dialysis session. Once thrombosis has occurred, however, it must be dealt with in a timely fashion, hopefully causing no more than one missed dialysis and definitely no more than one temporary catheterization for access. Surgical correction of graft thrombosis should not only employ thrombectomy, but should also effectively treat the presence of an arterial plug or inflow stenosis as well as the treatment of any venous stenosis in the circuit. Thus by mandate of K-DOQI guidelines, surgical thrombectomy must be accompanied or followed by angiography to confirm the return of unimpeded flow and resolution of all stenoses. In our experience venous stenoses are associated approximately 92% of the time and arterial inflow stenoses measured by the objective criterion of balloon deformity during the movement of a partially inflated balloon into the arterial anastomosis occur in 48% (780/1624) of thromboses. Because of these observed findings in our practice we have developed a technique, aspiration and extrusion thrombectomy with PTA (AET/PTA), to effectively treat the entire

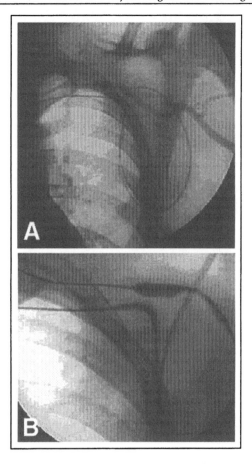

Fig. 7.28A,B. Chest wall loop declot. A) Inflow axillary angio for chest wall loop—note previous surgical removal of mid clavicle and an axillary jump graft from previous post op pseudoaneurysm. B) Arterial balloon deformity pre PTA.

access graft (Fig. 7.27, Appendix A). This technique was developed in response to a need to safely provide time- and cost-efficient delivery of care to these patents in an outpatient setting. It minimizes the potential for arterial and venous (pulmonary) embolism, dilates all stenoses in the dialysis circuit, provides excellent wall contact for total clot removal, has few complications, is very cost effective (supply cost including angio tray of $324), can be used in any graft setting (Figs. 7.28 and 7.29), and has shown excellent primary patency. Over 500 of these procedures were initially performed in a hospital outpatient setting followed by an additional 2,500 cases in a freestanding outpatient dialysis access center. This large series of declotting procedures showed primary patency rates that outperform the DOQI guidelines (58% patency at 3 months and 44% patency at 6 months) and have resulted in

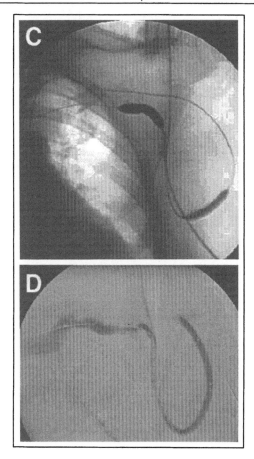

Fig. 7.28C,D. C) Venous PTA and closed system for extrusion. D) Final angio after percutaneous thrombectomy.

prolonged patency when compared with local surgical and pharmacomechanical techniques of dialysis access declotting.[35]

After successful endovascular dialysis access procedures, the method of hemostasis depends to a large degree on local practices. I suture almost all puncture sites because of time and safety related to a large and busy practice. Even if the patient is going to dialysis the same day, hemostasis is much easier to achieve for the dialysis personnel if it is from a needle puncture than from a 6 F sheath puncture. I use a Figure-of-eight suture rather than a purse-string[36] suture because it is much easier for the dialysis center nurses to remove in 2-5 days. Hemostatic sutures do not actually go through the access, but only the adjacent skin. The use of nonporous suture material (monofilament) also decreases the potential for suture infection. Some operators prefer direct compression hemostasis or a gelfoam compression dressing.

Fig. 7.29A-C. Thigh loop declot.
A) Venous stenosis after initial
thrombus aspiration. B) Arterial
limb after clearing with "modified
Fogarty" maneuver. C) Graft after
aspiration thrombectomy

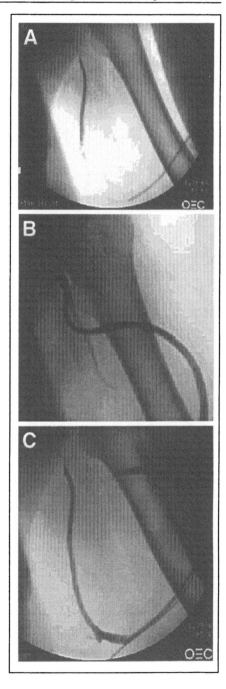

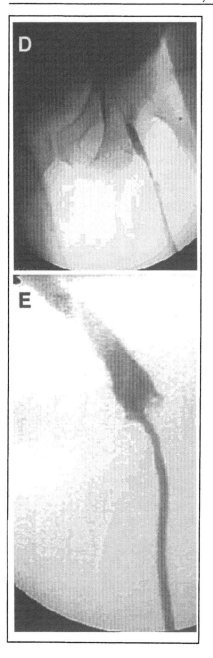

Fig. 7.29D,E. D) Venous balloon deformity-8 mm PTA. E) Final venous outflow.

7.

Patients with "Frequent Failure" (FF) of Dialysis Access Grafts

In the initial analysis of access graft thrombectomy outcomes, we noted that a small group of patients had a very high incidence of graft thrombosis and I coined the term "frequent failure" patients. We found that they represented 17% (54/301) of the thrombectomy patient population, with 41% (213/518) of graft failures and 77% (114/148) of first-month failures. Because of their high failure rate, this "frequent failure" patient population deserves close scrutiny. Using DOQI guidelines of two PTA procedures (extrapolated to include thrombosis) in 3 months as indicating a need for surgical revision and 40% 3-month patency after an endovascular declotting procedure (this would result in four thrombotic episodes per year) as a benchmark, I defined the "frequent failure" patient as one in whom there occurred three episodes of vascular access thrombosis in any 2-month period or five episodes of vascular access thrombosis in any 12-month period. These criteria are more stringent than the criteria for surgical referral in the K-DOQI guidelines.

Eventually 73 patients fit the criteria to be included in this group. These patients are a small group with excessive early access failure often associated with early use (< 4 weeks). Early use should be discouraged. In addition, they tend to be elderly, African-American, and female with grafts and multiple co-morbid conditions including diabetes, hypertension, and cardiac disease. Males can be affected also, but much less frequently. Early access failures can frequently (70%) be overcome with repeated declotting procedures in the first 6 months, resulting in patency rates similar to non-frequent failure patients after 6 months. Interestingly, when these patients are excluded from analysis with the other graft thrombosis patients, endovascular thrombectomy patency results are almost identical to the patency of non-clotted angio/PTA grafts.[37]

To Stent or Not to Stent

As previously mentioned, venous stenosis is the major cause of dialysis access complication. In some instances, patency cannot be maintained in an otherwise adequate access circuit and an intravascular stent is appropriate for use. One must remember that once a stent is in place, it will eventually present its own set of problems and complications. Indications for stent placement include:

- recurrent central stenosis (<3 months),
- flow impedance by elastic recoil in spite of overdilatation,
- recanalization of central vein occlusion,
- venous rupture not controlled by balloon and/or external tamponade, and
- sometimes compression of the brachial vein by obesity or abnormal venous anastomotic angle.

The FDA as yet approves no stent for venous use. I use the Wallstent exclusively because of its flexibility and self-expanding nature. The flared ends promote stability in intravascular placement, preventing not only stent migration but also providing a relatively smooth transition at both ends of the stent conduit. It also can be collapsed for removal if it migrates to the heart or pulmonary bed. There is extensive experience with this device in dialysis access. Wallstents are the treatment of choice for PTA failures and early recurrences in central veins. For peripheral lesions, stent patency is no better than that of angioplasty alone, as has been shown in several randomized studies to date.[38,39] Therefore, stents should be reserved for angioplasty failures due to rupture (Fig. 7.30), elastic recoil or early restenosis (Fig. 7.31). In the

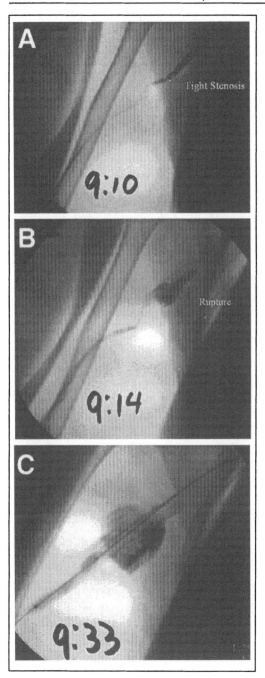

Fig. 7.30A-C. Venous rupture not responsive to balloon tamponade is an indication for stenting. A) 9:10 am—a tight venous stenosis (hypertrophic valve) is identified and dilated with an 8 mm balloon. B) 9:14 am—rupture with extravasation identified. C) 9:33 am—in spite of 20 minutes of balloon tamponade the leak is persistent and a stent was placed.

7

Fig. 7.30D-F. D)
9:48 am—follow-
ing completion of
forearm access
declotting proce-
dure, venous in-
tegrity has been
restored. E) Twenty
months   later
rethrombosis oc-
curred related to
stent occlusion. F)
Patency following
successful declot.

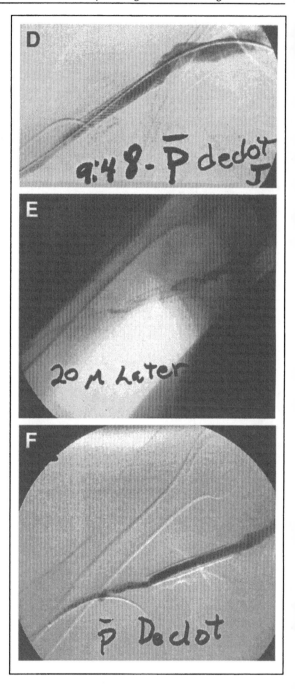

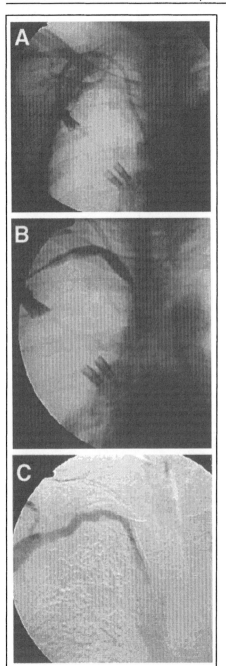

Fig. 7.31A-C. Some central lesions require stents. A) Angio shows jugular catheter and right subclavian occlusion. B) 10 mm PTA. C) Angio after 10 mm PTA.

7

Fig. 7.31D-F. D) Re-occlusion after 2 weeks. E) 12 mm PTA after catheter removal. F) Angio after subclavian stent placement.

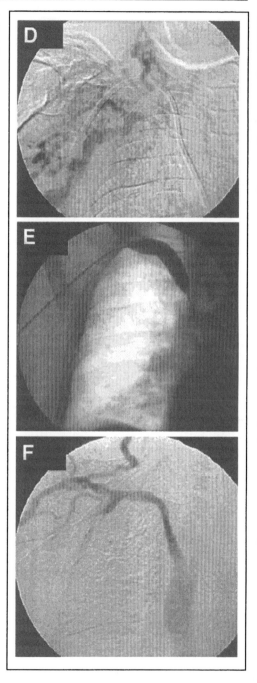

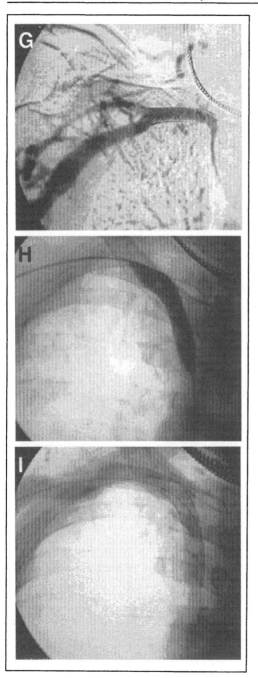

Fig. 7.31G-I. G) More stenosis central to stent after 3 weeks. H) 12 mm PTA after second stent placement. I) Restored flow through interlocking stents.

7

7

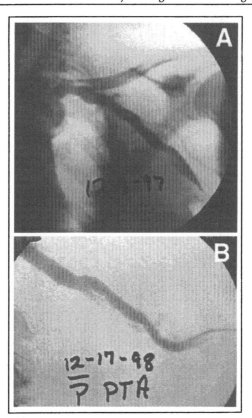

Fig. 7.32A,B. Recurrent stenosis in an obese patient may respond to stenting. A) Upper brachial vein stenosis was treated with 8 mm PTA. B) One year later the recurrent stenosis was again dilated to 8 mm.

event of an angioplasty failure, placement of a Wallstent more than doubles the cost compared to balloon angioplasty alone. The cost effectiveness of stent deployment versus surgical revision for angioplasty failures has not been determined. I use a stent for the occasional obese patient whose arm weight causes compression above an upper arm graft resulting in repeated thrombosis (Fig. 7.32). In this setting prolonged patency (> 6 months) has been observed by placing a stent through the area of compression. If a stent is used, it should be of the minimum length needed to treat the stenosis to avoid depleting the vein for future surgical interventions.

Central venous stenosis or occlusion can occur anywhere along the course of the draining veins from the axillary vein to the superior vena cava. These stenoses can be crossed from above using the access for entry; however sometimes a retrograde femoral approach is necessary to establish a through-and-through guide wire so that catheters, balloons, and stents can be passed through the lesion (Fig. 7.33). In our experience of dealing with more than 100 central stenoses, PTA alone was satisfactory to

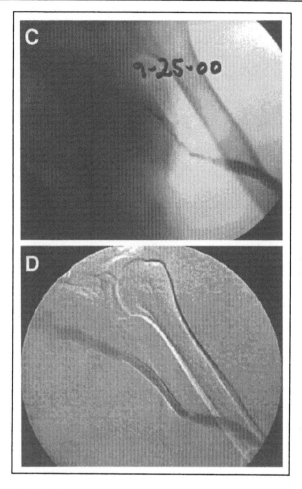

Fig. 7.32C,D. C) After three recurrences in nine months a stent was placed because the size of the patient made a high axillary anastomosis technically difficult for surgery. D) Final angio post stenting. This access is still functional after 16 months.

maintain prolonged patency approximately 50% of the time. However, stents can be a very helpful adjunct when dealing with central stenoses, particularly in the proximal cephalic, subclavian, and innominate veins. To repeat for emphasis, the placement of an intravascular stent in the venous tree is as yet not an approved FDA application and must be used as an off-label application. It must also be remembered that placement of a stent in the venous tree is a permanent application and will eventually be associated with either stenosis or thrombosis related to intimal hyperplasia (Fig. 7.34). In our practice, if venous patency of central stenoses or recurrent stenoses can be maintained for 3 months or more with a PTA alone, stents

Fig. 7.33A-C. Dual approach to central vein occlusion. A) Angio from left arm access shows central vein occlusions. B) Attempt at antegrade crossing was unsuccessful but showed a hint of left innominate vein. C) Selective central venogram from a femoral approach shows the left innominate vein junction.

7

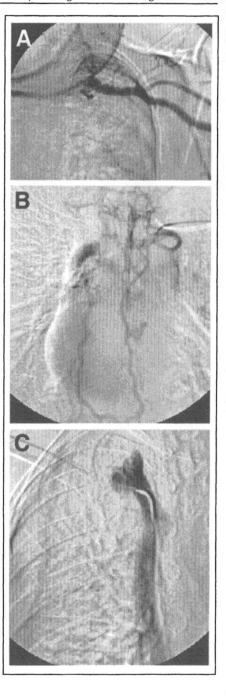

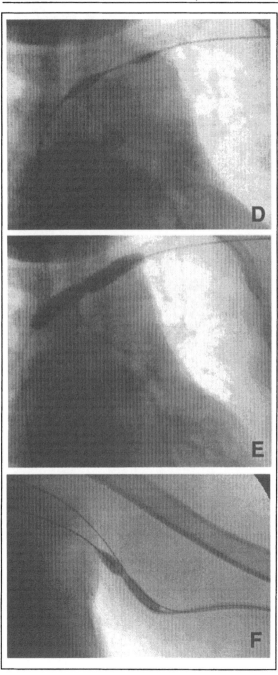

Fig. 7.33D-F. D) Retrograde catheterization was successful in establishing a through and through guide wire (stiff glide) allowing PTA. E) 10 mm PTA of central occlusion. F) Aspiration and extrusion thrombectomy concluded and flow restored.

7

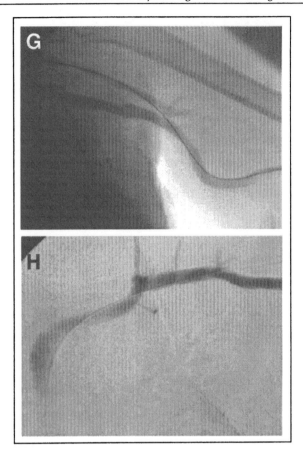

Fig. 7.33G,H. G) Aspiration and extrusion thrombectomy concluded and flow restored. H) Final angio shows patent central veins-now functional without stent for 5 months.

are not used. However, recurrence of stenosis for a third occasion with an interval of less than 2 months is an indication for stent placement. Other indications for stent placement are elastic central vein stenosis, particularly if associated with a complete venous occlusion and if overdilatation to as much as 12 mm will not maintain primary patency for more than 3 months.

Occasionally venous stents are placed to facilitate straight-line flow through a ruptured vessel following PTA in which response to both balloon inflation and external tamponade is unsuccessful (Fig. 7.35). The re-establishment of straight-line flow with the central vasculature being the path of least resistance will result in preservation of venous flow in an active native AVF or re-canalized access graft. In our practice an additional indication for stent placement is venous compromise in a very heavy patient whose arms cause extrinsic compression of a central draining vein above an upper arm graft when the patient's arms hang at their side. This can be

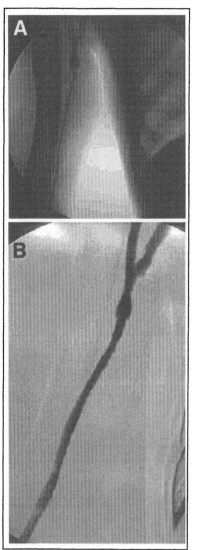

Fig. 7.34A,B. Peripheral stents cause problems and thus should be used only when necessary. A) Long brachial stent placed at another institution to treat a long stenosis presents with thrombus after 3 months. B) Flow restored after forearm graft declotting and PTA.

evaluated angiographically by positioning the arms appropriately during contrast angiography. If positional compression can be confirmed angiographically and recurrent stenosis/thrombosis occurs, the choices are either extension of the graft into a higher axillary anastomosis or placement of an intravascular stent (Wallstent) to facilitate patency. In very obese patients stent placement is technically much easier than surgical intervention.

7

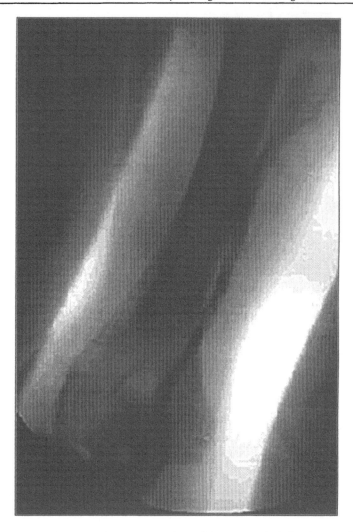

Fig. 7.34C. C) The venous drainage required PTA of almost 20 cm including the stent. Of note is that this access was abandoned and an upper arm graft placed two months later.

Stents are not recommended for resistant stenoses because they do not have enough radial force to further dilate the stenosis and they have the eventual problem of intimal overgrowth and occlusion. In order to treat resistant stenoses, I have developed a technique called infiltration and perforation PTA (Fig. 7.36). When a resistant stenosis is identified by rupture of a balloon at 22 atmospheres or more (AngioDynamics Workhorse or Meditech Blue Max), the lesion is defined, infiltrated locally with anesthetic, and the defined resistant tissue surrounding the stenosis is perforated multiple times with a 21 or 19 gauge hypodermic needle to weaken the

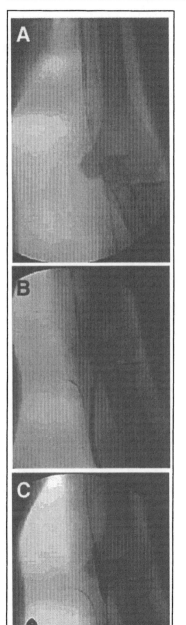

Fig. 7.35A-C. Venous rupture required stenting. A) Initial draining venogram shows long venous narrowing. B) Stenosis of draining vein extends 5-6 cm centrally from the anastomosis before the final step of declotting. C) 8 mm PTA of venous lesion.

7

Fig. 7.35D,E. D) Venous rupture with extravasation post PTA did not respond to 30 minutes of balloon and external tamponade. E) 68 mm Wallstent inserted.

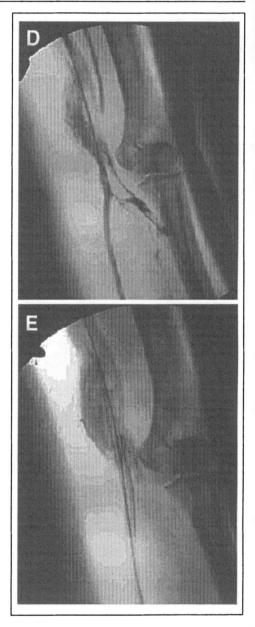

7

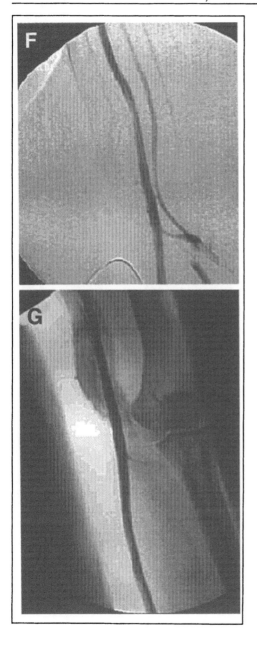

Fig. 7.35F,G. Final angio— no leak.

Fig. 7.36A-C. Persistent cephalic stenoses may require aggressive therapy. A,B,C) Angio of a one year old AVF with high pressure and poor clearance from central cephalic stenosis.

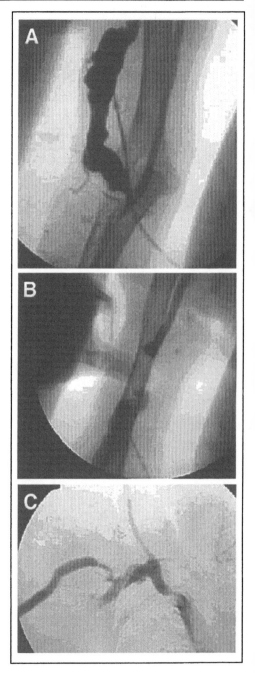

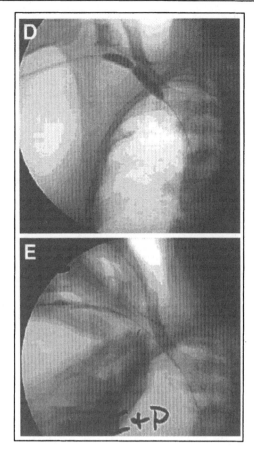

Fig. 7.36D,E. D) Resistant stenosis—balloon ruptured at 22+ atmospheres. E) Local infiltration with lidocaine and needle perforation with a 21 gauge hypodermic needle weakens the tissue for a controlled tear (infiltration and perforation PTA).

tissue so that a controlled-tear PTA (usually with a high pressure balloon) can be achieved to restore the desired caliber of the vein.[19]

Management of Endovascular Complications

Any diagnostic or therapeutic procedure that is invasive and is associated with injection of a substance foreign to the body or directed to correction of anatomic and physiologic defects presents an opportunity for a complication. These may be as simple as local discomfort or as severe as death. They will occur in spite of the talent and experience of the operator when using aggressive endovascular techniques, as well as in association with rather mundane procedures because ESRD patients are chronically ill with multiple co-morbid conditions. Complications are more frequent during the time of the learning curve associated with endovascular HVA care

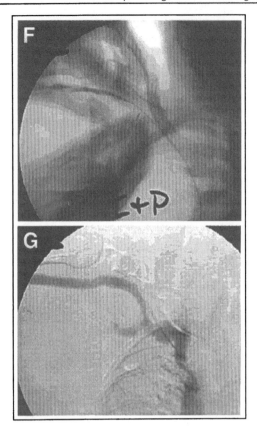

Fig. 7.36F,G. F) Full 8 mm PTA for 2 minutes. G) Final angio shows restored caliber and flow.

(usually about 150 cases) than after experience and confidence have been gained. However, experience brings a more aggressive approach into the equation because the practitioner has a feel for the resilience of the venous tree in these patients. The result is a continuing complication rate of about 2%. Thus it is necessary for all who perform these procedures to be trained to deal with iatrogenic complications.

If complications are collected using SCVIR criteria, many minor occurrences go unreported because they have no clinical sequelae. This is probably appropriate because usually only those problems that require additional treatment outside the interventional suite or that require hospitalization are consequential. Complications that prolong a procedure (brachial artery embolectomy) or result in additional cost (placement of a stent to preserve a ruptured vessel) should always be reported even though they require no additional treatment outside the interventional suite and have no clinical sequelae.

Catheter complications occur most commonly when performed blindly without sonographic and fluoroscopic imaging guidance. When using good sonographic technique the incidence of carotid or subclavian artery puncture should be 0%. With fluoroscopic observation there is no excuse for vascular rupture or malposition of a catheter tip. Infection can occur when a catheter is exchanged through the same tunnel over a guide wire; thus creation of a new tunnel can significantly decrease the potential for sepsis. Catheter exchange can be complicated by loss of a catheter fragment into the central venous tree or pulmonary artery (Fig. 7.37). The fragments should be retrieved to prevent pulmonary infarction or abscess formation. Another complication of catheter exchange and removal is loss of the tissue in-growth cuff (Fig. 7.14). It should always be retrieved to prevent foreign body reaction or abscess formation. The maneuver is simple: grasp the cuff with a hemostat through the tunnel and twist it like a key until the cuff is separated from the tract and easily removed. Multiple catheter placements over time and the use of central venous stents can also cause problems with catheter insertion. Attention to detail and innovative application of catheter/guide-wire techniques will usually result in success (Figs. 7.38, 7.39). In the apparent absence of neck/arm veins by sonogram, a combined approach from the femoral vein to selectively find a vein, and a supraclavicular puncture using a balloon as a target almost always allows placement of a catheter (Fig. 7.40).

Complications associated with AVFs and SBGs usually are related to vessel leaks following PTA, peripheral embolism of thrombus, and development of a true "steal syndrome" with poor perfusion of the distal extremity. Leaks usually respond to a combination of balloon and external pressure to achieve tamponade. If leakage persists, stent placement can restore appropriate function (Figs. 7.30, 7.35). Resistant arterial plugs, especially when adherent to the native artery wall may require deployment of both a balloon and ATPTD to facilitate clearing (Fig. 7.41). Peripheral embolism can be easily treated by "back bleeding"[40] (Fig. 7.42), aspiration, or retrieval with a balloon (Fig. 7.43). All of these techniques require a widely patent access to siphon away the embolus once it is mobilized. If these maneuvers fail, surgical embolectomy or prolonged enzymatic thrombolysis perfusion may be needed to restore flow. These additional treatments must be performed in a timely fashion in order to prevent nerve damage from ischemia.

Stents are associated with many complications ranging from migration requiring removal from the heart, to eventual tissue overgrowth and occlusion (Fig. 7.34). Sometimes flow continues through the interstices of the stent and can be crossed, dilated and used for a catheter access or stent placement to revascularize an extremity. Stent complications ultimately result in access failure.

Practical Observations: A Summary From My Experience

- Aspiration and extrusion thrombectomy with PTA is a very efficient method of recanalization of failed (thrombosed) hemodialysis accesses, both grafts and native fistulae. The procedure time is usually less than 1 hour. Procedural cost containment can be achieved without compromising quality of care. Improved outcomes are related to PTA of the entire graft and all stenoses.
- Prospective PTA for dysfunctional accesses without thrombosis will improve access usefulness and decrease the thrombosis rate.
- Oversizing the venous stenoses will improve patency without increasing the complication rate. Our standard is 8 mm. Incremental or sequential dilatations may be required to achieve this goal, especially in long stenoses (> 6 cm in length) of native AVF. A terrible-looking vein/native fistula has great recovery capabilities and can be redilated after 2 weeks to achieve more optimal results.

Fig. 7.37A-C. Catheter loss and retrieval. A) Catheter fragment loss during exchange (obese patient with coughing episode while guidewire being inserted). B) Unsuccessful attempt to snare fragment pushed it centrally. C) Catheter fragment displaced from fibrin sheath (guide wire) rebounded into right external jugular vein.

7

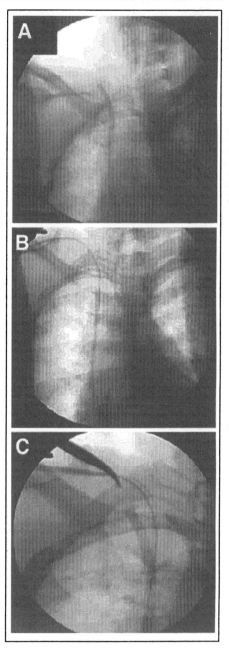

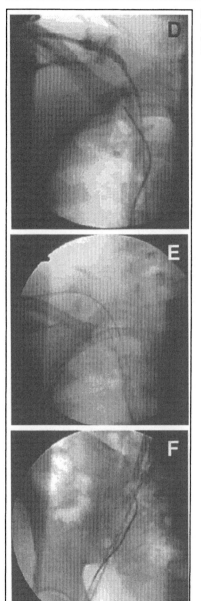

Fig. 7.37D-F. D) Femoral approach with selective catheter snare. E) Snare engaged on catheter fragment. F) Folded catheter in right iliac vein just prior to removal.

Fig. 7.37G-I. G) Removal system with catheter fragment. H) PTA of fibrin sheath. I) New catheter in place.

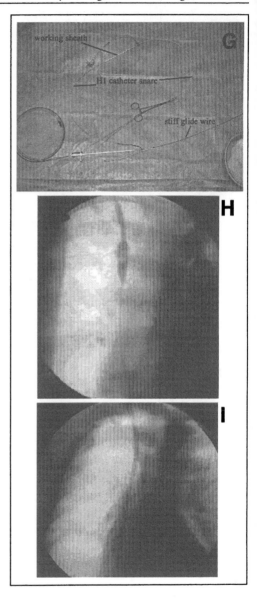

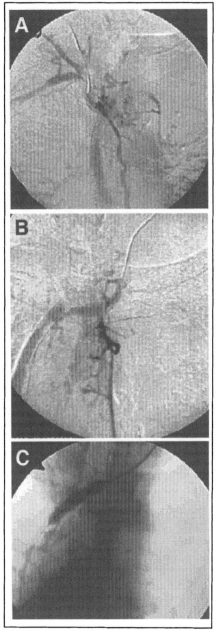

Fig. 7.38A-C. Catheter placement in central venous occlusion associated with a stent. A) Right IJ angio shows occlusion, venous cross over and minor flow through a central stent. B) Left IJ angio shows central flow through the interstices of the stent. C) PTA after passing a catheter and guide wire through the side wall of the stent.

Fig. 7.38D-F. D) Placement of a second stent through the sidewall of the existing stent. E) 10 mm PTA of new stent. F) New catheter in position for dialysis.

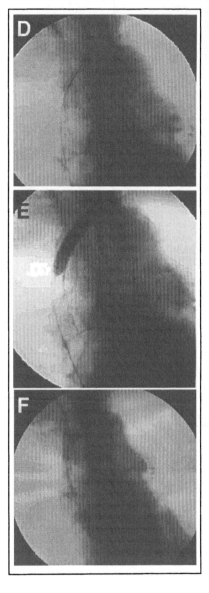

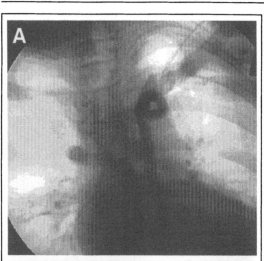

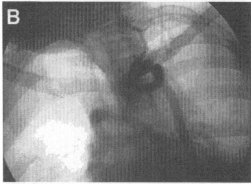

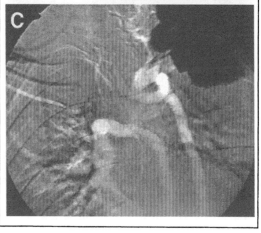

Fig. 7.39A-C. Pay attention to detail. A) Central venous occlusion caused placement of a femoral catheter at another facility. B) Oblique angiography shows the site of the occluded left innominate vein lumen. C) Road map with catheter into occluded lumen.

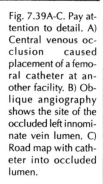

7

Fig. 7.39D-F. D) Angio shows patent central veins (faint opacification). E) 10 mm PTA of left innominate vein. F) Successful catheter placement for dialysis access.

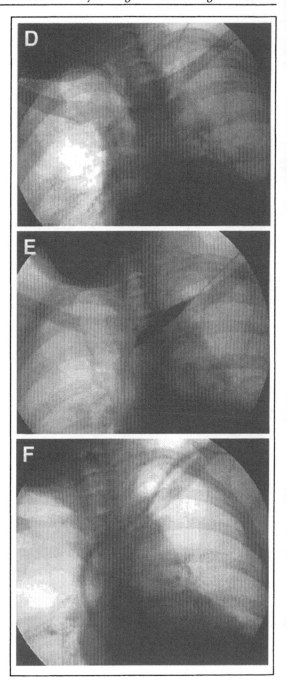

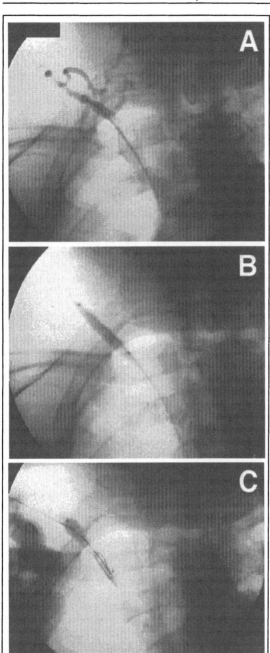

Fig. 7.40A-C. Dual approach for central catheter placement. A) Patient with no sonographically adequate vein for catheter access was entered from a femoral approach with selective venous catheter angio above the right clavicle. B) 8 mm PTA of vessel to SVC, now with inflated balloon parked as a target. C) Micropuncture of balloon with 0.018 wire placed inside the balloon.

7

7

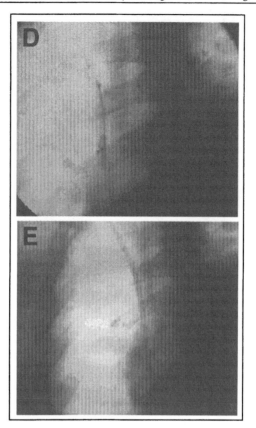

Fig. 7.40D,E. D) Balloon withdrawn to right atrium to place 0.018 guidewire. E) Central catheter placed by standard technique from above.

- PTA of the entire length of the graft will assure removal of all thrombotic material by good wall contact, thus improving patency by more nearly achieving laminar flow. We have observed arterial anastomotic stenosis related to tissue requiring PTA because it was not easily removed by a simple Fogarty catheter maneuver in 48% (780/1624) of graft declotting procedures. That finding is unusual in patent grafts but common in clotted grafts.
- The end point for adequate pressure in a balloon is successful dilatation of the lesion or balloon rupture. Minimum inflation time should be 1 minute. Most stenoses will respond to a regular balloon (Angiodynamics Workhorse) so that the need for a high-pressure balloon (Blue Max or Centurion) is less than 10%. Progressive dilatations should be in 2 mm increments except at the anastomosis of a native AVF. One millimeter sizing is not cost effective unless it involves the AV anastomosis of a wrist AVF or is a serial dilatation for maturation performed at a different session.

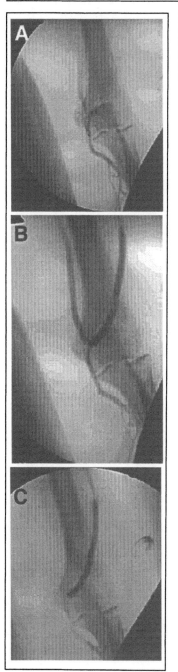

Fig. 7.41A-C. Arterial plug in the native artery required ATPTD for removal. A) Initial inflow angio shows deformity at the arterial anastomosis. B) Is the deformity an adherent arterial plug or surgical deformity? C) After PTA, inflow is occluded.

Fig. 7.41D-F. D) After both afferent and
efferent PTA there is an adherent triangular plug in the artery opposite the anastomosis with poor flow to the graft clinically.
E) Declot procedure is completed returning flow to the circuit. F) Repeated PTA of
both limbs of the artery would not remove
the plug.

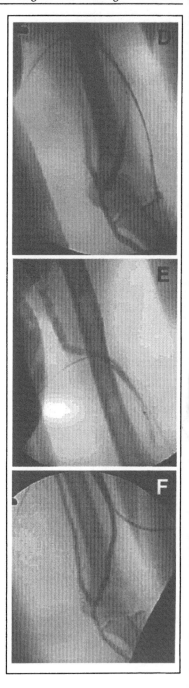

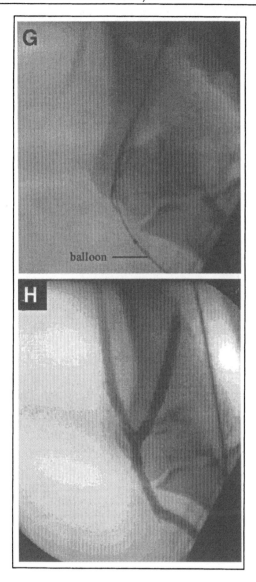

Fig. 7.41G,H. G) ATPTD applied to plug—note balloon is downstream to prevent emboli. H) Final angio shows good flow and no residual plug.

7

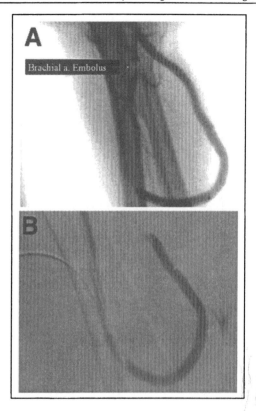

Fig. 7.42A,B. Brachial embolectomy by back-bleeding technique. A) Brachial embolus seen after arterial limb clearing. B) Percutaneous thrombectomy by AET/PTA completed to restore unrestricted flow. .

- These procedures are safe and ideally suited to performance in an outpatient setting. Most complications can be effectively handled at the table. Less than 2% require additional treatment after a procedure (major complication). One must be aware that if enough procedures are performed, severe complications (even death) will occur. Minor complications require increased procedure time, but usually have no sequelae. One must be prepared to handle any complication that arises and maintain a close working relationship with surgeons.
- There is a small cohort of patients (5-10%) who frequently fail both surgical creation/revision and declotting procedures of dialysis grafts. They tend to be female, African-American, and elderly (> 64 y.o.) with multiple co-morbid conditions. They need careful monitoring, but tend to do well after multiple endovascular interventions to maintain patency during the first 6 months following access graft creation. Our series shows that more than 70% can achieve prolonged patency/usefulness of the access after 6 months with patency curves similar to accesses that did not fail in the first 6 months. Primary patency of declotting procedures and angio/PTA are al-

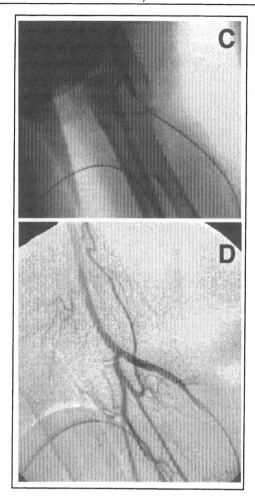

Fig. 7.42C,D. C) Occluding balloon inflated above the arterial anastomosis in the brachial artery for 3-5 minutes while exercising the hand and massaging the forearm. D) Final angio after balloon deflation shows clear arterial tree

most identical if frequent failure patients are excluded from the thrombectomy series.

- Tunneled catheters are used exclusively in our practice for non-permanent access. If we cannot restore function by percutaneous methods, the patient needs surgery and time to heal/mature. Our current choices are the Ash split-catheter and the Optiflow because of ease of insertion and high volume flow parameters. Manufacturing defects noted and reported with Optiflow catheters have been corrected. Because of current tip design, catheter function can be improved by explosive irrigation prior to each dialysis session.

Fig. 7.43A-C. Treatment of arterial ASVD and balloon embolectomy during AET/PTA. A) Inflow arterial angio shows ASVD of afferent and efferent limbs of the brachial artery. B) 6 mm PTA of arterial inflow afferent limb. C) Post PTA of inflow shows segmental occlusion of efferent brachial artery.

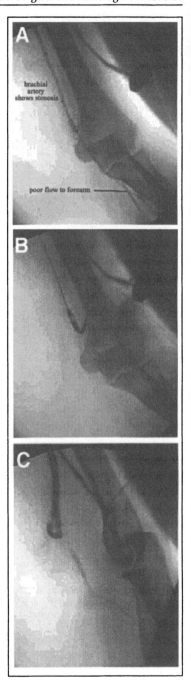

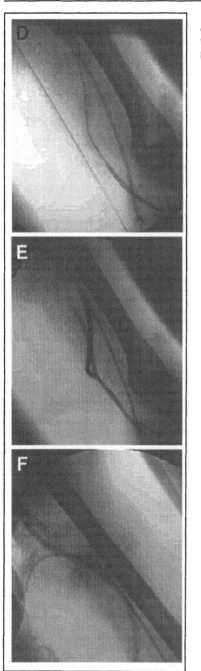

Fig. 7.43D-F. D) Balloon in position for down stream arterial PTA. E) Angio of forearm perfusion post PTA. F) Declot procedure is completed.

7

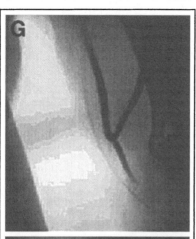

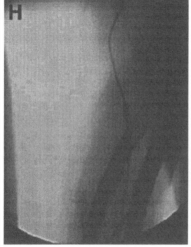

Fig. 7.43G,H. G) Repeat angio shows poor flow to forearm—spasm vs embolus. H) Downstream PTA.

- Native A-V fistulae require refined techniques and more innovative approaches to treatment than grafts. Many procedures will be for "failure to mature." Success can almost always be achieved if a single dominant vein can be defined for use as a dialysis conduit. Maturation can be enhanced after balloon PTA to dilate any flow-impeding stenoses and by occlusion of collateral "stealing" veins by either ligation or coil embolization if the side vein flow continues after adequate venous PTA. Long-term success may require serial dilatations at 2- to 4-week intervals to achieve optimal size for dialysis. "Watchful waiting" is often required after intervention to allow adequate flow to complete the maturation process.
- Thrombosis of a native forearm fistula is most often caused by a small plug of thrombus associated with a tight stenosis of the juxta-anastomotic inflow

Fig. 7.43I,J. I) Final forearm angio. J) 8 months follow up.

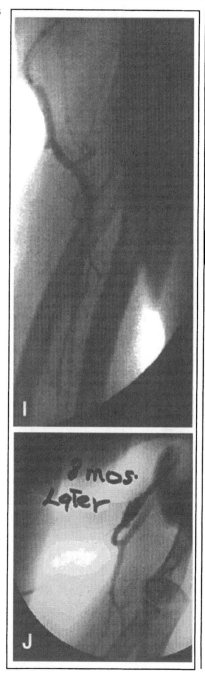

portion of the vein (first 5 cm of the fistula vein). Simple PTA, with or without aspiration, can usually restore adequate flow. In contradistinction, upper arm fistulae often present with total thrombosis to the subclavian level and must be treated by extensive aspiration with or without mechanical device application. Because the clot load is voluminous, great care is necessary to prevent massive pulmonary embolism. This can be accomplished by use of a central occlusion balloon while removing thrombus by aspiration through a 9 F sheath or use of a mechanical thrombolytic device. t-PA can enhance the success of native vein thrombectomy if the clot is more than 3 days old. Following percutaneous thrombectomy, attention is directed to correction of underlying stenoses. If thrombus has been present for more than three days, the final result may appear quite irregular at final angiography. We usually give additional subcutaneous heparin (10,000 u) and perform repeat angiography in 2 weeks. Follow-up angiography has routinely shown a smoothly contoured vein without detectable pathology in the lumen.

- Comparison of grafts with native AVFs has shown that AVFs are overall a much better access than a synthetic graft. Our local surgeons have responded well to the DOQI guidelines by placing increased numbers of AVFs, now approaching 70% for initial accesses attempted. Our data have shown that we can promote fistula maturation by endovascular means beginning at 4-8 weeks if maturation does not appear to be progressing well. Useful function has been achieved in more than 70% of immature fistulae. Thus, abandoning a fistula without a trial of endovascular therapy is a gross disservice to the patient. Our experience in dealing with native AVFs is currently more than 1,000 cases.

Conclusion

Maintenance of hemodialysis vascular access is a complex task requiring the focus, dedication, and communication of all caregivers. All members of the team play an important role in the care, quality of life, and survival of these patients. Monitoring and surveillance of the access can decrease the thrombosis rate and prolong the access life. Explosive irrigation of dialysis access catheters will improve function by clearing thrombus plugs from the tip. The small group of "frequent failure" patients can be identified that require persistent care to attain prolonged patency. Native fistulae are the most ideal access currently available and should be created and maintained aggressively. They should be abandoned only as a last resort.

Appendix A: Dialysis Access Recanalization Techniques

"Lyse and Wait"[28]

Despite early literature experience limited to a single technical note with 18 cases, this technique has become widely used for grafts. Now that urokinase is unavailable, some centers are now using t-PA (2-4 mg t-PA or 2-3 units of Retavase in association with 3-5000 U of systemic heparin). The mixture is injected into the graft 30 min to 2 hours before the patient enters the procedure room. The intra-access position of the sheath can be confirmed prior to injection by a small amount of blood return through the sheath or unrestricted passage of a small gauge wire (more reliable). During the injection, both ends of the graft are compressed manually in an effort to prevent peripheral arterial and central venous emboli. It is imperative that the patient is observed while "waiting" as significant bleeding through dialysis needle puncture sites can be encountered, especially with high-grade venous anastomotic lesions. There is some debate about how many clots are lysed vs. centrally embolized with this method, as for most pharmacomechanical and mechanical techniques. In addition, the status of the underlying anatomy is unknown, i.e. the venous outflow may not be amenable to PTA.

Once the patient enters the procedure room, preparation is similar to other declotting procedures. With initial angiography, partial or full thrombolysis may be found. Complete the procedure with Fogarty removal of the arterial plug and venous PTA. Any residual clot seen angiographically should be removed by Fogarty catheter or balloon maceration (my experience is that all enzymatic thrombolytic techniques have significant residual thrombus to be removed and flushed into the venous circulation). Although the technique is convenient and saves time in the angiography suite when successful, similar techniques were described in the early eighties mostly with streptokinase, but were abandoned due to high complication rates and poor results. Anecdotally, my experience, using only 2 mg of t-PA, has been that bleeding (not significant but bothersome none the less) from old puncture sites during the completion of the procedure using this technique has occurred in a high percentage of patients.

Infusion Thrombolysis

This technique is now used infrequently. It has been abandoned by most interventionalists due to high cost, long procedure time and associated complications. Today pulse spray protocol is commonly used; most published literature describes the techniques and results for urokinase. t-PA is actually faster and should have the same safety profile as urokinase if dose-equivalent amounts are used (1 mg t-PA has activity of approximately 100,000 units urokinase). Now available in small dose vials, dosage varies widely with some centers now using 2-5 mg of t-PA while others are using up to 20 mg t-PA. Whatever the dose, the manufacturer recommends that heparin not be mixed with the t-PA because it may cause the drug to precipitate out of solution. Heparin is therefore given into a central vein after the venous outflow has been assessed or, less optimally, through a peripheral IV.

Prep Common to All Declotting Procedures

1. Skin prep.
2. Obtain access to graft directed toward venous anastomosis.
3. Perform draining venogram to confirm venous patency to the SVC.
4. Abandon procedure if venous outflow unsalvageable or if unable to cross venous anastomosis or stenosis.
5. Administer 3-5000 units heparin IV.
6. Crossed catheter access is the standard approach to all current declotting techniques (See Figure 7.44) because it allows treatment of the entire access circuit.
7. Begin clot removal using any of the following procedures.

Pulse-Spray[26]

1. Position pulse-spray catheter(s) in graft. Most use two "crossing" catheters
2. Reconstitute t-PA. Doses vary from 2-20 mg .
3. Pulse t-PA.
4. Assess and treat venous outflow stenosis with PTA.
5. Treat arterial plug

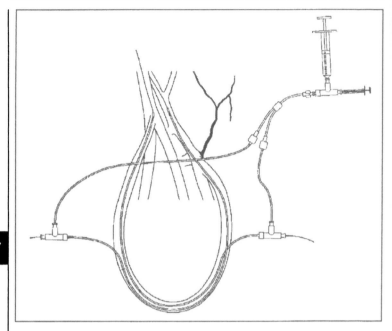

Fig. 7.44. The setup for graft (access) recanalization using crossed catheters (or sheaths) is standard procedure for endovascular declotting by any method. It allows infusions, pulsing, aspiration, Fogarty maneuvers and PTA (if needed or desired) to the entire access circuit.

6. Assess using physical exam of graft, thrill or pulse should be present (see endpoint determination above)
7. Treat residual clot in graft with maceration of balloon mobilization
8. Perform final fistulogram from arterial inflow to right atrium

Mechanical Devices (Generic, Modify According to Individual Device)[30,31]

1. Insert "venous" sheath and apply mechanical thrombolytic device to venous end of graft (some prefer this step before placing "arterial" sheath; others find it easier to place both wires and cross them before passing the sheaths).
2. Place "arterial" sheath.
3. Apply mechanical device to arterial end of graft.
4. If applicable (PTD) aspirate after each pass.
5. Treat arterial plug with balloon (ATD, Cragg, AngioJet, PTD) or device. With some devices (AngioJet, Gelbfish EndoVac) it is preferable to treat the arterial plug last.
6. Assess and treat venous outflow stenosis with PTA.
7. Assess using physical exam of graft, thrill or pulse should be present.
8. Treat any residual clot with device or balloon maceration as necessary.
9. Perform final fistulogram from arterial inflow to right atrium.

Balloon Assisted Thromboaspiration

1. Place 7 F removable hub "venous" sheath. Remove hub and aspirate clot if possible. Replace hub.
2. Place 6 F "arterial sheath".
3. Pass 5 F Fogarty catheter along arterial limb to near anastomosis. Inflate balloon and pull back toward sheath while another operator aspirates on 7F sheath with hub removed.
4. Repeat step 3, passing balloon beyond arterial anastomosis.
5. Assess graft for pulsatility to confirm inflow, if absent repeat step 4 until pulsatility is restored.
6. Pass 5 F Fogarty catheter through venous sheath beyond central limit of remaining clot. Inflate balloon and withdraw catheter while aspirating through sidearm of sheath(s).
7. Repeat steps 3-6 as necessary.
8. Assess residual clot burden; if large use mechanical device to complete procedure. If not, assess and treat venous outflow stenosis with PTA.
9. Assess graft with physical exam, thrill should be restored after successful PTA. Treat any residual clot with Fogarty balloon, pushing into lungs. Perform final fistulogram from arterial inflow to right atrium.

Aspiration and Extrusion Thrombectomy with PTA (Fig. 7.27)[35]

1. Place a 6 F removable hub sheath 10-12 cm from and directed toward the venous anastomosis. It should be positioned so that the sheath is within the graft and the 15 cm inner dilator extends beyond the venous anastomosis and usual site of stenosis.
2. Inject 3,000 units of heparin and perform a draining venogram to confirm venous size and patency to the SVC.
3. Aspirate the venous limb while applying external massage. Repeat several times. Empty the thrombus into a container for visual inspection.
4. Attach the sidearm and perform a gentle venous angio to confirm and size the venous stenosis. Do not inject forcefully enough to cause arterial reflux and embolization.
5. The arterial directed puncture is placed midway between the venous sheath and the venous anastomosis to establish crossed catheters. There is no need for an arterial sheath.
6. A guide wire and 6mm balloon catheter are advanced into the native artery where an inflow angio is performed to confirm arterial size and peripheral arterial patency.
7. The balloon is then partially inflated and drawn into the arterial anastomosis to displace soft clot and size any stenosis at the anastomosis.
8. Perform a 6 mm PTA of the inflow. Needless to say that judgment must be exercised if the inflow artery is small. This is followed by incremental dragging of the partially inflated balloon toward its puncture site and full dilatation about every 2 cm until its position comes to rest (inflated) at the venous puncture site (modified Fogarty maneuver). Perform a gentle retrograde angio to confirm clearing of the arterial limb and patency of run-off vessels to the forearm.

9. Perform a second series of aspirations of the venous limb. This maneuver (8 and 9) should be repeated several times to insure complete clearing of the arterial limb.

10. Attach a check-flow valve and place 8 mm PTA balloon through the sheath into the draining vein.

11. Partially inflate the balloon and draw it into the venous anastomosis where a full PTA completely effaces the balloon for a minimum of one minute. Note that there is now a closed system between the two inflated balloons.

12. Remove the sheath onto the venous catheter shaft and then incrementally withdraw and reinflate the venous balloon (modified Fogarty maneuver) until it "kisses" the inflated arterial balloon extruding the thrombus trapped between the balloons. Repeat this maneuver until no more thrombus can be extruded. This maneuver minimizes systemic venous (pulmonary) emboli.

13. Replace the sheath, deflate the balloons, and perform final angiography through the arterial balloon catheter and venous sheath. Any residual clot or stenosis should be treated by repeat PTA, modified Fogarty maneuver, or extrusion. Rarely is a mechanical device needed to clear adherent clot.

14. Hemostasis by figure-of-eight sutures.

Manual Catheter-Directed Thromboaspiration[24,25]

1. Aspirate clots starting from the most central with a 7 or 8 F slightly angled (vertebral or multipurpose type) aspiration catheter. Perform as many passes as necessary until no additional clots are aspirated.

2. Leave a guide wire through the "venous sheath" and place 7 or 8 F "arterial sheath."

3. Push very gently a slightly angled SF catheter over a hydrophilic guide wire into the artery and check the exact level of the arterial anastomosis.

4. Aspirate clots starting from the sheath. If the catheter is clogged, flush it into gauze and reintroduce it further toward the anastomosis. If blood is aspirated with no resistance, flush it through a gauze and ... (some will return this "filtered" blood to the patient).

5. Inject gently 2 to 5 ml of contrast medium through the arterial sheath during compression of the graft and look for residual clots.

6. Push the angled aspiration catheter to contact any residual clots and aspirate with quick back and forth movements to detach the thrombi. Stop only when absolutely no residual clot is visible.

7. Leave a guide wire through the "arterial" sheath. Aspirate again through the "venous" sheath, as clots may have migrated from the arterial side and can be blocked between the two introducers or at the venous anastomosis.

8. When absolutely no residual clot is visible, dilate the stenosis of the venous anastomosis. To dilate before aspiration of all the thrombi increases the risk of pulmonary embolism and is against the spirit of the technique.

9. Final angiogram from arterial inflow and outflow to right atrium

Declotting of Native Fistulae[21-25]

1. Skin prep.

2. Obtain access to fistula toward both (arterial) anastomosis and central veins in a crossed fashion.

3. Pass catheters down to the feeding artery and up to central veins. Do pull-back angiograms to assess the exact level of the anastomosis and the central

extension of the thrombus.

4. If the anastomosis cannot be reached from the fistula, inject contrast from the brachial artery at the elbow to provide mapping and use an antegrade brachial approach for selective catheterization of the feeding artery. If unable to cross the anastomosis or the venous outflow, abandon the procedure (<10% of cases).

5. Begin clot removal starting from the venous outflow, ending with the arterial inflow. Predilate stenoses (5 mm) and use safety guide wire if necessary to allow passage and safe use of the thrombectomy catheter.

6. When all clots are removed or lysed, dilate all stenoses sufficiently. Restoration of a good inflow is essential.

7. Final angiogram from the arterial inflow to the right atrium.

Appendix B: Billing and Coding

Component coding for billing as used in all interventional radiology procedures applies to dialysis vascular access. The principle of component coding /billing is that if a procedure is indicated as "medically necessary," it should be performed, documented in the report, and coded for what was done with full expectation for payment. At the present time, no national uniform coding scheme for dialysis access interventions is in place. Many carriers have local medical review policies (LMRPs) in place, which vary from state to state. Individual physicians should consult with their Medicare carrier for local coding requirements. This section is intended to share general guidelines that have been used with success.

1. Effective for 2001, code 36870 describes percutaneous declotting of permanent dialysis accesses using aspiration, mechanical, pharmacomechanical and pharmacological (infusion < 1 hour) methods. This is a national stand-alone code that has no bundled components; it is only for clot removal by any means. This new code replaces codes used in the past. Its one disadvantage is that it comes with a 90-day global restriction that does not consider complications that occur in dialysis over which we have no control and are not related to the efficacy of the procedure. For thrombolytic infusion of greater than 60 minutes, 37201 is an appropriate choice.

2. Although component-coding rules indicate that 36145 can be used for each graft catheterization, some carriers limit use of this code to just one use, despite the number of punctures actually necessary for graft revascularization. Code it as many times as it is performed and make the payers deny it.

3. For venous angioplasty, 35476 and 75978 can typically be used only once, even if several peri-anastomotic or intragraft stenoses are treated. It can be used with each anatomic venous name change again as a new code, especially if documented in the setting of a separate central venous stenosis.

4. Although some carriers have proposed denying reimbursement for the off-label use of vascular stents, most states with published policies now specifically allow reimbursement when performed in the setting of dialysis access.

5. Supervision and interpretation code 75790 should be used for dialysis access angiography. Whether an additional S&I code (75820) should be used for upper extremity venography is not specifically addressed in the SCVIR Coding Users' Guide; however, in the setting of multiple injections to visualize the entire venous drainage, it would seem appropriate. When specifically indicated and performed, central venography codes are also appropriate, superior vena cavagram 75827.

6. If a true arterial lesion remote from the graft is dilated, codes 35475 and 75962 may be used. However, appropriate justification (documentation both with images and in the report) must be given. It is inappropriate to use this code for treatment of clot or plug at the arterial anastomosis if it cannot be documented by images as an anatomic lesion with > 50% narrowing of the inflow. I find this to be present and code for it almost half of the time (48%). If dedicated arm arteriography is performed (such as in the identification of arterial embolism or for evaluation of steal syndrome), S&I code 75710 may be used in this setting.

7. Coil occlusion of significant stealing side veins is coded as non-neuro embolization, 37204 with S&I 75894 and can be coded for each vessel occluded. Ligation of side veins is 37607.

8. Evaluation and management codes are appropriate when seeing a new patient, 99241, and when following a previous patient for a chronic problem 99212. These must be documented by appropriate history, physical exam, and treatment decisions.

Selected References

1. Windus DW. Permanent vascular access: A nephrologist's view. Am J Kidney Dis 1993; 21:457-471.
2. United States Renal Data System 1997 Annual Report. US Dept of Health and Human Services; April 1997.
3. Zibari GB, Rohr MS, Landreneau MD et al. Complications From permanent hemodialysis access. Surgery 1988; 104:681-686.
4. Health Care Financing Administration End Stage Renal Disease Program Highlights 1995; August 31, 1996.
5. United States Renal data system 1999 annual data report. US Dept of Health and Human Services; April 2000.
6. Gray R. Percutaneous intervention for permanent hemodialysis access-a review; JVIR 1997; 8:313-327.
7. International Panel. Is the AV fistula the gold standard for dialysis vascular access? 2nd International Congress of the Vascular Access Society; London, UK, June 2000.
8. Fillinger M. Pathophysiology of dialysis access graft dysfunction; Second Annual Venous and Hemodialysis Access Symposium; Keystone, CO, Feb 1998.
9. Allon M, Bailey R, Ballard R et al. A multidisciplinary approach to hemodialysis access: Prospective Evaluation. Kidney Int 1998; 53:413-419.
10. Savader S et al. Treatment of hemodialysis catheter-associated fibrin sheaths by rt-PA infusion: Critical analysis of 124 procedures. JVIR 2001; 12:711-715.
11. Merport M, Murphy TP, Egglin TK et al. Fibrin Sheath Stripping versus catheter exchange for the treatment of failed Tunneled Hemodialysis Catheters: Randomized Clinical Trial. JVIR 2000; 11:1115-1120.
12. Gray RJ et al. Percutaneous Fibrin sheath stripping versus transcatheter urokinase infusion for malfunctioning well-positioned central venous catheters: A prospective randomized trial. JVIR 2000; 11:1121-1129.
13. Schemmer DC, Gray RR et al. Prospective analysis of central venous catheters after balloon angioplasty of fibrin sheath. SCVIR 2001: poster abstract no. 345: S137, JVIR Supplement 12:1, part 2, Jan 2001.
14. Beathard GA, Settle SM, Shields MW. Salvage of the nonfunctioning arteriovenous fistula. Am J. Kidney Dis 1999; 339: 910-916.
15. Turmel-Rodrigues L et al. Salvage of immature forearm fistulas for haemodialysis by interventional radiology. Nephrol Dial Transplant 2001; 16:2365-2371.

16. Arnold WP. Efficacy of endovascular interventions in promoting maturation of native arteriovenous fistulae. Abstract Presentation 4.7:32; 2nd International Congress of the Vascular Access Society, London, UK, June 2001.

17. Falk A. The management of hemodialysis arteriovenous fistulas. New Developments in Vascular Diseases 2001: 2:2:11-14.

18. Arnold WP, Replogle M. Native arteriovenous fistulae for hemodialysis access: three years experience with endovascular maintenance. JVIR 2001:12(2 pt 1):S55.

19. Arnold WP. Infiltration and Perforation PTA: A percutaneous endovascular technique for treatment of resistant venous stenosis in hemodialysis vascular access. Abstract presentation 8.6:56; 2nd International Congress of the Vascular Access Society, London, UK, June 2001.

20. Arnold WP, Shetye KR. The treatment of long stenoses in dysfunctional hemodialysis accesses. J Amer Soc Nephrol 2000; 11:A0955, SU570 (PS).

21. Turmel-Rodrigues L et al. Treatment of stenosis and thrombosis in haemodialysis fistulas and grafts by interventional radiology. Nephrol Dial Transplant 2000; 15:2029-2036.

22. Haage P et al. Percutaneous treatment of thrombosed primary arteriovenous hemodialysis fistulae. Kidney International 2000; 57:1169-1175.

23. Turmel-Rodrigues L, Pengloan J, Rodrigue U et al. Treatment of failed native arteriovenous fistulae for hemodialysis by interventional radiology. Kidney International 2000; 57:1124-1140.

24. Poulain F, Raynaud A, Bourquelot P et al. Local thrombolysis and thromboaspiration in the treatment of acutely thrombosed arteriovenous hemodialysis fistulas. Cardiovasc Intervent Radiol 1991; 14.

25. Tunnel-Rodrigues L, Sapoval M, Pengloan J et al. Manual thromboaspiration and dilation of thrombosed dialysis access: mid-'term results of a simple concept. JVIR 1997; 8:813-824.

26. Valji K. Transcatheter treatment of thrombosed hemodialysis access grafts. AJR 1995; 164:823-829.

27. Beathard GA, Welch BR, Maidment HJ. Mechanical thrombolysis for the treatment of thrombosed hemodialysis grafts. Radiology 1996; 200:711-716.

28. Cynamon, J, Lakritz, PS, Bakal, CW et al. Hemodialysis graft declotting: Description of the "lyse and wait" technique. JVIR 1997; 8:825-829.

29. Arnold WP. Aspiration and extrusion thrombectomy with percutaneous transluminal angioplasty: Technique, efficacy, and patency in hemodialysis access management. Abstract Presentation SCVIR 1998 Annual Meeting, SS 21-4, 3 Mar 1998.

30. Sharafuddin MJA, Hicks ME. Current status of percutaneous mechanical thrombectomy. Part II. Devices and mechanisms of action. JVIR 1998; 9:15-31.

31. Sharafuddin MJA, Hicks ME. Current status of percutaneous mechanical thrombectomy. Part III. Present and future applications. JVIR 1998; 9:209-224.

32. Arnold WP. The effect of venous outflow angioplasty balloon size on primary patency in hemodialysis vascular access graft recanalization. Abstract poster presentation 82:120; 2nd International Congress of the Vascular Access Society, London, UK, June 2001.

33. Ayus JC, Sheikh-Hamad D. Silent infection in clotted hemodialysis access grafts. J Am Soc Nephrol 1998; 9:1314-1317.

34. Schur I, Koh E. Resistant arterial anastomotic lesions in thrombosed hemodialysis conduits: Current methods of percutaneous treatment. Abstract presentation: Vascular Access for Hemodialysis VII. 2000:16.

35. Arnold WP. Aspiration and extrusion thrombectomy with PTA: Technique, efficacy, and patency in hemodialysis access management. Poster presentation 0424VI, Radiology (supplement) 1998; 209:569.

36. Vesely TM. Use of a purse string suture to close a percutaneous access site after hemodialysis graft interventions. JVIR 1998; 9:447-450.

37. Arnold WP. The Restoration of function in hemodialysis grafts with recurrent thrombosis. Supplement to Radiology 2000; 217 (P); Abstract Presentation 543:332.

38. Quinn SF, Schuman PS, Demlow TA et al, Percutaneous transluminal angioplasty versus endovascular stent placement in the treatment of venous stenoses in patients undergoing hemodialysis: Intermediate results. JVIR 1995; 6:851-855.

39. Funaki, B, Szymski, GX, Leef, JA et al. Wallstent deployment to salvage dialysis graft thrombolysis complicated by venous rupture: Early and intermediate results. AJR 1997; 169:1435-1437.

40. Trerotola SO, Johnson MS, Shah H et al. Technical note: 'backbleeding' technique for treatment of arterial emboli resulting from dialysis graft thrombolysis. JVIR 1998; 9:141-143.

41. Schwab S, Besarab A, Beathard G et al. NKF-DOQI clinical practice guidelines for vascular access, Am J Kid Dis 1997; 30(4, Supp 3), S150-S191.

7

CHAPTER 8

Common Dialysis Access Management Strategies

Ingemar J.A. Davidson

Men occasionally stumble over the truth, but none of them pick themselves up and hurry off as if nothing had happened — Sir Winston Churchill

Introduction

Thrombectomies with or without revisions currently constitute approximately half of all vascular access OR procedures (Appendix V), not counting dialysis catheter placements. With increasing dialysis access surveillance programs, revision prior to thrombectomy is a likely and highly desirable trend (Chapter 10). The benefits of access surveillance programs include: increased dialysis efficiency from prospective monitoring and routine troubleshooting, avoiding placement of temporary dialysis catheters associated with clotting episodes, thereby saving central veins for future access. Second, the surgery can be planned causing less interruption in the dialysis routine, as well as for the surgeon and the operating room scheduling. Also, the patient is less likely to be admitted for complications associated with thrombosis such as fluid overload, congestive heart failure and hyperkalemia. Many effective surveillance options are available, from straightforward dialysis machine-driven testing to invasive fistulography. Even though surveillance programs are associated with some cost, proactive interaction is likely to save dollars for society in addition to the improved quality of life for patients and dialysis access team members.

This chapter outlines the authors overall approach to some common access management problems. Specific examples are also illustrated in Appendix I case reports.

In general, the safest and correct approach is to carefully perform physical exam on each patient, then consider further tests such as duplex Doppler, fistulagram and central venograms before presenting to the patient his or her options, as well as a prediction of outcome.

Another basic principle is not to take the easy road, i.e., placing a central venous catheter just to avoid or delay a difficult access problem. The less traveled road in this case is to do what most likely will give the patient dialysis access long term and save central veins for emergencies.

Doing the right thing in this ESRD context is to prevent or delay renal disease. This starts with the patient and the primary care physician by treating hypertension, preventing infections, engaging in regular exercise and maintaining healthy weight. Then the primary physician and/or the nephrologist must follow the patients renal function decline closely and refer for access placement when the GFR approaches 20-25 ml/min. Early referral alone will increase the likelihood of native AV fistula placement, a highly desirable outcome.

Education to the patient and family is key to success (Appendix III). Patients who are proactive, knowledgeable about their disease process are more likely to undergo a preemptive kidney transplant from a living donor, leaving their professional

Access for Dialysis: Surgical and Radiologic Procedures, 2nd ed., edited by Ingemar J.A. Davidson. ©2002 Landes Bioscience.

life for only a brief period of time. However, there is much patient denial causing delays in the prevention and treatment. Compassionate teaching by health care personnel at all levels is needed and recommended, and will largely alleviate patient delay. Responsible family members, i.e., spouse, parent and children are encouraged to become active participants. They are more likely to hear the facts. Patient delay in access placement is a major problem and cause of suboptimal dialysis access and morbidity. For example, in the U.S. 40% of all initiation of dialysis is with a temporary dual lumen percutaneous or cuffed catheter.

Valuable Tools

Duplex Doppler
Duplex Doppler is of immense value for diagnosis, vein mapping, anatomy definition of vascular access cases.

Vascular Lab Testing
The author makes daily visits to the vascular laboratory, using the arsenal of testing procedures (Table 8.1). The quality of information is very operator (technician) dependent in terms of skills, knowledge and availability. Donna Nichols, vascular laboratory technician at Medical City Dallas Hospital superbly exemplifies all these qualities (Fig. 8.2).

For maximal outcome effectiveness, the surgeon must be present for most of these examinations to direct and determine the optimal anatomic sites and mark skin for surgical incisions. This takes time and causes inconveniences for all involved, but correctly done dramatically improves outcome. Often vein mapping is best done on the day of surgery while patient is in the day surgery or preop holding area.

Portable Ultrasound
The use of portable ultrasound, i.e., the Site Rite® (Fig. 8.1) device is highly recommended. Its indications and usefulness were discussed elsewhere (Chapter 1, Table 1.1; Chapter 5, Fig. 5.4). The author uses this device in all cases of central vein dialysis catheter placement including femoral vein percutaneous dual lumen dialysis catheters. The benefits include increased safety avoiding inadvertent arterial punctures, not to mention erroneous placement of the dialysis catheters in the carotid, subclavian and femoral arteries.

Table 8.1. Examples of vascular laboratory testing for vascular access procedures

---

1. Determine patency, anatomy and flow pattern of central veins (internal jugular, subclavian and femoral veins).
2. Diagnosing and characterizing stenosis, aneurysms and other complications of PTFE grafts and native AV fistulae.
3. Estimate flow rate (ml/min) in grafts and fistulae (flow (ml/min)= flow velocity (cm/sec) x r^2 (cm) x 3.14 x 60 sec).
4. Vein mapping for native AV fistulae in the forearm, demonstrating (cephalic) vein patency.
5. Vein mapping upper arm for optimal graft to vein anastomosis site.
6. Finger pressures in cases of arterial "steal" symptoms.
7. Arterial (i.e., brachial, ulnar, radial) mapping in cases of peripheral ischemia.

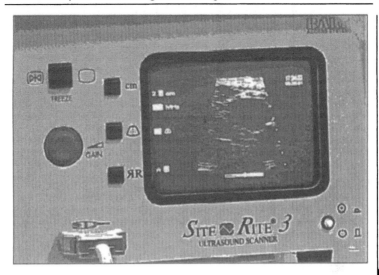

Fig. 8.1. Site Rite® portable ultrasound device.

8

Fig. 8.2. Donna Nichols, Vascular Laboratory, Medical City Hospital, Dallas, Texas; Has the three qualities for effectiveness, skills, knowledge and a great attitude.

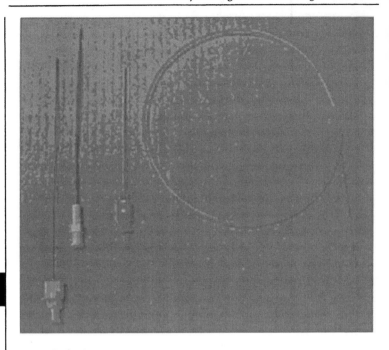

Fig. 8.3. Cook micropuncture set

Micropuncture Technique

The micropuncture technique (Fig. 8.3) for catheter placement was described in detail elsewhere (Chapter 5, Table 5.4, Fig. 5.10A-F). The micropuncture technique also increases safety. It is the author's impression that long term patency of central veins is enhanced, because of less perioperative trauma and hematoma causing less obstruction.

Obesity

Obesity in the US has reached epidemic proportions, and needs to be addressed as a national disaster. Recent statistics indicate that at least 60% of Americans are overweight. Twenty-five percent are morbidly obese, that is 30% above expected norms based on sex, height and body frame. The obesity problem is seldom addressed when the ratio of primary AV fistula to PTFE is compared to that in other countries. The lower native vein utilization in the U.S. is often perceived as a surgeons lack of interest or skills. Nephrology and industry pressure are also often cited. Most European countries have an obesity rate of 10% or less.

Clearly, access placement in obese people presents several challenges (Fig. 8.4). The patient must be made aware of these as well as the prognosis for future access problems and interventions and the diminished likelihood for both acceptance into a transplant program's waiting list and a successful transplant. Controlling the eating addiction is of course the only real long term solution. Having said all this, the reality is that the obese access population is going to affect us in about 70% of cases.

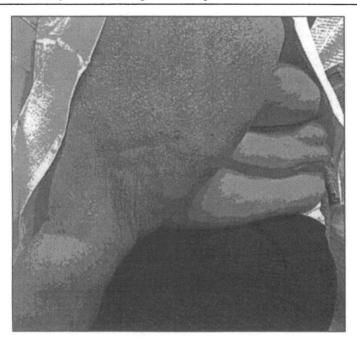

Fig. 8.4. One of the most unrewarding challenges in vascular access surgery is adipositis. Not only is the creation of native fistulae questionable or impossible, but PTFE grafts are also difficult to place at the appropriate depth below skin level.

Information on the relation of body habitus to access options must be presented to the patient by the nephrologist, surgeon and renal dietitian. The impetus should be placed on the patient to lose weight, but the dialysis system must make education and appropriate referral available.

Native Fistulae (AVF) in Obese People

The author has recently changed his approach in the indications for primary AV fistulae. Whenever patients come early, that is ideally 3-6 months before needing dialysis, a primary AV fistula is placed at the wrist on the side most likely to mature to a usable dialysis vein. Should this vein not become usable, a few weeks prior to need of dialysis a PTFE graft is placed using the dilated vein as venous outflow. In other words the fistula is converted to a PTFE AV graft (See Appendix I, Case #1,2,3). Often the graft to cephalic vein anastomosis can be placed in the distal third of the forearm, leaving room for future revisions below the antecubital fossa (Appendix I, Case #1). Of 24 AVFs so converted (12/31/01), 21 have been working for up to 1 year (94% predicted graft survival, Kaplan-Meier) (Appendix V, Fig. 5A). Three of the four that have had clotted episodes were declotted and revised. Two of the 24 have failed from repeated clottings. Although small numbers, this approach is quite encouraging.

8

PTFE in Obese People

Even with the expanded indications for primary AVFs in overweight people, most obese patients will eventually have PTFE placed. Again, a forearm PTFE in a loop formation is appropriate in the majority of cases. When there is long standing diabetes or fragile skin, an upper arm PTFE graft with the arterial anastomosis more proximal to avoid arterial steal or hand ischemia may be indicated (See Appendix I, Case #38).

The specific challenge in obese people is placing the graft at the appropriate depth under the skin. This was described in detail previously (Chapter 4, Fig. 4.10A-C). Planning and increased awareness on the surgeon's part will decrease the dialysis needle puncture complications. Again, the surgeon should mark the skin preoperatively for the intended course of the loop, choosing optimal veins in the antecubital fossa, possibly by preoperative duplex Doppler vein mapping. Sometimes, the basilic vein is a better outflow option, not just in the obese but in any patient. In obese patients especially, but ideally in all patients, the graft or fistula should be marked prior to cannulation, using duplex or hand held Doppler or portable ultrasound. The anastomoses, direction of flow and course of the access should be marked with indelible pen, and the information communicated to the patient and dialysis unit.

Dual Lumen Catheters in Obese People

Regarding catheters in overweight individuals, the use of the ultrasound (Site Rite®) (Chapter 5, Fig 5.4A-B, Chapter 1, Table 1.1) for internal jugular vein identification and puncture should be considered mandatory. This portable ultrasound device guidance and the micropuncture set markedly improves safety. The usual access problems are just magnified in obese people. Great care must be exercised with anesthetics and sedatives, especially if comorbidity is present, such as: fluid overload, diabetes (usually type II) and advanced age. Excess sedatives not only creates an uncooperative patient but may require intubation and time in the ICU for weaning off the respirator. As addressed in Chapter 5, special consideration should be given to the position of the catheter tip and exit site to avoid proximal catheter migration in the vein when tension is placed on the exit site by excess, pendulous tissue.

The Elderly

The average age for patients entering ESRD programs is now 65 years in 2001. Patients needing PTFE are ten years older than those receiving native fistulae in the author's experience (Appendix 5, Fig. 2A). In addition to many social issues in the elderly, this patient population uniformly has various degree of comorbidity, making even minor day surgery a stressful event. The proper level of anesthesia becomes a key to success. Choosing the right access also becomes more challenging because of the decreased life expectancy, life quality issues, support system availability, etc. In elderly with short life expectancy, major severe comorbidity, and decreased mentation, the author relies more heavily on cuffed, dual lumen tunneled catheters, especially initially (Appendix I, Case #48). Should the patient make significant recovery after initiation of dialysis treatments, permanent access is considered. Family involvement is of paramount importance in dealing with the elderly in order to put the disease process in perspective. High outcome expectation from access placement and dialysis must be moderated. However, being old is not a limitation or a contraindication for access placement and dialysis. Elderly without comorbidity can often have AVFs placed early and dialysis started without the use of catheters.

Paper Thin Skin

Fragile skin is commonplace, more so in females and in the elderly. Other patient categories include failed kidney transplant after years of steroid treatment. Also problematic are individuals with many years of outdoor sun exposure, where tendons are sometimes outlined under the paper thin skin. This kind of forearm does not lend itself to PTFE placement. Alternatives to forearm AV grafts include: cuffed tunneled catheters or upper arm native fistulae or PTFE grafts. The decision is made after thorough physical exam and consideration and discussion with the patient and family.

Arterial Steal

Hand ischemia has become increasingly common with PTFE grafts secondary to the changing ESRD population, most notably increasing age and diabetes. Also, with emphasis on primary AV fistulae, the elderly and high risk patients are left for PTFE placement. In the author's last consecutive 191 PTFE first time, fresh arm placements, 11 (6%) developed hand ischemia requiring surgical banding (Chapter 4, Fig 4.29). The graft salvage rate is about 60% at one year, making this procedure very worthwhile rather than abandoning the access (Appendix V Fig. 5B). In comparison, only 1 of the 163 (0.6%) primary consecutive wrist AVFs in the same time span developed symptomatic hand ischemia, likely reflecting the lower flow rate in AVFs.

Arterial steal with hand ischemia is a serious symptom and warrants close attention. Mild symptoms, i.e., noticeable coolness of the hand, some tingling but no pain or slightly impaired motor function can be watched by frequent examinations. Symptoms often dissipate over the next few days or weeks. In the authors experience, this situation represents finger pressures of 30-50 mm Hg. Partially compressing (banding) the graft will relieve symptoms within seconds to minutes. In cases of rest pain often increased during dialysis, blue or white fingertips, urgent banding is indicated. Ulcerations occur with chronic ischemia especially at fingertips under the nail Appendix I, Case #28) . There is an emergent need for banding when there is a cool bluish hand with pain and impaired movements. Rarely is there an indication for ligation of the graft.

The banding procedure with a gradual occlusion with a large hemoclip while measuring finger pressures or observing a pulse oximetry curve was described in Chapter 4, Fig. 4.29, Table 4.1.

Banding is performed in the operating room with a vascular lab technician available for finger pressures (waveforms) before and after the procedure for documentation as well as guiding the degree of banding. Pulse oximetry curves before and after are helpful.

Banding access with no vascular lab support (the emergent weekend situation) requires the surgeon's estimation of capillary refill (i.e., pulse oximetry recording) and patient (without sedation) reporting relief of symptoms is an acceptable approach. The banding procedure uniformly relieves the patient symptoms. Clotting may occur early or weeks or months after banding. At declotting of banded grafts months later, the hemoclip may be permanently removed with no new ischemia symptoms occurring.

Poor Arterial Inflow

This situation represents patients with preexisting, easily recognizable or suspected problems. These individuals may have positive Allen test or Doppler determination of insufficient palmar arch flow with radial or ulnar arteries sequentially (manually) occluded. Duplex Doppler is used to assess arterial quality from the subclavian artery to the wrist. Finger pressures are obtained as well. The presence of subclavian, axillary or brachial artery stenosis should be subjected to balloon angioplasty. Distal ulnar and radial artery angioplasty may be attempted, but is less likely to improve flow, and carries a risk for occlusion of remaining hand circulation.

In cases of known distal inflow impairment, (that is atherosclerosis, lack of peripheral pulses, history of smoking, diabetes), a very high arterial graft anastomosis may be justified in the extremity with the lesser estimated problems (Appendix I, Cases #33 and 38). This approach is based on the principle that a larger proximal artery used for graft anastomosis is less likely to cause distal (hand) ischemia.

Other options in these cases are, of course Peritoneal dialysis or dual lumen cuffed tunneled catheters. However, this patient population is likely to be elderly, diabetic, and obese.

Aneurysms

True Aneurysms

True aneurysms occur frequently in native AV fistulae and are exceedingly rarely in PTFE grafts. However, they occur often in bovine (5-10%) grafts, which are rarely used.

True aneurysms in primary AV fistulae should be surgically corrected when they become bothersome because of size or when the skin becomes shiny (Appendix I, Cases 9 and 10). There are several operative approaches which must be individualized, ranging from removal of the entire fistula and ligating at the arterial anastomosis to resection of the aneurysm with reanastomosis to healthy vein. An interposition vein or PTFE graft are other options, depending on local anatomy. Often a more proximal stenosis contributes to or causes aneurysms to expand. Therefore, the entire fistula from anastomosis to right heart must be examined, first by physical exam and duplex Doppler and possibly also radiologic fistulagram and balloon angioplasty performed as indicated.

Pseudoaneurysms

Pseudoaneurysms occur rarely at the PTFE anastomosis sites but more commonly along the PTFE graft from repeat dialysis needle puncture holes. Again, these should be corrected when the skin becomes shiny or the aneurysm becomes large, painful or is expanding rapidly, over weeks to a few months (Appendix I, Cases #30-32). The technique for repairing these varies with local anatomy. Also a more proximal outflow stenosis should be corrected, which in itself may decrease pressure in the aneurysm and prevent further surgery directly aimed at the aneurysm itself.

Pseudoaneurysms also occur in native AV fistulae, caused by dialysis needle punctures. Indication for surgical correction are similar to PTFE grafts.

Crossing the Antecubital Fossa

When clotting occurs in a PTFE graft a venous anastomosis obstruction can be expected in 90% of cases, most effectively addressed with a patch PTFE (i.e., Accuseal) angioplasty (Appendix I, Cases #20 and 21). In cases of repeat clottings or with more extensive fibrosis affecting a longer segment, other options include finding a new outflow vein or graft extension or interposition PTFE grafts. Sometimes the antecubital fossa has to be crossed to find an adequate outflow vein (i.e., basilic vein) in order to save the current access (Chapter 4, Figs. 4.22-4.25). A standard PTFE graft crossing the antecubital fossa may kink, inducing thrombosis. A new kink resistant graft (Intering W.L. Gore, Flagstaff, AZ) with reinforced (condensed PTFE) rings within the wall itself offers a distinct advantage over current external plastic ringed grafts. The author has used this graft for crossing the antecubital fossa on 15 occasions (Appendix I, Cases # 18 and 22). It can be cut and sutured through the rings with minimal resistance.

The System Factors

The overall success of vascular access and the effectiveness of dialysis treatment depend on many interfacing "system" factors (Table 8.2). When browsing through this table the reader must come to the conclusion that many of these are not immediately correctable, such as obesity. Skills and knowledge levels are what medical training and education are all about. Changing team members' attitudes may take time and requires in itself specific leadership skills.

Even though a "perfect" dialysis and vascular access system may not be attainable, the best situation for the patient and team members is acceptable quality of life. Improvement programs must be put in place in areas where they are most effective. These efforts are best initiated by combined efforts of the medical profession and institutions, for instance the Center for Medicare and Medicaid Services (CMS), the National Kidney Foundation (NKF), The American Society of Transplantation (AST) and other ESRD private and public institutions.

Doing the Right Thing

The management of the ESRD patient is a multidisciplinary endeavor. Key players are the nephrologist, the dialysis unit personnel, the vascular access surgeon and the interventional radiologist (Table 8.3). The most important team member is the patient. A coordinating team member, the vascular access coordinator, is

---

Table 8.2. Example of system factors contributing to dialysis access success or failure

---

1. Skill, knowledge and attitude of the access team members, including the patient.
2. Institutional, departmental, hospital support of the overall ESRD mission including transplantation, organ procurement and dialysis.
3. Advanced age.
4. Patient compliance.
5. Surveillance program effectiveness; mission and goals.
6. Level of diabetes and blood pressure control.
7. Life style habits (i.e., obesity, smoking, lipids/cholesterol control, diabetes glucose (Hba$_1$c) control.
8. Continuing educational objectives for all team members, including patients and families.
9. Proactive attitude of team members, initiating communication and feedback on all aspects of patient care.

Table 8.3. Team members

Nephrologist
Dialysis RN - Technician
Surgeon
Interventional Radiologist
Patient
Access Coordinator
Administrative Bodies (i.e., CMS, NKF, Hospital Administration)

often missing. The too common less effective dialysis access team structure as well as an ideal model is cartooned in Figure 8.5A-B.

Traditionally, the dialysis access surgery is put on the "back burner". Several factors contribute to this less than optimal management. The ESRD patient is typically elderly, often with several severe comorbidities and from the lower socio-economic status. Adding the tiring effect of dialysis and uremic symptoms makes this patient group less likely to stand up for themselves.

There is much delay in getting dialysis access placed in patients approaching uremic status. Forty percent of all patients need temporary catheters for starting dialysis. Patient delay relates to denial or inability to deal with a potentially deadly, chronic disease.

On a daily basis each dialysis patient's quality of life and survival depends on each team member's intention of Doing the Right Thing. The Right Thing may vary depending on local community's available expertise and technical support, patient and family desires. The attitude, that is willingness to do the right thing, must be the primary goal of each team member and cannot change by geography, equipment or scheduling.

Examples of What Every Team Member Must Do

- Educate patients, each other and self (i.e., read, attend seminars, inservices).
- Be proactive, take charge, don't wait to be told.
- Be nice and also firm with patients and each other. Don't accept an unacceptable practice.
- Expect and demand patient compliance with medication, appointment times, treatment schedule and objectives.
- Don't accept or use abusive language or behavior.
- Listen to and hear all sides in discussions and disputes.
- Preserve access sites by cooperation, communication and trust building.
  - Adjust schedules in order to facilitate declotting and revision of failing access without need for placement of central vein catheters.
  - "Go the extra mile" in gray areas between team members. Do not assume that "they" are doing it.
  - Make the patient/family part of decision/action taking.
  - Reinforce importance of maintaining access sites to patients and families.

What the Nephrologists Must Do

- Optimally treat patients to delay the need for dialysis initiation. This is especially important for blood pressure and glucose control and prevention of infections.

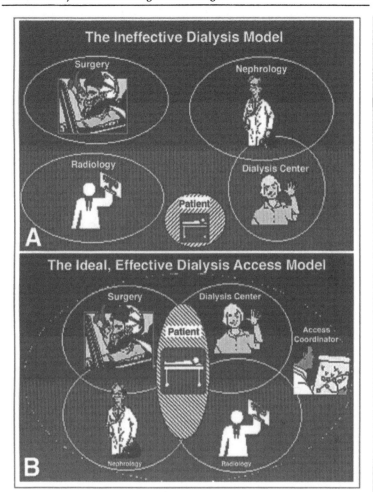

Fig. 8.5. The too common ineffective dialysis model (A) promotes ineffectiveness, and is characterized by late referrals, last surgery of the day ("add on" cases), high usage of temporary catheters and blaming other team members for causing the access failures. The "ideal" dialysis access model (B) in contrast places the patient's need in the center, has a vascular access coordinator, team members communicate effectively to minimize scheduling disruption; chronic catheter use in these centers is about 5% or less.

- Educate patients early regarding future need for dialysis.
- Advocate preemptive living donor transplantation when appropriate.
- Refer patients early for dialysis access placement (GFR = 15-20 ml/min).
- Regularly support and educate dialysis unit personnel through inservices.
- Communicate to the surgeon pertinent information and laboratories (Cr/S, BUN, GFR) and estimate time for dialysis initiation.

- Refer patients for continuing surgical and medical specialty care as needs arise, based on each patient's individual case.
- Review renal replacement therapy options including transplantation regularly as patient condition changes and adjustment to dialysis is made.

What the Surgeon Must Do

- Carefully examine each and every patient for dialysis options, that is native AV fistulae, peritoneal dialysis catheter, PTFE graft.
- Think "art", not "plumbing".
- Educate patient and family as needed about the options including transplantation (Appendix III).
- Communicate to dialysis unit and the patient's nephrologist after surgery (i.e., OR report and/or picture of access).
- Basic information to patient and family regarding post op management and exercise of arm for optimal access development (Appendix IV).
- Support dialysis units through educational programs (inservices, CME seminars, etc).
- Review access surveillance results as received to determine the need for prophylactic intervention of early problems.
- Work with dialysis unit staff and patients when cannulation or other access utilization difficulties arise.

What the Dialysis Director of Nursing (DON) Must Do

- Train and maintain high quality, state of the art dialysis staff.
- Encourage and support educational efforts.
- Bring nephrologists and surgeons to the unit for inservices.
- Encourage or demand patients and/or families to hold the puncture site after dialysis.
- Put patient issues/safety before the economics of the dialysis company.
- Implement and maintain an effective access surveillance program.
- Communicate regularly with surgeons and nephrologists on access function and utilization.
- Encourage "best teach the rest" mentality for training new staff on cannulation and patient care.
- Facilitate and participate in patient support groups and patient mentoring projects.

What the Patient and the Family Must Do

- Be willing to learn about the ESRD diagnosis (Appendix III).
- Work with the medical and dialysis staff to achieve treatment compliance.
- Ask questions when presented with new information or problems.
- Be willing to participate in their care and medical decision making.

Future Dialysis Access Issues

There is recently much emphasis on dialysis access. There are at least four areas receiving attention (Table 8.4).

The Access Center

In the US, several so called Vascular Access Centers are sprawling. It is the author's strong belief that the dialysis patient is better served in such a center as long as its

Table 8.4. Dialysis access concern areas

1. The need for optimal utilization for native vein for AV fistulae
2. Decreasing the (ab)use of central vein dialysis catheters
3. Improving dialysis access outcome after declotting and revision procedures
4. Dealing with the elderly

mission is to optimally manage dialysis access problems, and not a company's stock value.

Correctly done, the leadership for such a center would correct or significantly improve the first three issues listed in Table 8.4.

Who Does What and When

With rapidly changing and improving technology, interventional radiology has come to play an important role in many community hospitals and access centers (Chapter 7). This situation has created some turf issues with the surgeon. The recent addition of "interventional nephrologists" makes this a triangular drama.

This healthy competition is good, as long as it is driven by competence and outcome data and not by politics and dialysis companies. The issues are likely to continue for some time as the optimal or "best" treatment regimens continue to be defined. Again, attitude of the treatment team is a major factor in outcome, in addition to the obvious skill and knowledge requirements.

The Anatomy of the Troubled Vascular Access

The most common vascular PTFE anatomic problem is the venous anastomosis stenosis from intimal hyperplasia, occurring in 90% of clotted accesses. This is typically corrected by surgeons using a patch angioplasty (Appendix I, Case #20-21), or interposition grafts (Appendix I, Case #18-19). Therefore surgery, by preserving several venous outflow veins is the most effective corrective measures in this respect.

Arterial stenosis, although to a lesser degree and frequency is also common. It is the author's opinion that not only must the arterial plug be retrieved, but the anastomosis may benefit from balloon, angioplasty in the majority of cases. This is perhaps the interventional radiologist's territory, but can also be done in the operating room and requires fluoroscopy capacity.

Also stenoses in the outflow venous system in the upper arm, subclavian vein or superior vena cava are present in about 30% of clotted vascular access. Not only must the entire outflow venous system be visualized, but problems corrected with balloon angioplasty. Again, the interventional radiologist has the upper hand in this setting.

Finally, intragraft problems, stenoses from fibrosis and scarring from needle punctures and pseudoaneurysms may be corrected by a variety of procedures available to the surgeon, including balloon angioplasty.

Therefore, to completely treat the troubled access, the availability of both surgical and radiologic expertise is necessary, ideally at the same time and in the same setting to treat all problems. This represents the ideal access center. There seems to be several options. In the first model, a dedicated surgeon and radiologist work closely together to correct the problems in one setting. This requires much planning, cooperation and a fairly large volume of patients to be economically feasible.

The second model is represented by a surgical approach where the surgeon has attained the skills to perform angioplasty and intraoperative diagnostic venograms.

8

This requires a dedicated surgeon, technical support with fluoroscopy and trained personnel.

The most common and current model is a mixture of the above, with surgery and radiology sharing the access complications, depending on local referral patterns.

Whatever system is in place, stratified outcome follow-up data will indicate the most effective local model. Data must be shared and new technology skills taught between surgeons, radiologists and nephrologists. This is consistent with the mission of a vascular access team, namely doing the right thing (for the patient).

Selected References

1. Covey SR. Principle Centered Leadership. New York: Simon and Schuster, 1990.
2. Goleman D. Working with Emotional Intelligence. New York: Bantam Books, 1998.
3. Howard, P. The Death of Common Sense: How Law is Suffocating America. New York: Random House, 1994.
4. NKF-DOQI Clinical Practice Guidelines for Vascular Access. New York: National Kidney Foundation, 1999.
5. U.S. Renal Data System, USRDS 1999 Annual Data Report: Atlas of End-Stage Renal Disease in the United States, National Institutes of Health, National Institute of Diabetes and Digestive and Kidney Diseases, Bethesda: 1999.
6. Munschauer CD, Gable DR. Vascular access coordination at a large tertiary care hospital. Presented to Vascular Access for Hemodialysis VIII, Rancho Mirage, CA, May, 2002.

8

CHAPTER 9

Dialysis Technique and Access Management

Carolyn E. Munschauer, Sandra Hinton and Angela Kuhnel

Hemodialysis

The vascular access is the ESRD patient's "lifeline", providing the route for hemodialysis therapy. A well maintained and properly utilized access contributes to the efficacy of the dialysis prescription. Management of vascular access is multidisciplinary, involving the patient, nephrologist, surgeon, radiologist, dialysis nurses and technicians. The dialysis center has frequent, prolonged exposure to the patient. Therefore, the unit staff are key personnel to perform duties related to quality improvement, including access surveillance, identification of problems, timely referral for intervention and, most importantly, patient education.

The Dialysis Center Organization

300,000 patients are on chronic dialysis in the US. If the average number of chairs at a dialysis center is 20, with three shifts per day, six days per week, then 120 patients may receive dialysis each week at one center. The dialysis center is staffed with nurses, patient care technicians and technical staff responsible for water quality and dialyzer reuse. The director of nursing (DON) is the unit supervisor, responsible for compliance with patient care protocols and facility regulations, as well as unit staffing and administration. For each shift the charge nurse is responsible for overall patient care, including triage and referral of problems, medication administration and rounding with the medical staff. Staff nurses are responsible for initiating, monitoring and terminating treatments, medication administration and patient assessment. Patient care technician responsibilities vary by state or unit and may include pretreatment assessment, initiation and termination of treatments, but not administration of medications.

Assessment

Before treatment initiation the access site is assessed for patency and condition, including skin breakdown and pseudoaneurysm formation (Fig. 9.1). Complications should be presented to the charge nurse, who will use unit protocols and system resources to discuss the problem with the nephrologist on unit rounds, refer the patient to the surgeon or other available resources.

Direction of Flow

A patient with a new access or revision should have information on the type of access, blood flow direction, vessels used, date placed and surgeon. Ideally, each patient has an access data log with technical information including a picture indicating direction of flow, abnormalities, revisions. Copies of this log should be sent with the patient to the surgeon when referred for problems, and back to the dialysis

Access for Dialysis: Surgical and Radiologic Procedures, 2nd ed., edited by Ingemar J.A. Davidson. ©2002 Landes Bioscience.

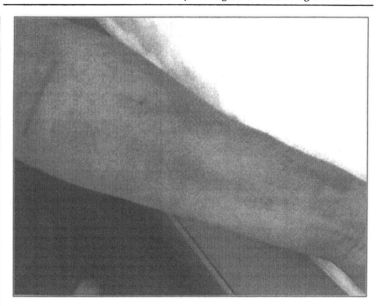

Fig. 9.1. Pretreatment access assessment.

unit after interventions. If this is not available, access type and direction of flow are easy to determine (Fig. 9.2). Native fistulae usually have a straight configuration, are generally soft to palpation, may move laterally or "roll" when gently compressed, and may have undulations or visible branching structures. Synthetic grafts are usually placed in loop or straight configurations, are firmer to touch, generally resist lateral rolling after tissue integration has occurred, and have no undulations or visible branches (Fig. 9.3).

Direction of flow in native vein fistulae is predominantly distal to proximal. However, flow direction should be confirmed with palpation for thrill and auscultation for bruit, both of which are stronger at the inflow end. Palpation of a fistula beyond the incision(s) may reveal the pulse around the arterial anastomosis. In synthetic grafts, thrill and bruit should be assessed over both ends, regardless of configuration (Fig. 9.3). The side with more pronounced bruit and thrill is presumed to be the arterial side. Next, the graft is gently compressed at the midpoint, while palpating each limb. The side with the remaining pulsation is the arterial side. Once determined, blood flow direction is charted, added to the patient log and communicated to the patient as "pull" and "return" sides

Effective Cannulation

Cannulation of an access is based on communication between the patient, dialysis staff and surgeon.

Selection of Puncture Site

Rotation of cannulation sites results in prolonged graft life, reduction in the instance of pseudoaneurysms and needle-stick abscesses, decreased incidence of perigraft

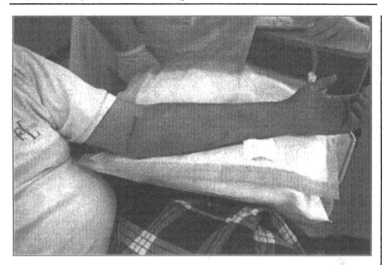

Fig. 9.2. Manual determination of flow direction.

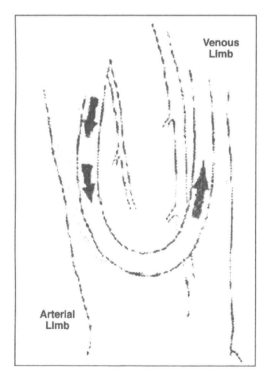

Venous Limb

Arterial Limb

Fig. 9.3. Direction of flow in a synthetic loop graft.

9

or perifistula bleeding, and more even maturation of native vein fistulae (Fig. 9.4). Patients may request repeat punctures at the same site to reduce discomfort, but they must be educated about potential complications and surgical procedures that may result from that practice. When planning cannulation pathways, needles should be rotated around the access to maintain constant needle distance (Fig. 9.5). For effective treatment without recirculation, the needle bevels should be at least 3 inches apart. For 16 gauge needles, this translates to a 1 inch separation between hubs when the bevels are in opposite directions (Fig. 9.6).

Preparation of Site

A common cause of access infection is contamination at cannulation. This is especially true in patients who are difficult to stick, as needles are pulled and reinserted, and skin repeatedly palpated. Skin preparation, cannulation technique and minimal exposure of the needles during treatment diminish infection risk. Before going to their station, all patients should wash their hands and arms with antibacterial soap, even those with leg grafts or catheters. Before contact with the access, dialysis staff and patient should use surgical masks. The National Kidney Foundation DOQI Guidelines makes recommendations for preparation of vascular access sites before cannulation (Table 9.1, Fig. 9.7).

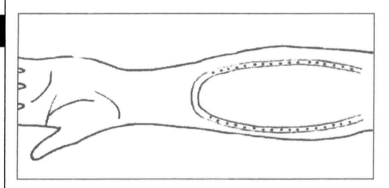

Fig. 9.4. Rotation of cannulation sites prolongs graft life.

Table 9.1. Preparation of vascular access

1. Locate and palpate cannulation sites prior to skin preparation.
2. Wash access site using an antibacterial soap or scrub (e.g., 2% chlorhexidine) and water.
3. Cleanse the skin with 70% alcohol and/or 10% povidone iodine using a circular rubbing motion.*

---

* Alcohol has a short bacteriostatic action time and should be applied in a rubbing motion for 1 minute immediately prior to needle cannulation, while povidone iodine needs to be applied for 2 to 3 minutes for its full bacteriostatic action to take effect and must be allowed to dry prior to needle cannulation.

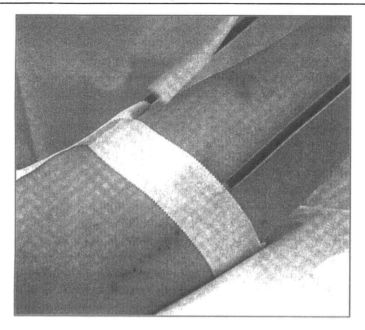

9

Fig. 9.5. Counterclockwise needle rotation pattern. The venous needle (foreground) rotates from anastomosis to apex, the arterial needle from apex to anastomosis. This assures consistent bevel spacing.

Whatever the unit policy, the cannulation area is prepped twice, with the final prep in a circular fashion from the cannulation site outward in one motion (Fig. 9.7). Catheters are also prepped twice, first with the end caps on, before draping. The catheter is then draped with the adaptors accessible, end caps removed and lumens and hubs prepped from center to edge. The author pours betadine into an open 4x4 package and wraps the package around the catheter hubs. This way, the patient can hold the package from the outside without contaminating the 4x4. Material adherent to the hubs can be removed with peroxide before applying the 4x4 package.

Cannulation

Local anesthesia use during cannulation is controversial. Routine use of lidocaine may lead to skin breakdown, especially over synthetic grafts. The author has recently begun using Emla® cream on patients requesting local anesthesia, with good results. Needles are placed under strict sterile technique, with needle bevels untouched, even with sterile gloves. If the access has to be palpated after prep, it should be done with new sterile gloves. Needle gauge and length are part of the dialysis prescription, mainly dictated by blood and dialysate flow rates and access depth. To avoid back wall punctures, the shortest needle providing flashback should be used. The needle is held bevel up by the wings at a 45 degree angle for synthetic grafts, 25-35 degrees for native fistulae (Fig. 9.8, 9.9). The venous needle is always placed

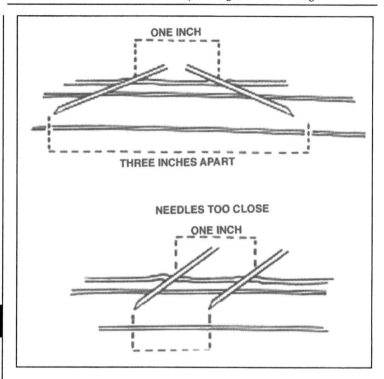

Fig. 9.6. Relation of hub distance to tip placement.

towards outflow, while the arterial needle can be directed toward outflow or inflow. The needle is advanced through the skin and subcutaneous tissue into the access in one motion, only until a flashback is noted. This motion is quick, without hesitation, and with a minimum of force applied. This helps the needle slide through surrounding tissue without spearing it. Needles should not be stopped, retracted or advanced in a new path through subcutaneous tissue. If the skin is scarred or thick, cannulation is a two-stick motion, the first motion advancing the needle through the skin into the subcutaneous tissue. The other hand stabilizes the puncture site while the needle is gently advanced into the access (Fig. 9.10). This helps prevent back wall sticks from excessive needle pressure through the skin and no resistance once the needle enters the access. After flashback, the needle is advanced another one-eighth inch (one-third cm). At this point, the angle between the needle and skin is decreased until the needle is flat, then advanced to the hub (Fig. 9.11). The bevel should not be rotated to avoid reinforcing the needle hole. The needles are secured with tape, covered with a dry sterile dressing and left alone. If needles must be repositioned, sterile technique should again be followed, and a new sterile dressing used for cover.

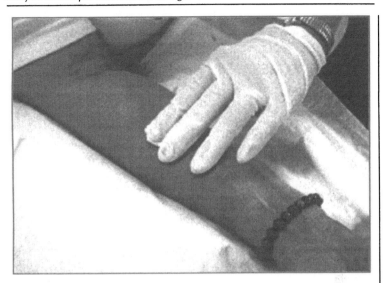

Fig. 9.7. Skin preparation.

9

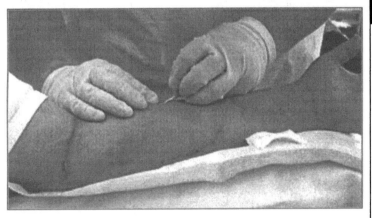

Fig. 9.8. Cannulation technique and needle angle for synthetic grafts.

Treatment

The patient should be checked often and kept within direct vision of the staff at all times. If marked extremital edema was noted at the beginning of the treatment, the needles should be visually inspected for change in position as edema diminishes. If movement does occur, needle position can be stabilized with folded, sterile 2x2 dressings under the hubs.

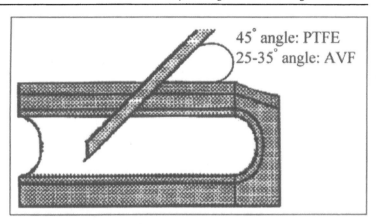

45° angle: PTFE
25-35° angle: AVF

Fig. 9.9. Needle position when flashback is seen.

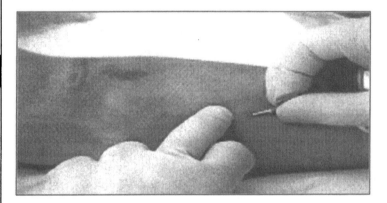

Fig. 9.10. Two step cannulation motion for scarred skin.

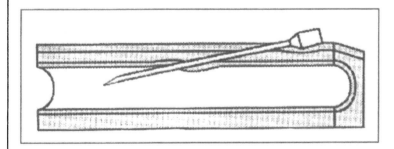

Fig. 9.11. Needle position after flattening and advancing. The needle position within the graft lumen is stabilized by the length of travel through graft and skin, as well as by the hub.

End of Treatment

At the end of treatment needle position is maintained while preparing for withdrawal. The needle is lifted and withdrawn at a slight angle, about 20 degrees. No pressure is applied to the site until the needle is completely out (Fig. 9.12) to avoid driving it through the back wall of the access, resulting in a large hematoma in the presence of systemic heparin. After needle removal, light, even pressure is placed on the needle sites with the index and middle fingers over a gauze dressing for 10 minutes without looking. The thumb is held on the back of the limb to stabilize the fingers, so that pressure is applied directly to the access (Fig. 9.13). The bruit and thrill along the access are assessed often, with pressure adjusted to maintain a balance between thrill and bleeding. Coagulating compression buttons or spring loaded so-called "C" clamps must not be used, as it is difficult to gauge the amount of pressure or respond to any episodes of transient hypotension. Some staff have been trained to "challenge" clot formation by rubbing the visible fibrin scab with gauze. In the author's experience, a scab without evidence of subcutaneous or active bleeding following 10-15 minutes of continuous pressure is stable, and should not be tested. A dry, sterile dressing is applied to the puncture sites with noncircumferential tape. This dressing is not hemostatic, just protective of the cannulation sites. Hemostasis must be assured before the dressing is applied. The combination of a relatively loose dressing and minimal tape prevents excess pressure to the access if post treatment swelling occurs. The patient may be discharged from the unit when hemostasis is assured and blood pressure is stable.

When to Refer for Access Consultation

Needle Related Complications

Complications related to cannulation include back wall sticks or lacerations (Fig. 9.14), needle placement outside or against the lumen, cannulation of dilated collaterals

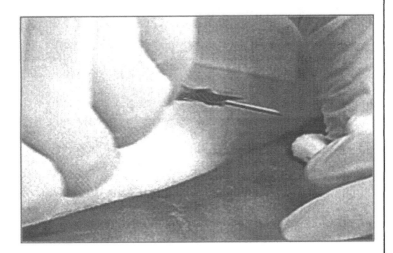

Fig. 9.12. Needle removal and hemostasis.

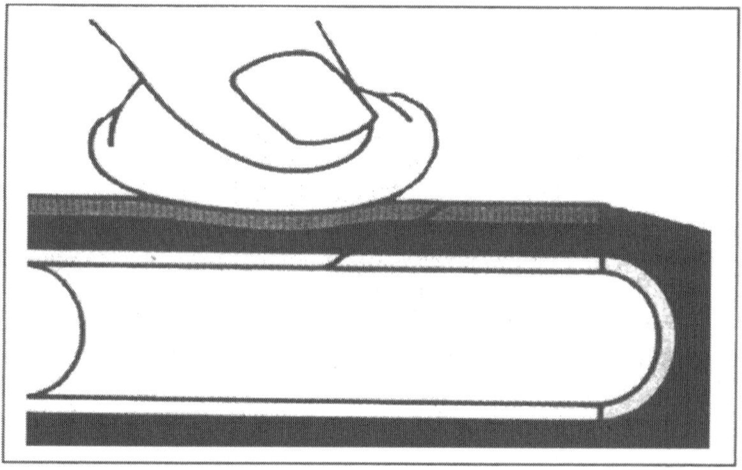

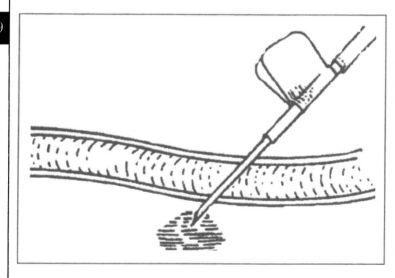

Fig. 9.13. Hemostatic pressure should be applied to a spot in front of the needle path and in the opposite direction to assure sealing of the graft at the needle entry site.

Fig. 9.14. Back wall needle stick.

and aspiration of clots. Needle placement complications can be avoided by attention to the flashback. If flashback is diminished or absent until the needles are moved, they should be repositioned before starting treatment. Cannulating dilated collaterals is prevented by pretreatment access assessment, including patient interview. If the site feels different than adjacent areas, or points away from the access path, it should be avoided. If there is no flashback after cannulation in an access with palpable thrill, the needle may be outside the graft, against the wall, or clotted from passing through subcutaneous fat or old clot (Fig. 9.15). Clots are aspirated for a variety of reasons, but usually do not indicate a clotted or failing access. If the access has recently been infiltrated, clots may come from the resultant hematoma outside the vein/graft.

Rising Venous Pressure

If venous pressure is noted to rise during treatment, with all other variables stable, then there is a presumption of outflow resistance. This includes patient or needle position, kinked return lines to the dialyzer, clotted or clotting dialyzer or an outflow problem with the access. This rise should be reassessed at the next treatment. If increased pressure continues, or baseline venous pressure increases over several treatments, the access needs evaluation, i.e., Doppler, fistulogram for venous outflow stenosis. The nephrologist or surgeon should be notified, depending on unit algorithm.

Aneurysms and Skin Complications

Skin problems over the access may result from repeat cannulations, breakdown of sterile procedure or overuse of local anesthesia. Rotation of cannulation sites

9

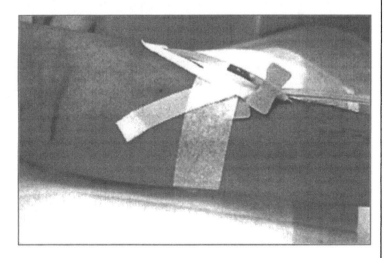

Fig. 9.15. Cannulation of primary AV fistula with crossed needle ("one up, one down") technique. Note absence of blood flash in venous needle (background).

prolongs access life and reduces skin complications. Precannulation assessment should include recent cannulation sites and their condition. Any redness, swelling or discharge from old stick sites should be reported to the surgeon or nephrologist, based on unit algorithm. Surgical repair is recommended for pseudoaneurysms when overlying skin is thin and shiny. Literature suggests less graft resection and faster healing if aneurysms are referred early.

Arterial Steal

Patient complaints of temperature or sensation changes in the access hand must be evaluated by the dialysis nurse for timing, duration and physical evidence. Patients with a new access may complain of hand pain and cramping, often related to immobilization and edema. Pain associated with the dialysis treatment accompanied by decreased temperature, longer capillary refill or discoloration, that improves quickly when dialysis stops is indicative of arterial steal, and should be referred to the surgeon (Fig. 9.16). Hand pain occurring before or after dialysis should be reported to the surgeon to rule out inflow obstruction or distal embolization (rare).

Clotted Access

Every center and hospital system has its own protocol for treating clotted vascular access. The authors' institution utilizes a vascular access coordinator as the triage person for acute access events, including thrombosis. The dialysis unit calls the

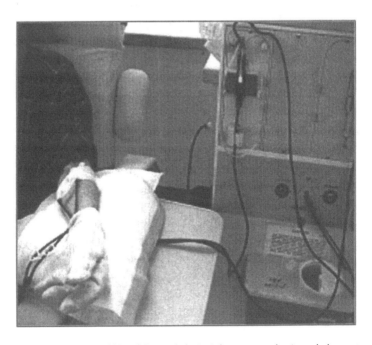

Fig. 9.16. Patient with cold hand during dialysis. Subsequent evaluation ruled out arterial steal as the cause.

coordinator with the patient's last treatment date, interdialytic weight gain and fluid status, as well as availability of transportation and NPO status. While the coordinator contacts an available surgeon, the dialysis nurse contacts the shift nephrologist for medical clearance. Any concerns the nephrologist has are relayed to the coordinator and the surgeon. A post-thrombectomy outpatient dialysis treatment is arranged and all pertinent records are forwarded before the patient is discharged. The overall goal of all involved should be to declot the access before the next dialysis treatment, without the need for temporary central vein catheters. In the ideal world, the consulting services—dialysis, nephrology, surgery and interventional radiology should work together to ensure that the patient can safely undergo surgical repair or radiologic intervention, with preprocedure dialysis done via a temporary femoral catheter.

Prolonged Puncture Site Bleeding
If post treatment hemostasis is not achieved after 10-15 minutes despite direct pressure, the charge nurse may notify the nephrologist, based on unit protocols. Patients may be hypocoagulable, related to late heparin bolus or metabolic disorder, or have a proximal venous stenosis increasing pressure in the access (Table 9.2).

Cannulation Related Problems
Problems with cannulation related to access anatomy are often difficult to quantify and eliminate. Often the location of a patient's access is based on the proximity of vessels or saving future sites. If a new access cannot be successfully cannulated by experienced dialysis nurses, the surgeon should be contacted for advice or evaluation (Table 9.3). Many times, an access requires time to heal and relieve edema before becoming manageable.

9

Table 9.2. Causes of prolonged post dialysis bleeding

1. Heparin
2. Drugs—coumadin, Plavix, aspirin
3. Venous outflow obstruction
4. Cannulation problems predialysis with multiple needle sticks
5. Underlying coagulopathies

Table 9.3. When to refer an access for surgery

1. Thrombosed access*
2. Abnormal duplex Doppler or fistulogram (i.e., stenosis >50%) with symptoms and poor dialysis (low BFR or kt/v)*
3. Aneurysms or pseudoaneurysms with skin changes
4. Access complication not amenable to interventional radiology repair*
5. Access thrombosis or other complications within 2-3 weeks of open repair*
6. Steal, rest pain or finger ulcerations

* Depending on local protocols and availability of interventional services including chemical thrombolysis

Peritoneal Dialysis (PD)

Peritoneal dialysis may be a viable dialysis option for up to 50% of ESRD patients. The percentage of patients offered peritoneal dialysis as a treatment modality in the United States is low (16%) compared to that in other countries.

Training

After surgery, the PD catheter is covered with gauze dressing, which the patient is instructed not to change until seen in the PD unit one week after placement (Chapter 6). The catheter is then flushed with 500 cc of peritoneal dialysate with heparin through a transfer set and allowed to drain to gravity. Exit site care is taught during this visit (Table 9.4). After this initial flush, the catheter is flushed daily, again with 500 ml dialysate.

The exception to this algorithm is the Moncrief-Popovich catheter, which is a swan neck Tenckhoff catheter (Fig. 9.17) placed as usual, except the external portion is placed in a subcutaneous tunnel to allow complete healing of the preperitoneal cuff. After 4-6 weeks, the external portion is exteriorized through a small incision and the patient is referred to the PD unit. This catheter type has been advocated for use in diabetic patients or others who are slow to heal or have had previous exit site infections.

Peritoneal Dialysis Training

Chronic peritoneal dialysis requires a gradual expansion of the peritoneal cavity. Small volumes of 1000 ml are infused for several days, gradually increasing to

9

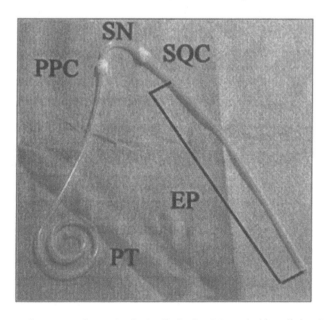

Fig. 9.17. The swan neck (proximal), pigtail (distal), 62.5 cm double cuffed peritoneal diaysis catheter. PT: Pigtail; PPC: Preperitoneal Cuff; SN: Swan Neck; SQC: Subcutaneous Cuff; EP: Externalized Portion.

2000 ml over a period of 2 weeks. During this "break in" period, in-depth training is provided to the patient (Table 9.5).

Table 9.4. PD catheter exit site care

1. Stabilize the catheter before removing dressing
2. Examine site for redness, swelling, pain and drainage
3. Showers only, no tub baths
4. Wash site with soap and water, rinse
5. Wash again with antibacterial cleanser (Hibiclens, povidone scrub)
6. Dry area gently
7. No lotions or powders around exit site
8. Stabilize the catheter with tape
9. Apply new dressing

Table 9.5. Training objectives for the peritoneal dialysis patient

1. Understand renal function and renal failure: uremia, anemia and the relationship between diet, medication and dialysis on bone disease
2. Learn to measure blood pressure
3. Understand dialysis diet
4. Understand peritoneal dialysis principles—equipment required for exchange, safe exchange procedure, the process of fill, dwell and drain, the use of different strength solutions for control of weight, blood pressure and edema.
5. Learn environment and hygiene principles
6. Understand laboratory tests and interpretations
7. Trouble shooting—How to identify and solve potential problems, when to call for assistance, common scenarios that can be solved quickly.
8. How to administer intraperitoneal medications

9

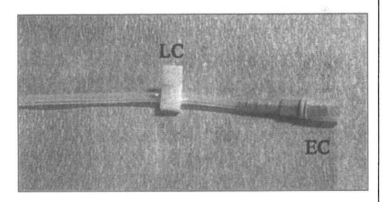

Fig. 9.18. The peritoneal dialysis catheter after insertion. If the dialysis system system is not known at the time of surgery, the two piece sterile end cap (EC) and line clamp (LC) packaged with the catheter are attached.

Assembly and Exchange

Following implantation in the operating room, all catheters are fitted with an end cap or transfer adaptor to the dialysis exchange system. The choice of dialysis exchange system varies among units and physicians. If the system is not known at the time of surgery, the two piece sterile end cap and line clamp packaged with the catheter is attached until training begins at the dialysis center (Fig. 9.18). In the case of the swan neck Moncrief catheter, the catheter can be capped with a standard syringe luer lock at the time of exteriorization.

The Baxter peritoneal dialysis system (Figs. 9.19A-F) (Baxter Healthcare Corporation, Deerfield, IL 60015) consists of a two piece titanium catheter adaptor and an integrated transfer set. The second peritoneal dialysis exchange system is the Fresenius Freedom™ Set (Figs. 9.20A-J) (Fresenius USA, Inc., Ogden, Utah 84404). There is no integrated transfer system to the plastic end cap with the Fresenius system. A new transfer apparatus comes on each exchange system aimed at minimizing catheter adaptor manipulation.

Complications

Patients are seen every two months in clinic, or as symptoms develop, for solution or volume adjustment and contamination prophylaxis.

Peritoneal catheter complications include: exit site and tunnel infections, marked by redness, swelling, pain and drainage around the exit site. Contributing factors include poor exit site care, cuff erosion, bacterial formation around granulation tissue, chronic irritation of the catheter and use of occlusive dressings. The surgeon should be consulted for a suspected exit site or tunnel infection. Peritonitis is a serious complication, often resulting in catheter removal (Table 9.6). Peritonitis can largely be avoided with proper exchange technique.

Leaks

Exit site leaks, most common early after placement, may occur any time. Signs include clear exit site drainage and damp dressings. Causes include poor surgical technique, nonhealed exit site, chronic exit site infections, trauma to the catheter or increased intra-abdominal pressure. Treatment includes decreasing intra-abdominal pressure through supine exchanges and decreased infusion volume. The most effective management is to stop exchanges for several weeks (3-4) then slowly increase volume. Subcutaneous and scrotal leaks occur due to inguinal hernia. Dialysate flows along the course of the hernia defect, terminating in the scrotal area. Symptoms include subcutaneous edema in the abdominal or genital area, dimpled, "orange peel" skin appearance, or patient complaint of incomplete dialysate return. Hernia repair with the catheter in place is possible, but exchanges must be discontinued for 4-6 weeks while the patient receives hemodialysis.

Exchange Complications

Inflow and outflow problems occur for many reasons and can be difficult to reproduce. First, the transfer set should be checked for kinks or mechanical malfunctions at all levels, while an x-ray can evaluate catheter position. Patients are trained to change position during exchanges, and laxatives may be used to decrease intra-abdominal volume. The effluent should be examined for fibrin, which may respond to heparin or tissue plasminogen activator (t-PA) administration. If these treatments are unsuccessful, surgical intervention, i.e., laparoscopy may be necessary to evaluate the catheter for omental obstruction or malposition.

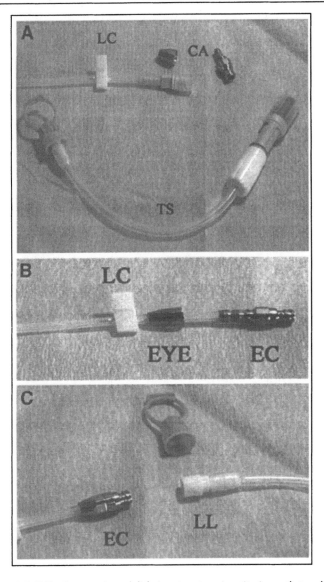

9

Fig. 9.19A-C. A) The Baxter peritoneal dialysis system: two piece titanium catheter adaptor (CA) and integrated transfer set (TS). The transfer set and adaptor are applied in a sterile fashion with the line clamp (LC) closed. B) The titanium end cap (EC) is inserted after the eyelet (EYE) is placed over the catheter. The eyelet is then screwed onto the end cap. To insure sterility, the line clamp (LC) is left in place during adaptor insertion. C) The blue plastic cap on the transfer set snaps off by pulling the ring, leaving the luer lock adaptor exposed (LL). The transfer set is then attached to the assembled titanium end cap (EC).

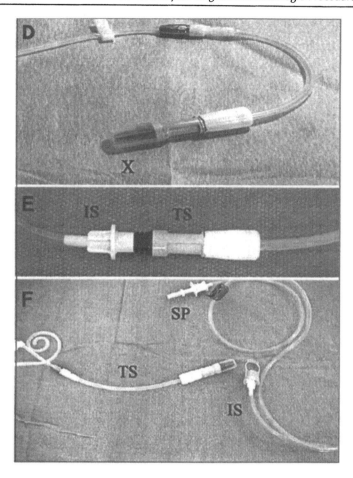

Fig. 9.19D-F. D) The completed transfer system. The distal end (X) is threaded to attach to the the infusion set. E) The transfer set (TS) is finally attached to the infusion set (IS) for the initiation of dialysis. F) The remaining line of the infusion set is used to spike (SP) the dialysate bag. At the conclusion of the treatment, the infusion set (IS) is removed and a povidone iodine containing end cap is applied to the transfer set (TS).

Table 9.6. Factors associated with peritonitis

1. Poor personal hygiene
2. Nonsterile exchange technique
3. Failure to wear face mask
4. Break of transfer system
5. Break in tubing
6. Exit site infection

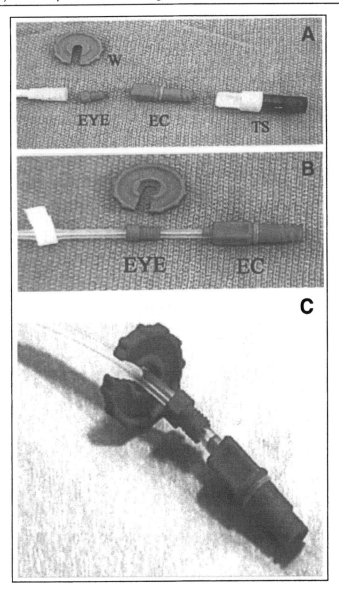

Fig. 9.20A-C. A) The catheter adaptor on this system consists of an eyelet (EYE), a luer lock end cap (EC), a wrench/guide (W), and a povidone iodine transfer set adaptor (TS). B) The eyelet (EYE) is applied to the catheter first, followed by the insertion of the luer lock end cap (EC) into the catheter. C) The wrench / guide applied to the eyelet to screw into the luer lock end cap.

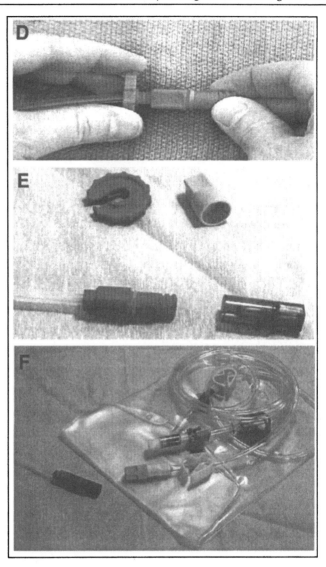

Fig. 9.20D-F. D) The eyelet is screwed in place using the enclosed plastic wrench/guard. E) The completed end cap luer lock system is attached to a protective povidone iodine transfer set adaptor. F) The Fresenius disposable Freedom™ set peritoneal dialysis exchange system.

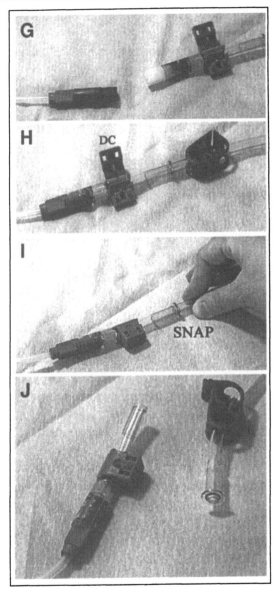

9

Fig. 9.20G-J. G) The distal end of the Fresenius exchange system is povidone-iodine gel containing Safe-Lock®. H) Note the open square clamp (DC) (Del-Clamp closure) just proximal to the Safe-Lock®. I) At the completion of the exchange this Del-Clamp is locked, and the redundant section of tubing (SNAP) (Snap™ Disconnect) grasped and J) And broken, leaving the povidone-iodine Safe-Lock® connector attached to the catheter adaptor.

CHAPTER 10

Tools and Methods for Vascular Access Surveillance

Aamer Ar'Rajab and Mitchell Henry

Introduction

This chapter presents an overview of current available tests for preoperative evaluation as well as surveillance methods for vascular access. All of these interventions are designed to improve the longevity of dialysis access and reduce access failures.

More than 300,000 patients are treated for chronic renal failure yearly. While about 5% of them will undergo renal transplantation each year, dialysis remains the only option for the majority. Because of the continuously increasing number of patients on hemodialysis, vascular access surgery has become part of the daily work of many surgeons.

The main complication of all types of vascular access is thrombosis. Many centers report one and three year patency rates of arteriovenous (AV) grafts of 60 and 20%, respectively. Fifteen percent of hospitalizations of patients with end-stage renal disease are caused by vascular access complications. This makes it a major cost as well as a health issue. Accordingly, a significant number of dialysis patients run out of sites for access. Measures to improve the longevity of vascular accesses are needed. Preoperative evaluation to select the most appropriate site and type of access could play a role. Access surveillance to predict and prevent access failure and consequently to correct the lesion is likely to prevent access thrombosis or loss.

Preoperative Evaluation—Duplex Doppler

Most centers do not routinely use radiologic and vascular laboratory methods before vascular access placement. Physical examination is thought to be adequate (Table 10.1). Koksy et al[1] abnormal findings in 5 of 17 patients had studied by duplex scan preoperatively. Two had an insufficient brachial artery flow rate and three had obstruction of the subclavian vein. The opposite extremity was therefore used, and no early graft failure occurred in this group. In contrast, 3 out of 22 patients that underwent access placement without prior study developed early thrombosis. Their work-up revealed proximal venous stenosis in two patients and atherosclerotic narrowing of the brachial artery in the third patient. In fact, ultrasound correctly predicts abnormalities in 100% versus 50% for clinical exam.

Preoperative duplex ultrasonography to identify arteries and veins suitable for hemodialysis access significantly increases the number of primary fistulas from 14% to 63%. One-year patency rate improved also from 48% to 83% for primary fistulas and from 63% to 74% for grafts.

Magnetic resonance arteriography (MRA), although useful in rare cases, is not an appropriate screening tool (Table 10.2).

Access for Dialysis: Surgical and Radiologic Procedures, 2nd ed.,
edited by Ingemar J.A. Davidson. ©2002 Landes Bioscience.

Table 10.1. Preoperative AV access: factors in the history and physical examination

History:
- previous vascular access
- previous central venous or peripheral venous or arterial catheters
- pacemaker placement
- heart valve disease or prosthesis
- previous arm, neck, or chest surgery/trauma
- severe congestive heart failure
- diabetes mellitus
- anticoagulant therapy or any coagulation disorder
- limited life expectancy due to comorbid conditions
- imminent renal transplant from living donor

Examination:
- dominant arm
- bilateral upper extremity blood pressures
- results of Allen test
- arterial system: character of peripheral pulses, with Doppler evaluation when indicated
- venous system: presence of edema or collaterals; palpation/vein mapping
- comparability of arm size

Table 10.2. Preoperative AV access: pertinent diagnostic studies

Duplex Doppler and/or venography are indicated:
- edema in the planned access extremity
- size difference of planned access extremity
- previous access placement in planned access extremity
- development of collateral veins in planned access extremity or on neck or chest
- diminished arterial pulses in planned access extremity
- history of subclavian catheter placement in planned access side
- history of transvenous pacemaker in planned access side
- arm, neck, or chest trauma or surgery on planned access side

10

Surveillance

While impending graft thrombosis may be heralded by rising venous pressure or poor flow characteristics on dialysis, prompting intervention, many such episodes occur without warning. The ability to predict imminent graft failure more accurately would allow elective revision in a greater number of patients. In fact, Sands and Miranda[2] followed 153 hemodialysis access patients (56 fistulas and 97 PTFE grafts), findings that elective access revision prior to thrombosis improved longevity of the access in both primary fistulas (999 days vs. 358 days) and PTFE grafts (1023 days vs. 689 days). In addition, early revision prior to thrombosis significantly decreased the number of clotting episodes for primary fistula (0.5 clots per patient year vs. 4.8) and PTFE grafts (1.1 clots per patient years vs. 3.6).

Physical exam and simple clinical parameters such as difficult needle insertion are of questionable value. Physical exam is noninvasive and subjective. It evaluates 1) a thrill changing to a pulse (venous stenosis) or 2) "full" graft changing to empty one (arterial abnormalities). Although angiography is considered the gold standard

Table 10.3. Methods of dialysis access surveillance

1. Intra-access flow
2. Static venous pressures
3. Dynamic venous pressures
4. Measurement of access recirculation using urea concentrations
5. Measurement of recirculation using dilution techniques (nonurea-based)
6. Unexplained decreases in the measured amount of hemodialysis delivered (URR, Kt/V)
7. Physical findings of persistent swelling of the arm, clotting of the graft, prolonged bleeding after needle withdrawal, or altered characteristics of pulse or thrill in a graft
8. Elevated negative arterial prepump pressures that prevent increasing to acceptable blood flow
9. Doppler ultrasound

for vascular assessment, it is expensive, invasive and demonstrates hemodynamic changes poorly. These factors limit its use. Less invasive studies such as venous pressure monitoring, access flow measurement, access recirculation and duplex ultrasonography are needed (Table 10.3).

Venous Pressure Monitoring

Venous pressure monitoring is based on the premise that as stenosis develops in the venous outflow, the resistance to flow will increase. This will result in increased pressure in the access proximal to stenosis.

Venous pressure measured from the dialysis machine has been a useful screening test for vascular access dysfunction, although it is a crude reflection of intragraft hemodynamics. Dynamic venous pressure monitoring utilizes pressure measured at a standard pump speed (Table 10.5). Schwab et al[3] measured venous pressures during dialysis at blood flows of 200-255 cc/min. They found that 73 patients had pressure >150 mm Hg. Fifty of 58 patients (86%) who agreed to angiography had venous stenosis of > 50%. This technique requires standardization for different dialysis machines. It represents the sum of the actual intra-access pressure, the hydrostatic pressure between the needle and the measuring site, and the pressure gradients through the external venous return tubing and the venous needle.

Static pressure measures intra-access venous pressure at zero blood flow (Table 10.4A). Besarab et al[4] used this technique in 133 patients using an "in-line" three-way stopcock adjacent to the venous return needle. Patients with venous pressure/systolic blood pressure > 0.4 were referred for angiography. On 80 occasions, accesses had significant stenosis with overall sensitivity and specificity of 91%.

Increased pressure in the access proximal to stenosis is a result of increased resistance to flow (Table 10.4B). Van Stone et al[5] evaluated access outlet stenosis by measuring the relative resistance of the outflow segment. Graft resistance is determined by comparing the pressure in the dialysis arterial and venous lines with the graft open to these pressures with the graft occluded by digital compression between the needles. With the graft occluded, the arterial line pressure is equal to systemic arterial pressure and the venous line pressure is equal to peripheral venous pressure. Since there is no flow through the graft, these pressures are not affected by any stenosis present. The difference between the occluded and nonoccluded line

---

Table 10.4A. Steps in measuring static intra-access pressure (IAP)

1. Establish baseline pressure values after the access has matured and shortly after it is first used.
2. Make sure the pressure transducers of the dialysis machine are calibrated to within ±5 mm Hg (P_0).
   - If not sure about calibration:
     - clamp the tubing from the drip chambers to the pressure transducer protectors
     - pull pressure protectors off the nipples, record the zero values (P_0)
     - replace the pressure transducer protectors
     - unclamp the lines
3. Measure mean arterial blood pressure (MAP) in the arm contralateral to the access.
4. Make machine ready for pressure measurement
   - stop the blood pump
   - cross clamp the venous line proximal to the venous drip chamber with hemostat (avoids having to stop ultrafiltration (UF) during measurement).
   - no hemostat is needed on arterial line (with occlusive roller pump).
5. Measure intra-access pressure (IAP)
   - wait 30 seconds (stabilize venous pressure)
   - record arterial and venous IAP values (arterial pressure available only if there is a prepump drip chamber and if the dialysis machine can measure absolute pressures greater than 40 mm Hg)
6. Unclamp the venous return line
   - return blood pump to its previous value
7. Determine the offset pressure (P_{offset}) between the access and the drip chambers one of two ways:
   - **Direct measurement**: measure height from venous or arterial needle to top of blood in venous drip chamber in cm. The offset in Hg = height (cm) x 0.76.
   - **Formula**: based on height difference between top of drip chamber and top of dialysis chair arm rest (Δ).
   The offset in mm Hg = 3.6 + 0.35 x Δ.
8. Calculate normalized arterial and venous static intra-access pressure ratios (PIA).
   Arterial PIA: (arterial IAP + arterial P_{offset}—arterial P_0)/MAP
   Venous PIA: (venous IAP + venous P_{offset}—venous P_0)/MAP

Suspected venous anastomosis stenosis can be detected with venous PIA alone. The higher the degree of anastomotic stenosis, the greater the venous (PIA) pressure ratio. Intra-access stenosis detection between arterial and venous cannulation sites requires simultaneous measurement arterial and venous PIA.

---

pressure is equal to the pressure decrease caused by graft blood flow across the graft segment and venous system distal to the venous needle. The resistance of the outflow segment relative to total graft resistance is equal to the difference between systemic arterial pressure and peripheral venous pressure. The relative outflow resistance is thus calculated by dividing the difference between nonoccluded and occluded arterial line pressures by occluded venous line pressure. The investigators found that a relative resistance of 0.4 has a sensitivity of 90% and a specificity of 53% for detecting hemodialysis access outlet obstruction (Table 10.4B).

Table 10.4B. Assessment of PIA ratios from arterial or venous needles

Access Type	Graft	Graft	Native	Native
Normalized PIA	Arterial Ratio	Venous Ratio	Arterial Ratio	Venous Ratio
Normal	0.35-0.74	0.15-0.49	0.13-0.43	0.08-0.34
Stenosis				
Venous outlet	>0.75	**>0.5**	>0.43 or	**>0.35**
Intra-access	**>0.75 and**	<0.5	>0.43 and	<0.35
Arterial Inflow	**<0.3**	NA	**<0.13**	NA
				+ clinical findings

Abnormal readings over two consecutive weeks represent a likely 50% diameter lesion. Criteria in bold type is the primary criteria for the location of the stenosis, nonbold is supportive.

Table 10.5. Steps in measuring dynamic venous pressure

1. Establish baseline pressure measurement when the access is first used.
2. Measure venous pressure from the dialysis machine at Qb 200 ml/min during the first 2-5 minutes of every treatment.
3. Use 15 gauge needles (or establish own protocol for different needle size).
   - most important variable affecting the dynamic pressure at a blood flow of 200 ml/min is the needle gauge
4. Assure that the venous needle is in the lumen of the vessel and not partially occluded by the vessel wall.
5. Pressure must exceed the threshold three times in succession to be significant.
   - eliminate the effect of variation caused by needle placement
   - threshold for different dialysis machines:
     - Cobe Centry 3: 125 mm Hg
     - Gambro AK 10: 150 mm Hg.
     - Baxter, Fresenius, Althin: data not available, but likely similar to Cobe Centry 3 with 15 gauge needles.
6. Assess at the same level relative to hemodialysis machine for all measurements.
7. Patients with progressively increasing pressures or pressures above threshold at three consecutive treatments should be referred for venography.

Physical Indications of Decreasing Access Function

Physical examination of an access graft should be performed weekly and should include, but not be limited to, inspection and palpation for pulse and thrill at the arterial, mid, and venous sections of the graft, and/or unexplained decreases in the measured amount of hemodialysis delivered (URR, Kt/V). Physical examination is a useful screening tool to exclude low flow (<450 ml/min) in grafts with impending failure. A palpable thrill at the arterial, mid-graft, and venous segments is likely associated with flow >450 ml/min. Conversion of thrill to pulse indicates lower flows. Intensification of bruit (higher pitch) indicates a stenosis. Therefore, in the context of proper needle position, an elevated negative arterial prepump pressure that prevents increasing the blood flow rate to the prescribed level is also predictive of arterial inflow stenoses. Physical findings of persistent swelling of the arm, clotting of the graft, prolonged bleeding after needle withdrawal, or altered characteristics of pulse or thrill in a graft may indicate outflow restriction. Persistent abnormalities in any of these parameters should prompt referral for venography.

Measurement of Dialysis Access Recirculation

Recirculation is defined as immediate return of venous (dialyzed) blood to the dialyzer, effectively short-circuiting the patients. When blood is pumped out of the access into the dialyzer, a low-resistance circuit is created that should increase total access blood flow. As a result, the venous drainage of the access is increased during dialysis. Should venous outflow be restricted, the likelihood of back-flow (recirculation) will be increased. Recirculation also will be facilitated by an increase in negative pressure at the arterial needle. Therefore, the measurement of dialysis access recirculation has important diagnostic implications. The presence of recirculation is confirmed by demonstrating that the concentration of a dialyzable solute in dialyzer afferent (arterial line) blood is lower than that in systemic blood. Recirculation can be measured directly using classical solute dilution technique or indicator dilution methods provided by a variety of newly developed devices (Table 10.6). The blood flow entering the dialyzer (Q_a) is composed of a mixture of true systemic blood (Q_s) and recirculated blood (Q_r); $Q_a = Q_s + Q_r$. The rate of solute (such as BUN) delivery to the dialyzer can be expressed $C_a \times Q_a = C_s \times Q_s + C_v \times Q_r$. C_a is dialyzer arterial solute concentration, C_s systemic concentration, C_v dialyzer venous concentration. The fraction of arterial flow that consists of recirculated blood is calculated by the equation $F_r = Q_r / Q_a = (C_s - C_a) / (C_s - C_v)$. Thus recirculation can be calculated by measuring BUN concentration in three blood samples drawn simultaneously. The arterial and venous samples are obtained from blood samples drawn simultaneously from the arterial and venous ports in the bloodline.

Recirculation may also be detected indirectly from the results of urea modeling. Urea is the most widely used solute for the measurement of recirculation because it is easily measured and highly extracted by the dialyzer. The difference between modeled and expected urea clearance is a measure of recirculation provided no other error (e.g., blood flow) contributes to the difference. Recently it has been suggested that dialysis access recirculation measurement is fraught with the potential for substantially overestimating recirculation.[6] Most of the potential error in the measurement can be overcome by using an arterial rather than a venous specimen for the systemic sample.

10

Table 10.6. Techniques for measuring recirculation in AV access

1. Urea concentrations
2. Nonurea-based dilutional method
3. Two-needle urea-based method.
4. Three-needle peripheral vein method of measuring recirculation should not be used.
   - overestimates access recirculation in an unpredictable manner
   - requires unnecessary venipuncture
5. Any access recirculation is abnormal.
   - recirculation exceeding 10% using two-needle urea-based method, or 5% using a nonurea-based dilutional method, should prompt investigation of its cause.
   - access recirculation values exceeding 20%, correct placement of needles should be confirmed before conducting further studies.
6. Elevated levels of access recirculation should be followed with fistulography.

Table 10.7. Protocol for urea-based measurement of recirculation. Perform test after approximately 30 minutes of treatment and after turning off ultrafiltration

1. Draw arterial (A) and venous (V) line samples.
2. Immediately reduce blood flow rate (BFR) to 120 ml/min.
3. Turn blood pump off exactly 10 seconds after reducing BFR.
4. Clamp arterial line immediately above sampling port.
5. Draw systemic arterial sample (S) from arterial line port.
6. Unclamp line and resume dialysis.
7. Measure BUN in A, V, and S samples and calculate percent recirculation (R).
8. Recirculation Formula: $R = ((S-A)/(S-V)) \times 100$

The urea-based method uses two needles and avoids overestimation of recirculation (Table 10.7). The recommended approach for this method is simple and is based on two considerations. The dead space of arterial lines to the sampling port is less than 12 ml. Access recirculation generally does not occur (except for reversed needles) unless access blood flow rates are less than dialyzer blood pump flow rates. This conclusion is supported by studies using nonurea-based methods that show that recirculation is absent (0%) in a properly cannulated, well-functioning access. A blood flow rate of 120 ml/min for 10 seconds will clear the arterial line dead space. Sampling at this time will provide arterial blood prior to onset of rebound. The results using the recommended two-needle method should average zero (-5% to +5%) in patients with unimpaired accesses.

Although nonurea-based dilutional methods are more accurate and avoid problems with cardiopulmonary recirculation, they require specialized devices which limit their applicability.

Measurement of recirculation is a more useful screening tool in AV fistulae compared to grafts because fistula flow can decrease to a level less than the prescribed blood pump flow while the fistula stays patent.

Duplex Ultrasonography

Only Doppler ultrasound has proven effective in evaluating both anatomic vascular features and blood flow parameters. Its value in the follow-up of the vascular bypass graft is well established. This technique has also been successfully adopted for patients with vascular accesses.

Strauch et al[7] used color Doppler flow imaging to study the predictive value for future episodes of thrombosis in vascular access graft patients, in relation to the degree of stenosis as well as access volume flow. The authors found 57% of patients with stenosis of greater than 50% had clotting episodes within six months, in contrast to 11% of patients with stenosis less than 50%. Patients with low access graft flow (500-800 ml/min) had more clotting episodes than those with higher flow.

Shackleton et al[8] evaluated flow characteristics using duplex scan in 18 patients with forearm PTFE vascular grafts. They found that mean Doppler flow in grafts that subsequently thrombosed was significantly lower than in those that did not thrombose (544±218 ml/min versus 843±391 ml/min, P<0.001). The interval from exam to thrombosis ranged from 13 to 58 days.

Sands et. al[9] used ultrasonography to evaluate access flow rate in 253 patients (177 PTFE grafts and 76 arteriovenous fistulas). They found that patients with flow rate < 800 ml/min had 92.9% incidence of thrombosis within six months compared to a 25% thrombosis rate in those with higher flows. Elective revision of low-flow grafts with > 50% stenosis had a thrombosis rate of 6% over six months vs 42% in those that were not revised.

Bay et al[10] found that quantifying blood flow in access grafts using color Doppler ultrasound could predict graft failure. The relative risk of graft failure increased 40% when the blood flow in the graft decreased to less than 500 ml/min and the relative risk doubled when the blood flow was less than 300 ml/min. Duplex ultrasound flow measurement in native fistulae has less or no predictive value.

Preoperative ultrasound evaluation is better than clinical assessment in predicting both successful and unsuccessful outcomes. Moreover, intraoperative duplex ultrasound flow measurement is widely variable due to vascular spasm.

Factors limiting the use of Doppler ultrasound in vascular access screening include: quality data is highly operator (technician) dependent, high capital cost of the equipment coupled with technician/physician costs and lack of reimbursement for testing.

Access Flow Measurement

Low or falling access blood flow rates are predictive of access dysfunction (Table 10.8). The dilution method has been used to measure access blood flow. In 1995, Krivitski[11] introduced the reversed line approach for measuring access flow during hemodialysis. This technology is based on reversing the delivery of the dialyzer outflow and placing it upstream from the arterial line with respect to the direction of flow in the vascular access. In this arrangement, the indicator introduced through the venous line into the vascular access will mix with the incoming access flow. After mixing, a portion of this mixed blood will reenter the dialyzer via the arterial inlet, which has been placed downstream from the mixing site by virtue of the line reversal. Blood flow measurement is based on Stewart-Hamilton principle. The most commonly used formula for flow measurement is Qa= Qb (1/R-1), where Qb is blood flow in the venous line; R is access recirculation measured with reversed lines, Qa is access blood flow. Several methods are available to measure access blood flow based on this approach including ultrasound dilution, thermal dilution, and conductivity dilution. Lindsay et al[12] compared access flow rates measured by several indicator dilution methods including ultrasound dilution, optical dilution and conductivity dilution. They found that flow rates measured by ultrasound dilution

10

Table 10.8. Measurement of access flow

1. Measure access flow on a monthly basis.
2. Methods used include:
   - ultrasound dilution
   - conductance dilution
   - thermal dilution
   - duplex Doppler
3. Perform flow measurement early in the treatment
   - effect of reduced cardiac output is reduced due to ultrafiltration.
4. Access flow in a single treatment is the mean of 3 independent measurements.
5. Access Flow < 600 ml/min
   - refer for fistulogram.
6. Access Flow decreasing over 4 month period by 25% or more to < 1 L/min
   - refer for fistulogram.

and conductivity dilution were essentially identical. Measurement with optical dilution correlated with both other techniques but consistently measured higher access flow.

May et al[13] studied flow rates in 172 PTFE grafts for 12 weeks using ultrasound dilution technique. There were 34 episodes of thrombosis. They found that accesses that thrombosed had significantly lower flow rates than those that remained open (875 ml/min versus 1193 ml/min). Similarly, Depner et al[14] prospectively followed PTFE grafts after obtaining baseline flow volume measurement. They found a 77% failure rate over 6 months in grafts with base line access flow < 600 ml/min.

Conclusions

Early detection of dialysis access dysfunction and timely intervention is likely to result in prolongation of access function. Several strategies are available for the diagnosis of failing access. Doppler ultrasound has the distinct advantage of evaluating both anatomic and flow characteristics. Meaningful ultrasound information requires skilled technicians and presence of the surgeon to direct and interpret images. While static venous pressure monitoring is easy, it does not represent intra-access pressure only. The evolving technology of access flow measurement has important prognostic implications to predict failure as well as evaluate the outcome of intervention.

Selected References

1.   Koksoy C, Kuzu A, Erden I et al. Predictive value of color Doppler ultrasonography in detecting failure of vascular access grafts. Br J Surg 1995; 82:50-52.
2.   Sands JJ, Miranda CL. Prolongation of hemodialysis access survival with elective revision. Clin Nephrol 1995; 44:329-333.
3.   Schwab SJ, Raymond JR, Saeed M. Prevention of hemodialysis fistula thrombosis: early detection of venous stenoses. Kidney Intern 1998; 36:707.
4.   Besarab A, Moritz M, Sullivan K. Venous access pressures and the detection of intra-access stenosis. Trans Am Soc Artif Intern Organs 1992; 38:519.
5.   Van Stone JC, Jones M, Van StoneJ. Detection of hemodialysis access outlet stenosis by measuring outlet resistance. Am J Kid Dis 1994; 23:562-568.
6.   Sherman RA. The measurement of dialysis access recirculation. Am J Kid Dis 1993; 22:616-621.
7.   Strauch BS, O'Connell RS, Geoly KL et al. Forecasting thrombosis of vascular access with Doppler color flow imaging. Am J Kid Dis 1992; 19:554-557.
8.   Shackleton CR, Taylor DC, Buckley AR et al. Predicting failure in Polytetrafluoroethylene vascular access grafts for hemodialysis: a pilot study. Can J Surg 1987; 30:442-444.
9.   Sands J, Young S, Miranda C. The effect of Doppler flow screening studies and elective revisions on dialysis access failure. Trans Am Soc Artif Intern Organs 1992; 38:524-531.
10. Bay WH, Henry ML, Lazarus JM et al. Predicting hemodialysis access failure with color flow Doppler ultrasound. Am J Nephrol 1998; 18:296-304.
11. Krivitski NM. Vascular access flow measurement by dilution during hemodialysis: Overview of first four years' experience. In: Henry M, ed. Vascular Access for Hemodialysis-VI. 1999:79-89.
12. Lindsay RM, Blake PG, Malek P. Hemodialysis access blood flow rates can be measured by a differential conductivity technique and are predictive of access clotting. Am J Kid Dis 1997; 30:475-482.
13. May RE, Himmelfarb J, Yenicesu M. Predictive measures of vascular access thrombosis: a prospective study. Kidney Int 1997; 52:1656-1662.
14. Depner TA, Rizwan S, Cheer AY. High venous urea concentration in the opposite arm: a consequence of hemodialysis- induced compartment dysequilibrium. Trans Am Soc Artif Intern Org 1991; 37:141.

10

CHAPTER 11

CPT and ICD-9 Coding for Dialysis Access: A Practical Guide

Diana J. Adams, Ingemar J.A. Davidson and Carolyn E. Munschauer

Money doesn't always make people happier. People with ten million dollars are no happier than people with nine million dollars. — Herbert Brower

Introduction

Proper CPT (Current Procedural Terminology) and ICD-9 CM (International Classification of Diseases, Ninth Revision, Clinical Modification) coding of vascular access cases can reflect the complexity of the diagnosis, as well as the procedures. The codes allow for delineation of sound surgical judgment, the sequence of often-staged access procedures, and the level of urgency by which decisions and surgeries are performed. Reimbursement from Federally funded programs, at present, involves three areas of concern for both hospitals and physicians, in regards to coding and documentation requirements. The Federal reimbursement cost areas are: the inpatient prospective payment system known as diagnostic related groupings (DRGs, which involves only hospital level reimbursement), reimbursement based relative value scale (RBRVS) for physician reimbursement and the outpatient prospective payment system (OPPS) that HCFA introduced in August, 2000 for hospital outpatient departments. The OPPS system can be complicated and coding/billing will depend on provider-based status and/or facility-based status. Anyone performing the coding and billing for vascular access cases must have a basic knowledge in all three of these areas, so that coding and documentation guidelines can be followed appropriately.

Basic coding knowledge is as follows: CPT is a statistical coding system that is utilized for reimbursement purposes. First developed in 1966, it is used as a statistical tracking system for physician practices. Insurance companies, including Medicare, are not mandated to utilize the CPT system, that is maintained by the American Medical Association (AMA). They do recognize the CPT codes as a tool to record physician encounters and have utilized the CPT system in various degrees; but, Medicare and other third-party payers have also developed their own guidelines which can often differ from the guidance in the AMA's CPT text. This can be seen with Medicare's Level II coding (which are codes with an alphanumeric quality). However, ICD 9 diagnostic coding system is required by Medicare since a Federal mandate in the late 1980's.[1] The lack of clearly defined rules of "how to code" for both CPT and ICD9 often can lead to a confusing "second guessing" methodology set of guidelines developed by those attempting to interpret Medicare and third-party payer information. Coding training seminars have become an industry of often unqualified individuals who have influenced the following: scaring doctors into being threatened by audits and fines for not following the rulings, that they did not realize existed or were recently changed. This current situation has created unnecessary

Access for Dialysis: Surgical and Radiologic Procedures, 2nd ed., edited by Ingemar J.A. Davidson. ©2002 Landes Bioscience.

tension between the medical community and payers and can hurt patients and the public in terms of treatment delays and increased cost. Measures to improve clarity and understanding are badly needed and would benefit all concerned—physicians, patients and payers. Staying informed of upcoming Federal legislation (HIPAA-Health Insurance Portability and Accountability Act of 1997) as well as the changes in the AMA's CPT system with the introduction of codes for experimental and/or investigational procedures that may affect the coding and reimbursement area should remain a priority for all.

Compliance is another subject of concern when developing the appropriate coding and documentation policies and procedures. Unintentional erroneous coding is commonplace. Under-coding guarantees under reimbursement. Improper or "over-coding" results in one of the following: delays, over-payment, or, more likely denials from insurance companies. Only proper coding may result in appropriate reimbursement. Internal auditing of coding, billing and documentation for services rendered to patients is a key to the knowledge of practice compliance and quality of care for patients.

There is no insurance system standard definition of "global surgical package". In the 2001 CPT, the surgical package definition is not clearly delineated per the text and is described through several different venues of "Reporting more than one procedure/service"; "add-on codes"; "Separate procedures" and/or "Starred procedures or items". The following is a basic standard to the surgical package definition, which is maintained by the National Correct Coding Initiative (CCI), that are national guidelines per HCFA regulations: To report postoperative follow-up visits for documentation purposes only, use CPT 99024. Such visits include suture removal, evaluation of the outcome of surgery and check for complications.[2] Note that initial preoperative services are not included in this "package" guideline per Medicare, but this is not always the case with third party payers that through their own global package definition may bundle visits prior to the procedure. (Internal tracking of such denials is recommended for any physician's practice in this avenue). Verification of insurance policies per provider contract is a must when filing claims for these services. Each insurance payer has defined global time periods, for example Medicare has a 0 day global period for scope procedures, 10 days for minor (usually office) procedures and 90 days for major surgeries (Table 11.1). Again, verification of insurance contract policies for their defined global period is a must when filing claims for these services.

With all of this in mind, this chapter is developed to present a commonplace understanding of key documentation elements to the effective management of vascular access. The NKF-DOQI Clinical Practice Guidelines are utilized as the bases for these standards of documentation, which should lead to appropriate coding. Patient evaluation through history and physical examination must be performed "to determine the type of access most suitable for an ESRD patient". The history and physical exam are most often billed under the evaluation and management services (E/M) of the CPT coding system. E/M requires the following documentation: Chief complaint (CC), history of present illness (HPI), review of systems (ROS) and/or past, family, social history (PFSH), physical examination (PE) and medical decision making (MDM).

With any documentation, the pertinent positives and negatives per the nature of the presenting problem(s) should be documented. The following is an example of how the history and physical exam, per NKF-DOQI guidelines, can be documented:

Table 11.1. Medicare 90 day major global package summary

Reminder: with insurance companies other than Medicare ask for written definition of their global package prior to any coding/billing of services rendered to patients. Included in the global package definition, per the National Correct Coding Initiative, are the following:
1. All normal preoperative visits beginning with the day prior to a major surgery. (NOTE: some Medicare Carriers may have local medical review policies that set a different guideline per the global package definition.)
2. Intraoperative services are a necessary part of the surgical procedure.
3. All medical or surgical services required of the operating surgeon due to complications that do not require additional trips to the operating room.
4. Services such as: dressing changes; wound incision care and removal of packings, drains, sutures, staples, lines, wires; irrigation, removal of urinary catheters, IV lines, nasogastric tubes, rectal tubes, changes and removal of tracheostomy tubes.
5. Postoperative visits for 90 days for all settings related to recovery from surgery.
6. Same day visit is not paid separate from the procedure.
7. Postoperative pain management by the operating surgeon.
8. Supplies except for surgical tray for specified procedures in an office setting.
9. Reminder that if a starred procedure is performed (which are noted in the tables by the following XXXXX,* then refer to your local Medicare Carrier for any local medical review policies, prior to billing for services rendered.

Chief Complaint:
  Written referral from attending (medical physician) to a surgeon for access device due to failure of treatment with meds, or diet of chronic renal failure therapy. CC can be stated in the patient's own words, by ancillary staff or by the physician.

HPI:
  Consideration of documentation of the following: history of previous central venous catheter, impact of quality of life, and planned duration of ESRD therapy. Presence of any comorbid conditions that limit patient's life expectancy should be documented.

ROS:
  Cardiovascular: Pacemaker use, congestive heart failure, any heart valve disease or prosthesis, any coagulation therapy or disorder
  Endocrine: diabetes mellitus
  HEENT: trauma/ vascular damage
  Respiratory/chest: trauma/vascular damage

PFSH:
  Previous vascular access history, any transplant history

PE:
  Key physical exam areas are: arterial system in regards of character of peripheral pulses, Allen test results and bilateral upper extremity blood pressures; venous system in regards to edema evaluation, assessment of arm size comparability, exam of collateral veins, palpation and mapping of ideal veins for access, exam of previous catherizations and any vascular damage to arm, chest or neck.
  Also documentation of any associated systems, such as skin, respiratory, constitutional, neurological, etc. should be performed.

11

MDM:

Medical decision making involves the physician's assessment of these areas: diagnosis(es), data (such as Doppler evaluation) and risk (morbidity and/or mortality) to the patient. Per the physician's clinical assessment MDM should be either assessed at low, moderate or high complexity.

With all of this in mind, the rest of this chapter describes the most common vascular access case scenarios. These case scenarios reflect the multiplicity that goes into surgical judgment, preoperative medical morbidity, urgency, patient referral pattern and facility,[3] and may not always be reimbursed by insurance companies. The outcome standards outlined in this chapter reflect or exceed the NKF-DOQI (National Kidney Foundation Dialysis Outcomes Quality Initiative) guidelines.

Tables 11.2 and 11.3 show commonly used CPT codes for procedures and surgical management, as well as modifiers. Table 11.4 lists the most commonly used ICD-9 codes associated with vascular access surgical cases. They are used to support the degree of complexity, severity of illness and the use of CPT codes and modifiers. Table 11.5 shows the most commonly used V and E codes. Table 11.6 depicts a decision driven algorithm for the common vascular access procedure.

The authors realize that because of differences in opinions, practice styles between individual surgeons as well as geographical variations, the coding can and will vary. This publication should be looked upon as a guide or an example of one approach. Also, because of the changing practices with outpatient access centers where radiology and surgery procedures are performed in one setting, proper CPT coding related to vascular access becomes even more crucial, not just for reimbursement, but also for statistical and patient management purposes.

Scenario #1 Elective Outpatient

This case represents the ideal "straightforward" patients with known renal disease, at serum creatinine of 4-5 mg/dl (GFR 20-25 ml/min) in a diabetic, or less than 7 mg/dl (GFR 15-20 ml/min) in a renal failure patient with no other morbidity. The access is placed before anticipated need for dialysis to allow for healing and maturing. The office visit includes pertinent history and physical examination elements. Often, patients have no or little knowledge and understanding of options and benefits, requiring extended time for discussions. In a young patient not previously hospitalized or exposed to IV infusions, a primary AV fistula is a more likely possibility in this scenario.

Procedure	Suggested CPT / ICD-9 Code
Office visit	99203-99205[1]
Operative procedure	
Primary fistula	36821-sep proc. Per CCI bundled
OR	into 35860, 36825 and 36834
PTFE	36830
OR	
Peritoneal catheter	49421*
OR	
Moncrief peritoneal catheter	49421*-58 insertion
	49999 externalization (w/op note)
Postoperative visits	99024[2]
Examples of ICD-9 Coding:	Chronic renal failure 585
For a more complete list of ICD-9 codes,	secondary to (examples), i.e.,
see Table 11.4.	diabetes, hypertension, etc.

Table 11.2. Common CPT Codes in vascular access surgery- CPT codes are updated every January 1st

Category	Procedure	CPT	Comment
Primary AVF	Create	36821	
	De-clot no revise	36831	balloon
	De-clot w/revise	36833	balloon
	Revise only	36832	no thrombectomy
	Interposition	36834	for aneurysm-sep proc
	Ligate/Band	37607	
	Angioplasty	35460	modifier may be needed
PTFE	Create	36830	
	De-clot no revise	36831°	balloon
	De-clot w/revise	36833°	balloon
	Revise only	36832°	no thrombectomy
	Interposition	36834	for aneurysm- sep proc
	Ligate/Band	37607	
	Angioplasty	36005 diag 37201 lytic	
	Remove (infection)	35903	
Tenckhoff	Insert	49420 temp: w/99025 if NP 49421* perm	verify temporary coding policy
	Manipulate	49400 inject contrast/air	
	Remove	49422 perm (E/M If temp)	
Catheters	Cuff, indep lumens In	36533 first 36533 second	modifier 51 Exempt
	Cuff, indep lumens manipulate	36534	requires fluoro
	Cuff, indep lumens Out	36535	
	Permcath In	36491cut down, 36489perc	
	Permcath manipulate	36493	requires fluoro
	Permcath Out	E/M	
	Percutaneous In	36489	
	Percutaneous Out	E/M	
	Declot Catheter	36550	
Angiography	Thrombolysis	37201	
Transcatheter	Retrieval FB (Perc)	37203	Noncovered by Medicare- not approved for AV shunt
	Stent Place (Perc)	37205 one vessel 37206 add	
	Stent Place (Open)	37207 one vessel 37208 add	
Vasc Injection	AV Shunt	36145	inc: cath place/needle
Vasc Lab Studies	AV Acx study	93990-26	inflow, body-hard copy
	Veins (compression)	93971-26 unilat 93970 bilat	

° if two separate dialysis grafts are addressed use modifier 59

11

Table 11.3. **Common modifiers that can be used with the vascular access surgery and/or postoperative time frames. This table is a summary of basic modifier billing policy. Please note that not all insurance companies recognize modifiers.**

Physician Billing Modifiers

Modifier	Description
22	Unusual procedural service (time is not a factor)
24	Unrelated E&M service by same physician during postop period (ICD-9 code is different)
25	Significant, separate E&M by same MD on same day as procedure or other service (ICD-9 code does not have to be different, but documentation must support a separate visit.)
26	Professional component (separate from technical component) (To be utilized with radiology codes when that service is performed by the surgeon and the surgeon follows the guidelines of a separate x-ray report of the service performed)
50	Bilateral procedure at same operative session(to be used when CPT code description does not have bilateral in description. –LT (left) and/or –RT (right) may be more appropriate to use pending supporting documentation.
52	Reduced services—elimination or reduction of service at MD discretion.
53	D/C procedure after starting. (documentation must state why procedure was canceled)
54	Surgical care only—one MD does procedure, another provides pre and post op management
55	Postop management only—one MD does procedure, another pre and post op management
56	Preop management only—one MD does procedure, another pre and post op management
57	Decision for surgery—E&M service resulting in initial decision to perform surgery(may have to be utilized in association with the global package concept)
58	Staged/related procedure same physician in postop period—prospective, extensive (may be utilized when the procedure requires stages to be performed before completion)
59	Distinct proc svc—(refers to different sites, operative sessions, etc. per NCCI guidelines)
62	Two surgeons—primary surgeons performing distinct parts of total proc. Must have two operative reports are required demonstrating what each surgeon did
76	Repeat procedure by same physician—same procedure twice or more by same MD
77	Repeat procedure by another physician—same procedure twice or more by different MDs
78	Return to OR for related procedure during postop period—(ICD9 code must reflect complication)
79	Return to OR for unrelated procedure during postop period—if on same day as original, use -76
80	Assistant surgeon
82	Assistant surgeon when no qualified resident is available

continued on next page

Table 11.3, cont'd

Ambulatory surgery (ASC) facility only modifiers—Not for inpatient or physician use

These modifiers are to be used by the facility when billing for services performed in a free standing ambulatory surgery center. They are not to be used by the surgeon.

Modifier	Description
50	Bilateral procedure at same operative session
52	Reduced services—elimination or reduction of service at MD discretion.
59	Distinct proc svc—independent from other svc same day
73	D/C procedure prior to administration of anesthesia—NOT elective cancellation
74	D/C procedure after administration of anesthesia, and/or after starting procedure
76	Repeat procedure by same physician—same procedure twice or more by same MD
77	Repeat procedure by another physician—same procedure twice or more by different MDs

[1] Code level depends on the nature of the presenting problem and how the history, physical exam and medical decision making are documented. The decision on E&M level is not time dependent—it is based on the number and complexity of systems reviewed. Consult the E&M guidelines in the CPT manual for detailed description; *Third-party payers may also have specific guidelines concerning starred procedures. Medicare does not recognize the starred procedure concept. Refer to local Carrier guidelines for billing purposes. Starred procedures may be bundled into the more comprehensive procedure per insurance company's surgical package.
[2] Is a statistical postoperative code only with a no charge value attached. Subsequent visits related to the procedure have already been reimbursed per the global package concept and are tracked through the statistical code of 99024 only. All access categories will have postoperative visits coded 99024

11

Scenario #2 Elective Inpatient

This patient is similar to scenario #1, however, this patient has a higher medical risk because of morbidity requiring hospital stay. This morbidity may be advanced age, heart disease, social or mental factors, patient with amputations or other physical disability.

Procedure	Suggested CPT / ICD-9 Code
Office visit	99203-99205[1]
2. Admit to hospital	observation or 23 hr 99218-99220 [1]
(Days or weeks later)	If inpatient without time specified
	99221-99223 [1] first visit
	99221-99233 each add day [1]
	Once surgery is done, visits bundled into global
Operative Procedure	
Primary AV fistula	36821
OR	
PTFE	36830
OR	
Peritoneal catheter	49421*

Table 11.4.   Common ICD-9 codes in vascular access surgery—ICD9 is updated every October 1st

ICD-9 Code	Explanation
042	HIV—also list manifestations, i.e., 585
585	Chronic renal failure-inc nausea, edema, anuria (often primary code)
276.6	Fluid overload (secondary code)
276.7	Hyperkalemia (secondary code)
250.01	Diabetes, I, controlled
250.03	Diabetes, I, uncontrolled
250.00	Diabetes, II, controlled
250.02	Diabetes, II, uncontrolled
	It is key with the following codes to remember to also code the specific manifestation
250.40	Diabetes, II, with renal manifestations, controlled
250.41	Diabetes, I, with renal manifestations, controlled
250.42	Diabetes, II, with renal manifestations, uncontrolled
250.43	Diabetes, I, with renal manifestations, uncontrolled
403	Category code only (do not bill this code): Per Coding Clinic Assign codes from Category 403, hypertensive renal disease, when conditions classified to categories 585-587 are present. The linkage between these renal diseases and hypertension is presumed to exist per the classification system. Acute renal failure (584.5-584.9) does NOT make this assumption.
403.01	HTN w/ESRD, malignant—also use 585
403.11	HTN w/ESRD, benign—also use 585
403.91	HTN w/ESRD, unspecified—also use 585
278.00	obesity, unspecified (secondary code only)
278.01	obesity, morbid (secondary code only)
278.1	obesity w/fat pad (localized) (secondary code only)
451.89	central venous thrombosis, inc IJ, EJ, SCV
451.82	venous thrombosis, UE superficial, inc antecubital, cephalic, basilic
451.83	venous thrombosis, UE deep, inc brachial, radial, ulnar
459.2	edema D/T venous compression
787.01	nausea w/vomiting (secondary code)
787.02	nausea only (secondary code)
787.03	vomiting only (secondary code)
996	Category code only (do not bill this code): Complications peculiar to certain specified procedures (ie, internal anastomoses, patch grafts)
996.1	Mechanical Comps of vasc devices, implant, graft—not embolus or atherosclerosis
996.62	Infection & inflammatory reaction secondary to internal cath, graft, shunt
996.73	Other comps (NOS) renal dialysis dev, imp-embol, fibros, hemorrhage, pain, stenosis, thrombosis

NOTE: Some of these codes are listed as secondary codes, referring to the billing methodology where they cannot be coded in the primary or principal diagnosis position on the claim form for inpatient encounters.

11

Table 11.5. V and E codes for vascular access

ICD-9 Code	Explanation
V09	Infection w/drug resistant microorganisms: use 4th and 5th digits to identify resistance
V12.51	Pt. w/history of venous thrombosis, may impact current care
V42.0	Renal transplant status
V42.83	Pancreatic transplant status
V44.6	Artificial opening status, nephrostomy, ureterostomy, urethrostomy
V45.1	Renal dialysis status—presence of dialysis shunt
V45.73	Acquired absence of kidney
V56.1	Fitting and adjustment of dialysis extracorporeal cath, inc remove/ replace, clean
V56.2	Fitting and adjustment of peritoneal dialysis cath, inc remove/replace, clean
V49.6	Upper limb amputation status-use 4th digit to identify level
V49.61	Thumb
V49.62	Other finger(s)
V49.63	Hand
V49.64	Wrist
V49.65	Below elbow
V49.66	Above elbow
V49.67	Shoulder
V49.7	Lower limb amputation status-use 4th digit to identify level
V49.71	Great toe
V49.72	Other toe(s)
V49.73	Foot
V49.74	Ankle
V49.75	BKA
V49.76	AKA
V49.77	Hip
V58.3	Attention to surgical dressings and sutures, inc change and removal
V58.61	Long term use of anticoagulants, still in use
V67.0	Follow up examination: surveillance of pt after surgery completed
E870.2	Accidental cut, perf, hemorrhage during dialysis or perfusion
E871.2	FB left in during dialysis or perfusion
E872.2	Failure of sterile precautions during dialysis or perfusion
E874.2	Mechanical failure of instrument or apparatus during dialysis/perfusion
E878.2	Op w/anast,bypass/graft, natur/artif, cause abn rxn or late comp, w/o prob at procedure
E879.1	Kidney dialysis as cause of abn rxn or late comp, w/o prob at time of procedure

11

These ICD-9 codes are for secondary diagnoses only. It is key to remember that many V codes describe the patient's status but not necessarily medical necessity for the procedure being performed. E codes refer to external causes and are never primary/principal codes when coding/billing for services rendered. Medical necessity must be proven and documented and most often will be classified to categories of "malfunction of an internal device" or "complications of ESRD".

Table 11.6. Suggested algorithm for various vascular access settings

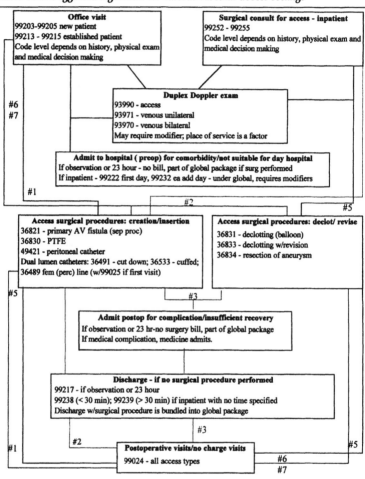

OR

Moncrief peritoneal catheter	49421*-58 insertion
	49999 externalization (w/op note)
Discharge if no surgery is performed.	If observation: 99217 (48 hr observation
Once surgery is done,	status only) If inpatient without time
discharge is bundled into global package.	specified 99238 (< 30 min), 99239 (>30 min)
Postoperative visits	99024 [2]
Examples of ICD-9 Coding:	Chronic renal failure 585
For a more complete list of ICD 9 codes,	secondary to i.e., diabetes, hypertension
see Table 11.4	Morbidities requiring admit for usual
	outpatient proc[3] BKA V49.75; AKA V49.76

¹ Code level depends on the nature of the presenting problem and how the history, physical exam and medical decision making are documented. The decision on E&M level is not time dependent—it is based on the number and complexity of systems reviewed. Consult the E&M guidelines in the CPT manual for detailed description.
*Third-party payors may also have specific guidelines concerning starred procedures. Medicare does not recognize the starred procedure concept. Refer to local Carrier guidelines for billing purposes. Starred procedures may be bundled into the more comprehensive procedure per insurance company's surgical package. For hospital billing, observation status will be bundled into the inpatient or the outpatient stay.
² Is a statistical postoperative code only with a no charge value attached. Subsequent visits related to the procedure have already been reimbursed per the global package concept and are tracked through the statistical code of 99024 only. These diagnoses represent secondary ICD-9 codes, and are for billing purposes only in developing the severity of illness profiles/databases. These are key to identify those patients who may run a higher risk due to these secondary conditions.

Scenario #3 Outpatient Becoming Inpatient after Surgery

This scenario is very similar to scenario #2. The procedure is planned as outpatient. For unforeseen reasons, such as insufficient recovery after surgery, vomiting, medical problems, chest pain, EKG changes, etc., requires admission, often for observation for 24 hours.

Procedure	Suggested CPT / ICD-9 Code
Office visit	99203-99205¹
Operative Procedure	
Primary AV fistula	36821
OR	
PTFE	36830
OR	
Peritoneal catheter	49421*
OR	
Moncrief peritoneal catheter	49421*-58 insertion
OR	49999 externalization (w/op note)
Dual lumen catheter	36489* percutaneous
	36491* cut down
	36533*,36533* independent lumen
Admit same day after surgery	Involves the policy of bundling the admit into the global package of the surgery and may not be billed for separately. Admission usually occurs because of a complication that may be related to the surgery. If admission is medical, i.e., patient requires dialysis for hyperkalemia, medicine service will admit and consult.
Discharge	If observation 99217
if no surgery is performed.	If inpatient without time specified
Once surgery is done, discharge is bundled into global package.	99238 (< 30 min), 99239 (> 30 min)
Postoperative visits	99024 ²
Examples of ICD-9 Coding:	Chronic renal failure 585

11

For a more complete list	secondary to–i.e., diabetes, hypertension
of ICD-9 codes,	Second category for reason for admit:
see Table 11.4	EKG change (postop arrhythmia) 997.1
	nausea and vomiting 787.01
	insufficient anesthesia recovery 995.2

[1] Code level depends on the nature of the presenting problem and how the history, physical exam and medical decision making are documented. The decision on E&M level is not time dependent—it is based on the number and complexity of systems reviewed. Consult the E&M guidelines in the CPT manual for detailed description.
*Third-party payers may also have specific guidelines concerning starred procedures. Medicare does not recognize the starred procedure concept. Refer to local Carrier guidelines for billing purposes. Starred procedures may be bundled into the more comprehensive procedure per insurance company's surgical package.
[2] Is a statistical postoperative code only with a no charge value attached. Subsequent visits related to the procedure have already been reimbursed per the global package concept and are tracked through the statistical code of 99024 only.

Scenario #4 Inpatient Staged Procedures

This represents an inpatient on medical service for ESRD, congestive heart failure, fluid overload, IDDM, hyperkalemia, ketoacidosis, shortness of breath, HIV, etc. This is usually a patient with known ESRD with rapid worsening uremic symptoms. Variations of case scenario #4 represent perhaps the most common. The staged procedures (type and timing) depends on the patient's condition and the medical judgment, facility sophistication, patient comorbidity and social factors; it is the author's experience (opinion) that sometimes stabilizing a patient with a dual lumen catheter for 2-4 weeks as an outpatient improves the outcome of permanent access. For example, a bruised forearm from IV's and blood draws will improve, increasing the likelihood of a primary AV fistula or successful PTFE placement.

Procedure	Suggested CPT / ICD-9 Code
Surgical consult for access	99251-99255 [1]
	If initial consult leads to decision for surgery the same day, use modifier -57 with consult code.
Operative Option 1:	
Patient too sick for OR	
1a. Placement of femoral line	36489*
(Emergent dialysis) OR	
Percutaneous dual lumen central line	36489*
(not recommended by author)	
Patient to option 2 or 3 when stable	
Operative Option 2: patient stable for OR	
2a. Dual lumen cuffed catheter	36491* cut down, 36533,36533*
(RIJ preferable)	independent lumens
Patient to option 4 when stable	
Operative Option 3: patient stable for OR	
3a. Placement of dual lumen	36491*cut down, 36533, 36533*
cuffed catheter	independent lumens
AND	
Primary AV fistula	36821
OR	

PTFE	36830
OR	
PD catheter	49421*
OR	
Moncrief peritoneal catheter	49421*-58 insertion
	49999 externalization (w/op note)

Operative Procedure 4:

4a. Elective placement of primary AV fistula or PTFE or PD or Moncrief	Now repeating scenario #1 or #2 or #3
Examples of ICD-9 Coding:	Chronic renal failure 585
For a more complete list	secondary to i.e., diabetes, hypertension
of ICD-9 codes, see Table 11.4.	Shortness of breath 786.09
The medical service, the admitting	Fluid overload 276.6
service, codes for diagnosis.	Uremia 586
The surgeon should also do this,	Bacteremia (unspec) 790.7
especially for the catheter placement	Sepsis (generalized) 038.9
for emergent dialysis, and to	
substantiate staged surgical procedures,	
patient transfer to the surgical service,	
or multiple operative procedures	
during the admission.	

[1]Code level depends on the nature of the presenting problem and how the history, physical exam and medical decision making are documented. The decision on E&M level is not time dependent—it is based on the number and complexity of systems reviewed. Consult the E&M guidelines in the CPT manual for detailed description.

*Third-party payers may also have specific guidelines concerning starred procedures. Medicare does not recognize the starred procedure concept. Refer to local Carrier guidelines for billing purposes. Starred procedures may be bundled into the more comprehensive procedure per insurance company's surgical package.

11

Scenario #5 Infected Dual Lumen Catheter

This is a fairly common scenario with some clinical urgency involved. Depending on patient's status the decision algorithm will vary. Prompt removal of an infected catheter and I&D of an infected PTFE graft is often warranted. Temporary femoral line for 24-48 h will bridge the patient to a cuffed dual lumen catheter 48-72 h later under antibiotic coverage, when patient is afebrile, white cell count down and clinically improved. If at all possible, the author would place a primary AV fistula at this time; but PTFE graft or PD catheter should probably wait 7-10 days; therefore, the patient may be discharged and revert to scenario 1, 2, 3 or 4 (Operative option 4).

Procedure	Suggested CPT / ICD-9 Code
5a. Admit for fever or R/O sepsis.	If admitted to surgery, 99221-99223 first visit 99231- 99233 each additional day until surgical procedure is performed. [1]
5b. Surgical consult for sepsis.	If admitted to medical service, 99253-99255 [1] If initial consult leads to decision for surgery the same day, use modifier -57 with consult code.

5c. Removal of dual lumen cath	36535 when surgical proc is required for removal. Op report must be submitted w/ code
5d. Femoral line placement	36489*
5e. New access placement	Dual lumen catheter 36489* percutaneous, 36491* cut down, 36533, 36533* independent lumens possible primary AVF 36821
5f. Discharge only if surgery not performed.Otherwise, bundled into global package	99238 or 99239
5g. Revert to scenario #1, 2, 3 or 4 (Operative option 4) while on antibiotics for permanent access placement.	Now repeating scenario #1 or #2 or #3 or #4 (option 4)
Examples of ICD-9 Coding:	Chronic renal failure 585
For a more complete list of ICD 9 codes, see Table 11.4	secondary to i.e., diabetes, hypertension
	Fever 780.6
The medical service, the admitting service, codes for diagnosis. The surgeon should also, to substantiate staged surgical procedures or patient transfer to the surgical service.	Sepsis
	Catheter 996.62
	Generalized 038.9
	Renal dialysis status V45.1

[1] Code level depends on the nature of the presenting problem and how the history, physical exam and medical decision making are documented. The decision on E&M level is not time dependent—it is based on the number and complexity of systems reviewed. Consult the E&M guidelines in the CPT manual for detailed description.
*Third-party payers may also have specific guidelines concerning starred procedures. Medicare does not recognize the starred procedure concept. Refer to local Carrier guidelines for billing purposes. Starred procedures may be bundled into the more comprehensive procedure per insurance company's surgical package.

Scenario #6 Clotted AV Access

This scenario represents perhaps up to 50% of all access procedures and is characterized by the unpredictable cause of access failure, as well as the outcome; the complexity becomes even greater when patients are treated by combined efforts of radiology and surgery; in the ideal world surgeons and radiologists work in an interdependent nature where procedures best benefiting the patient are performed. This could be the case in vascular surgicenters which are being established around the nation under various forms, missions and arrangements. Patients may be referred between surgery and radiology when either department has failed an attempted intervention or for problems discovered during the procedure.

Procedure	Suggested CPT / ICD-9 Code
Office visit (referred from dialysis, nephrology, radiology or self-referred)	Established pt 99212-99215 [1] New pt 99202-99205 [1]
Duplex Doppler exam	93990 (access), 93971 (venous unilateral), 93970 (venous, bilateral) are codeable
separately if performed in the office setting with appropriate modifiers per insurance requirements.	

6b. Operative procedure, same day
or next outpatient
6b1. Declotting only (balloon)   36831-(separate procedure)
OR
6b2. Declotting plus revision   36833
6b3. Failed declotting plus   36831/36833; 36489*percutaneous,
dual lumen catheter.   36491*cut down, 36533, 36533*
                                      independent lumens

Discharge to home, revert
to scenario #4, Operative option #4.
6b4. Failed declotting plus dual lumen   36831/36833; 36489*percutaneous,
catheter plus new PTFE   36491*cut down
or other permanent access.   36533, 36533*independent lumens
                                         36830 (PTFE) 36821 (Primary)

6c. Refer to radiology.   Office visit, Doppler codes only
Radiology: declot (mechanical),
declot (chemical, i.e., t-PA),
angioplasty
6c1. Successful: discharge to home   Radiology D/C
6c2. Failed - dual lumen catheter   Radiology codes
placed—radiology
6c3. Failed—dual lumen catheter   Surgery codes: 36489*percutaneous,
placed—surgery   36491*cut down, 36533, 36533*
6c4. Failed, revert to scenario #4,   independent lumens.
option #4.
Examples of ICD-9 Coding: Chronic renal failure 585
For a more complete list   Renal dialysis status V45.1
of ICD-9 codes, see Table 11.4   Complications of renal dialysis device 996.73
                                      Accidental cut during dialysis E870.2
                                      Operation w/anastomosis, etc as cause of
                                      abnormal reaction or later complication,
                                      without mention of problem at time of
                                      procedure E878.2 (i.e., intimal hyperplasia)
                                      Kidney dialysis as a cause of abnormal
                                      reaction or later complication, w/o mention
                                      of problem at time of procedure E879.1,
                                      i.e., dialysis induced metabolic hyperkalemia,
                                      hypovolemia.

11

[1] Code level depends on the nature of the presenting problem and how the history, physical exam and medical decision making are documented. The decision on E&M level is not time dependent—it is based on the number and complexity of systems reviewed. Consult the E&M guidelines in the CPT manual for detailed description.
*Third-party payors may also have specific guidelines concerning starred procedures. Medicare does not recognize the starred procedure concept. Refer to local Carrier guidelines for billing purposes. Starred procedures may be bundled into the more comprehensive procedure per insurance company's surgical package.

Scenario #7: Miscellaneous Vascular Access Problems

Procedure	Suggested CPT / ICD 9 Code
Office visit	Established pt 99212-99215 [1]
	New pt 99202-99205 [1]
Duplex Doppler exam	93990 (access), 93971 (venous unilateral), 93970 (venous, bilateral) are codeable separately if performed in the office setting with appropriate modifiers per insurance requirements.
Fistulogram	36145 (on table)
7b. Outpatient OR procedure	
Revision of vascular access	36832 sep proc
Primary AV fistulas	
Stenosis	
7b1. Reanastomosis	36832 (revision)
7b2. Patch angioplasty	36832 (revision)
7b3. Interposition graft	36832 (revision); 36834 (aneurysm)
7b4. Resection of aneurysm	36834 sep proc if performed alone, otherwise bundled
fistula- arterial steal	
7b5. Ligation/banding	37607
7c. PTFE graft problems	
infection	10140 (sep proc). Use modifiers 78 or 79 to identify cause
7c1. I&D	36832 (revision)
7c2. Bypass graft	35903
7c3. Removal	36834
pseudoaneurysm	36832
7c4. Resection and bypass	37607
7c5. Ligation and bypass	
Examples of ICD-9 Coding:	Examples of ICD-9 coding
For a more complete list	Chronic renal failure 585
of ICD-9 codes, see Table 11.4	Renal dialysis status V45.1
	Code problem as exists:
	996 Complications peculiar to specified procedures (ie, internal anastomoses, patch grafts)
	996.1 Mechanical Comps of vascular devices, implant, graft- not embolus or atherosclerosis
	996.62 Infection & inflammatory reaction secondary to internal cath, graft, shunt
	996.73 Other comps (NOS) of renal dialysis device, implant-inc embolus, fibrosis, hemorrhage, pain, stenosis, thrombosis
	E870.2 Accidental cut, hemorrhage during dialysis or perfusion
	E871.2 Foreign body left in during dialysis or perfusion
	E872.2 Failure of sterile precautions during dialysis or perfusion
	E874.2 mechanical failure of instrument or apparatus during dialysis/perfusion
	E878.2 Operation w/anastomosis, bypass/graft, natural/artificial, as cause of abnormal

11

rxn or later comp, w/o mention of problem at
procedure, i.e intimal hyperplasia
E879.1 Kidney dialysis as cause of abnormal
rxn or later complication, w/o mention of
problem at time of procedure, i.e., stenosis
from needle punctures

---

[1] Code level depends on the nature of the presenting problem and how the history, physical exam and medical decision making are documented. The decision on E&M level is not time dependent—it is based on the number and complexity of systems reviewed. Consult the E&M guidelines in the CPT manual for detailed description.

Scenario #8 Failed Thrombectomy with Return to OR

Failed de-clotting (or rethrombosis) for no obvious technical reason is common-place. National statistics indicate 50% of grafts fail within 3-6 months after de-clotting; similar both for radiology and surgery. Repeat procedures with appropriate coding are billable and must be reimbursed. Only when no attempts were made to correct a technical problem should no bill be issued as indicated below.

Procedure	Suggested CPT / ICD-9 Code
Office visit	99212-99215 [1]
Duplex Doppler exam	93990 (access), 93971 (venous unilateral), 93970 (venous, bilateral) are codeable separately if performed in the office setting with appropriate modifiers per insurance requirements.
Operative Procedure: same or next day outpatient	
8b1. Declotting (Balloon)	36831
Fistulogram	36145 (on table)
8b2. Declotting plus revision	36833
Fistulogram	36145 (on table)
8b3. Failed declotting w/out revision and return to OR same day	
a). If fistulogram done in first procedure w/out evidence of lesion and second exploration = hyperplasia or vasc anomaly	36833 (separate procedure)
b). If fistulogram done in first procedure w/out evidence of lesion and second exploration = no hyperplasia or vasc anomaly	36833
c). If fistulogram done in first procedure w/evidence of lesion and no revision attempted	no bill
d). If second exploration reveals technical problem	no bill
8b4. Failed de-clotting with revision and return to OR same day	
a). If second exploration reveals hyperplasia or vascular anomaly	36833
b). If second exploration reveals no hyperplasia or vascular anomaly	36833

c). If second exploration reveals technical problem

no bill

8b5. Failed de-clotting plus dual lumen catheter. Discharge home, revert to scenario #4, option #4

Bill de-clot as appropriate from above, add 36489* percutaneous, 36491* cut down, 36533, 36533* independent lumens.

8b6. Failed de-clotting plus dual lumen catheter. plus new PTFE or other permanent access.

Add 36830 (PTFE); 36821 (Primary AVF) to 8b5

Examples of ICD-9 Coding: For a more complete list of ICD-9 codes, see Table 11.4 Documentation of the differences in the outcome of the first procedure in this scenario may determine level of reimbursement for the second procedure, if any. All of these outcomes assume that a fistulogram was performed as part of the thrombectomy, and that appropriate action was taken based on these results, i.e., revision of the access. Failure of a thrombectomy and/or revision in the immediate postoperative period can be the result of many factors. The "no bill" designation is used here to reflect possible factors that could reasonably have been addressed during the initial procedure, i.e., failure to correct a stenosis as demonstrated by fistulogram.

Chronic renal failure 585
Renal dialysis status V45.1
Code problem as exists:
996 Complications peculiar to specified procedures (ie, internal anastomoses, patch grafts)
996.1 Mechanical Comps of vasc devices, dialysis device, implant, graft- not embolus or atherosclerosis
996.73 Other comps (NOS) of renal implant-inc embolus, fibrosis hemorrhage, pain, stenosis, thrombosis
E870.2 Accidental cut, hemorrhage during dialysis or perfusion
E878.2 Operation w/anastomosis, bypass/ graft, natural/artificial, as cause of abnormal rxn or later comp, w/o mention of problem at procedure
E879.1 Kidney dialysis as cause of abnormal rxn or later complication, w/o mention of problem at time of procedure
458.2 Postoperative hypotension (iatrogenic)

Selected References

1. The American Medical Association ICD Editorial Panel, eds. The International Classification of Diseases, Ninth Revision, Clinical Modification, 2000. Chicago: American Medical Association, ISBN 1-57947-150-1 (spiral), 2001.
2. The American Medical Association CPT Editorial Panel, eds. Physicians' Current Procedural Terminology 2000. Chicago: American Medical Association, ISBN 1-57947-106-4, 2001.
3. The Medicare Physician Self-Referral Improvement Act (HR 2650 IH). Stark, Pete (Rep), July 29, 1999.
4. American Hospital Association, Coding Clinic on CD.

Recommended Websites for Continuing Updated Information

www.hcfa.gov (for Medicare and Medicaid program memos and program transmittals regarding coding/billing policies.)

www.scvir.org (for coding involving interventional radiology procedures)

www.lmrp.net (for local Medicare Part B policies regarding vascular access coding).

Fifty Case Reports—Work in Progress

Ingemar J.A. Davidson
Photographers: Dwight Shewshuk, Carolyn E. Munschauer

We can chart our future clearly and wisely only when we know the path which has led to the present — Adlai E. Stevenson

Abbreviations

AV   arterio-venous
CV   cephalic vein
BA   brachial artery
RA   radial artery
BV   basilic vein
H   hematoma
IJ   internal jugular vein
R   right
L   left
PTFE   polytetrafluoroethylene
SVC   superior vena cava

The Antecubital Fossa Vascular Anatomy

The Cephalic Vein Patch Wrist AV Fistula

Cases

1. Conversion Primary AV to PTFE Graft
2. Conversion of Primary AV Fistula to PTFE Graft
3. Worn Out Forearm AV Fistula
4. Native Fistula Stenosis
5. Careful Exam Pays Off
6. Defective Forearm Cephalic Vein
7. Brachiocephalic AV Fistula
8. Brachiocephalic Fistula Rescue
9. Giant Primary AV Fistula
10. Bothersome AV Fistula
11. Vein abuse
12. More Vein abuse
13. Cephalic Vein I.V.
14. Traumatic Amputation
15. Cephalic Vein Branching
16. Clotted Primary AV Fistula with Stenosis
17. Upper Arm AVF with Central Stenosis
18. Intering PTFE Interposition graft

19. Outflow Stenosis
20. PTFE Patch angioplasty
21. Accuseal Patch angioplasty
22. Outflow Vein Occlusion
23. Exposed Sutures
24. Cutaneous Fistula
25. Infected Pseudoaneurysm
26. AV Graft Anastomosis Seroma
27. Seroma
28. Chronic Steal
29. Shiny Pseudoaneurysm
30. Multiple Shiny Aneurysms
31. Multiple Pseudoaneurysms
32. Forearm PTFE Pseudoaneurysm
33. Upper Arm PTFE AV Graft
34. No Dialysis Access
35. A Case of System Problems
36. Superior vena cava occlusion
37. Upper Arm Pseudoaneurysm
38. Distal Ischemia in Diabetic Patients
39. Loose Cuffed Tunneled Catheters
40. Subclavian Vein Catheter
41. Subclavian Vein Occlusion
42. Kinked Split ash Catheter
43. Subclavian Catheter
44. Poor Catheter Care
45. Mal-placed Dual lumen catheter
46. Catheter Infection
47. Subclavian Vein Catheter
48. The Elderly
49. Osteomyelitis of Clavicle-Sternal Joint
50. Malplaced PD Catheter

Background

This section represents work in progress. In fact, it will never be finished. I have collated a number of case reports with images. To be included several factors must come into play. The case scenario must have teaching value. Secondly, I did not forget to bring my camera (Sony Digital Mavica FD95) to the operating room. Third, the photographer in most cases has been Dr. Shewshuk, who besides being a superb anesthesiologist is an excellent photographer (judge for yourself). Also thanks to Cary Munschauer who took many images, but more importantly used Photoshop software to highlight the main points of the case. Because of printing cost the images appear in black and white; also, many times there are missing pictures for the purposes of obtaining important aspects in the case. Then often times the pictures shown here do not represent the perfect shot. As this case report section will expand, refine and perhaps appear in color, future editions will be more usable for our readers.

Scenarios are generally presented with a problem and an image followed by probable solutions, what I actually did, and sometimes further comments. This section (appendix) will likely be expanded in future editions, and perhaps made available on the internet.

The Antecubital Fossa Vascular Anatomy

The superficial venous (A) and arterial (B) anatomy in a typical case. The venotomy is made towards the graft, i.e., slightly on the side of the vein and includes the median antecubital vein (MAV), cephalic vein (CV) and the diving vena anastomotica (VA). There is an extra diving branch in this case at arrow (Fig. A.1.1A). These veins on this image will dilate markedly under slight pressure with saline (see Chapter 4).

The arteries also go into spasm. The arteriotomy is placed on the brachial artery (BA), sometimes extended into the radial artery (RA) or less commonly into the larger ulnar artery (UA). Muscular branches are seen (arrow) (Fig. A.1.1B). The surgical technique is described in detail in Chapter 4.

The Cephalic Vein Patch Wrist AV Fistula

The detailed surgical technique of using the dorsal branch of cephalic vein for a patch to radial artery anastomosis was discussed in detail in chapter 3. This case demonstrates some of the details. Fig. A.1.2A shows the two branches clamped with mosquito hemostats. At this point an 11 blade is used to partially divide the vein branches and the corner stitches Prolene® 7.0 BV-1 are placed. Fig. A.1.2B shows the corner stitches at trimmed "patch", matching the radial arteriotomy. Heifets clips are placed on the vein and artery. The small sized Alm retractor is quite useful in these small incisions.

Case Scenario #1: Conversion Primary AV to PTFE Graft

Patient is a 75-year-old woman just initiated on hemodialysis as evidenced by the large peri-fistula hematoma (H) (Fig. A.1.1) of the magnitude that renders this site nonusable for several weeks.

Solution

1. A right IJ split ash catheter was placed.
2. Primary AV fistula converted to a PTFE.

Fig. A.1.1A,B.

Comments

In cases like this the author connects the PTFE graft distally (Fig. A.1.2) to the cephalic vein in an end to end fashion (Fig. A.1.3). In this case the cephalic vein (CV) is quite large at this site. Contributing factor to the primary AV failure in this case was a stenosis (S) of the CV visible in image A.1.2. The AV fistula was also ligated at the radial artery cephalic vein anastomosis (arrow, Fig. A.1.1).

The benefits of placing the graft to vein anastomosis distally is that a long segment of native vein is available for cannulation. Secondly, when anastomotic hyperplasia develops, revisions can be made along the vein without interfering with the antecubital fossa.

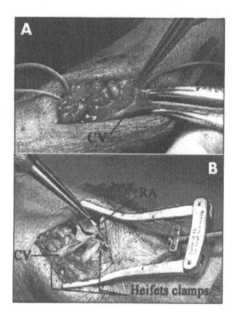

Fig. A.1.2A,B.

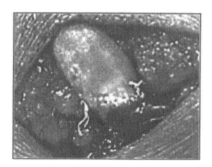

Fig. A.1.3.

Case Scenario #2: Conversion of Primary AV Fistula to PTFE Graft

Patient is a 60 year old female, predialysis, with a failing cadaveric transplant (1989). The primary AV fistula from 1985 clotted at the time of transplant (common), however, the cephalic vein (CV) remained patent (Fig. A.2.1)

Options

1. Move the anastomosis between cephalic vein (CV) and radial artery (RA) to site A in Fig. A.2.1.
2. Convert primary AV fistula to PTFE graft using the already enlarged cephalic vein.
3. Create new dialysis access at another site.

Solution

We chose option #2 (Fig. A.2.2). Need for dialysis was urgent and the already enlarged vein can be used for dialysis since the CV vein segment is long enough for cannulation with two dialysis needles (A). Figure B shows the arm 4 days after surgery, with some perigraft (PTFE) edema. The "old" cephalic vein is quite tortuous. The CV segment can be used for immediate dialysis needle punctures. The new PTFE graft was being cannulated 3 weeks after placement (Fig. A.2.2).

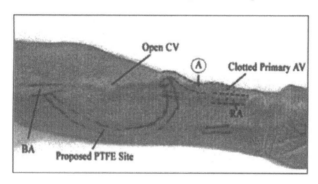

Fig. A.2.1.

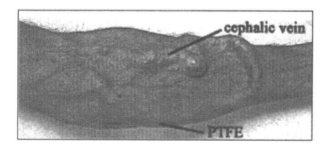

Fig. A.2.2.

Case Scenario #3: Worn-Out Forearm AV Fistula

Sixty-five year old man with a partly "worn out" native fistula, most notably a tight stenosis (between arrows). The dialysis unit also reports poor inflow (needle puncture below stenosis suggesting anastomotic stenosis as well). Note areas of resolving past dialysis hematomas.

Solutions

At surgery a PTFE loop was placed from brachial artery (BA) to the dilated cephalic vein (B); the remaining cephalic vein segment was too short for dialysis needle punctures. A temporary right IJ catheter placed. Also, the primary AV fistula was ligated at (A) (Fig. A.3.1).

Comments

This case represents one example of a failing native AV fistula conversion to PTFE. Features and benefits include increased longevity of this combined PTFE and native vein access (Appendix V, Fig. 5A); also, often the remaining segment of the native vein can be used for dialysis while the PTFE segment matures, thereby avoiding central vein catheters (Case #1-2).

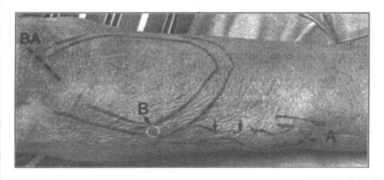

Fig. A.3.1.

Case Scenario #4: Native Fistula Stenosis

Another first time upper arm native fistula stenosis (between arrows) (Fig. A.4.1). The stenosis is easily palpable. Dialysis unit is experiencing poor inflow.

Solution

At surgery the tight stenotic area is exposed (Fig. A.4.2) and a short interposition 6mm PTFE graft is placed (Fig. A.4.3).

Comments

In this case the remaining graft segment was used for dialysis, and no temporary dialysis catheter was placed. As indicated in Figure A.4.1, the author discussed with the patient the possibility of a new PTFE graft around the old, in which case a dialysis catheter would be necessary.

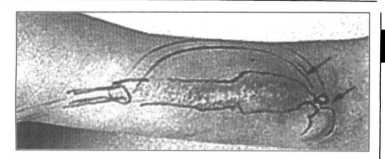

Fig. A.4.1.

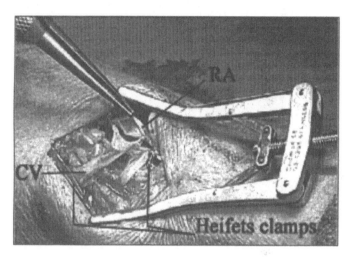

Fig. A.4.2.

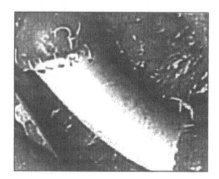

Fig. A.4.3.

Case Scenario #5: Careful Exam Pays Off

This 30 year old man has had a fairly recent failed forearm loop graft (Fig. A.5.1). Also, there is a past failed upper arm graft (UA-PTFE) at the medial aspect (does not show well, indicated by dotted line). There was a cephalic vein patent from the antecubital fossa to the shoulder (confirmed by duplex Doppler).

Solution
A brachial artery-cephalic vein (CV) native vein fistula was placed (incision at arrow).

Comments
Every patient, initial consultation as well as repeat access cases, must be carefully examined with duplex Doppler as indicated. Surprisingly there often are native veins, even in cases with previous PTFE grafts. It cannot be overemphasized to use the obvious principle to do the right thing the first time, which in this case likely would have prevented unnecessary procedures and much suffering and societal expenses.

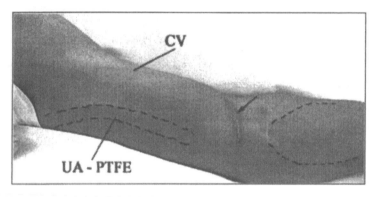

Fig. A.5.1.

Case Scenario #6: Defective Forearm Cephalic Vein

Patient is a 48 year old male being evaluated for permanent access. Creatinine clearance is 14.6 ml/min. Careful clinical exam (by palpation) reveals missing portion of cephalic vein (CV) midportion at the forearm (arrow). There is also a branch (perhaps the main CV) going up on the dorsal aspect of the forearm (Fig. A.6.1). There is a fairly large CV upper arm open up to the entrance into the subclavian vein (Fig. A.6.2). These findings were confirmed with color Duplex Doppler.

Solution

Placement of upper arm primary fistula between CV and distal brachial artery (BA) (Fig. A.6.2 and Fig. A.6.3). At surgery the CV is mobilized deep into the antecubital fossa (arrow, Fig. A.6.2) and brought over to the BA into an end CV to side BA anastomosis (Fig. A.6.3).

Comments

By placing the steri strips (1/2 inch) along the wound the risk of blisters is decreased (Fig. A.6.4), and wound edges are less likely to come apart at removal. This native fistula was successfully used for hemodialysis 8 weeks after placement.

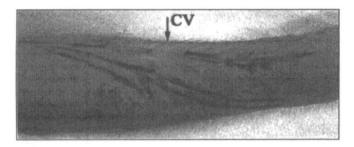

Fig. A.6.1.

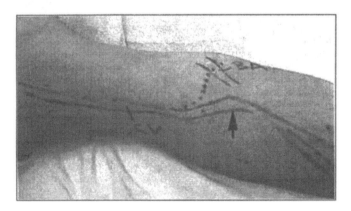

Fig. A.6.2.

AI

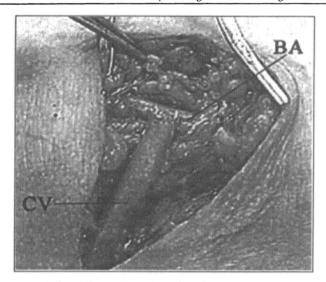

Fig. A.6.3.

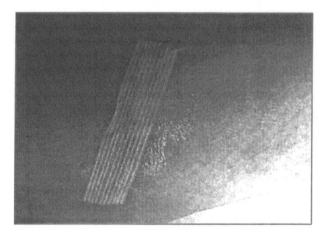

Fig. A.6.4.

Case Scenario #7: Brachiocephalic AV Fistula

This case fooled me for a while. First, a primary AV fistula (AV) had been placed at the wrist. When it failed, the cephalic vein was re-anastomosed (RAV) midarm, leaving a short segment of CV available for dialysis needle punctures; at the second surgery a short segment of PTFE was placed because of a stenosis in the vein; the CV also divides into two branches (CV_1 and CV_2).

Solution

Cephalic vein upper arm is fairly large by palpation and by Doppler patent up into the subclavian vein. The antecubital vein (ACV) was used to create a primary AV fistula for upper arm cephalic vein (UCV). Cephalic vein was divided at (CV₁X) to further mobilize the vein and prevent "backflow" into the forearm.

Comments

Again, this case illustrates the importance of thorough exam and duplex Doppler vein mapping. Preliminary vein mapping (vein, presence and patency) can be effectively determined by a handheld portable Doppler (Chapter 8).

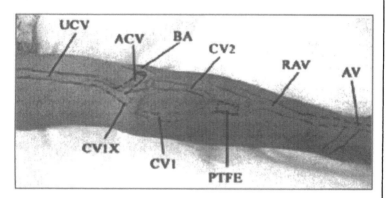

Fig. A.7.1.

Case Scenario #8: Brachiocephalic Fistula Rescue

A man in his thirties with a right large wrist AV fistula (AV) with a likely stenosis or occlusion of cephalic vein leading to numerous large collateral varicose looking veins (VV) in the forearm (not showing well in this image) (Fig. A.8.1). Dialysis unit nurses have a hard time cannulating; the veins reconstitute into normal antecubital venous anatomy (Fig. A.8.2) with an upper arm cephalic vein (CV), median antecubital vein (MAC).

Options

1. Create an upper arm CV to brachial artery primary AV fistula and ligate the CV at the antecubital fossa (arrow). Keep the distal anastomosis intact until upper arm is being used.
2. Ligate fistula at the wrist, create the upper arm primary AV fistula and place IJ dual lumen catheter until fistula is ready.
3. Place a contralateral arm PTFE or primary AV fistula, and ligate right arm fistula when new access is ready.

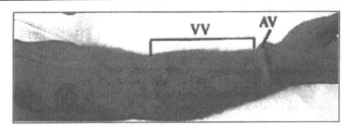

Fig. A.8.1.

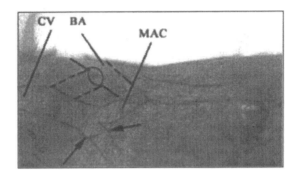

Fig. A.8.2.

Solution

The author recommended option #1, but patient did not return for surgery.

Comments

Creating an upper arm brachiocephalic fistula is an attractive solution for many different forearm access problems. A previous forearm access as in this case has dilated (matured) the antecubital veins and upper arm cephalic vein to the point that immediate use is possible after AV fistula creation.

Case Scenario #9: Giant Primary AV Fistula

This is an enormous brachiocephalic native fistula, left upper arm (Fig. A.9.1). Patient has pain at the lower end close to the anastomosis; the site often with calcification in the fistula wall. There is a stenosis at the cephalo-subclavian (SC) inlet. Patient is also practicing weightlifting. He wants "surgical" correction of this fistula. Due to the language barrier, unable to identify additional symptoms or complaints.

Options

1. Balloon angioplasty of cephalo-subclavian vein stenosis. Had been tried unsuccessfully by referring institution. May be attempted again by a more experienced interventional radiologist.

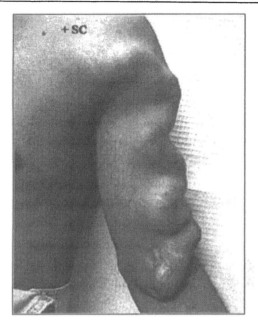

Fig. A.9.1.

2. "Banding" at the vein close to the brachial artery. This might be tricky because of large diameter and calcified walls. With this option the author would band with a 6-8 mm PTFE graft sewn around the vein.
3. Create a new primary AV fistula, right wrist where the patient has an excellent cephalic vein with a dorsal branch. When this new fistula is being used the upper arm fistula would be excised. (A forearm wrist AV fistula should have been placed in the first place; which would have been the "right" thing to do at that time).

Solution
Patient was offered option #3, but did not return for surgery.

Case Scenario #10: Bothersome AV Fistula

This young male with a giant right forearm AV fistula, similar to that of Case #9. In addition to some pain at the wrist the size of this had become bothersome. It occupies about a third of his forearm (Figs. A.10.1, A.10.2).

Solution
A new AV fistula was created in the left forearm (Fig. A.10.3). After the left AV fistula had been used 3 times, the right AV fistula was ligated and the radial artery reconstructed (no images were taken). The arm was wrapped in the hope the fistula

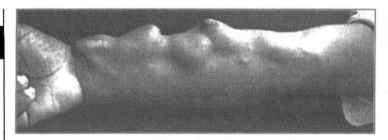

Fig. A.10.1.

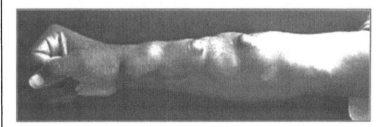

Fig. A.10.2.

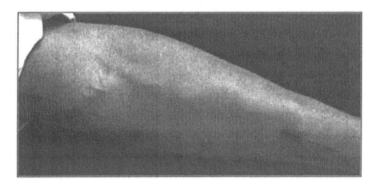

Fig. A.10.3.

would collapse. However, after 10 days with ace bandage binding the entire forearm fistula had formed a painful clot. The entire forearm thrombosed CV was excised through two skin incisions (Fig. A.10.4).

Comments
It is probably wise to excise these large veins at the time of ligation.

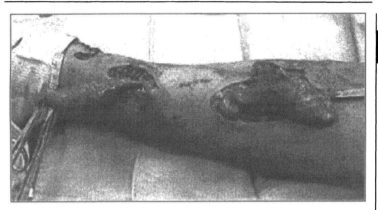

Fig. A.10.4.

Case Scenario #11: Vein Abuse

The author was asked to evaluate this confused, elderly (78 year old) gentleman for vascular access. He had been started on hemodialysis 3 days ago using a left subclavian (!!) percutaneous Quinton catheter. There is also a right subclavian triple lumen catheter. Both cephalic veins in the forearm have I.V. lines (Fig. A.11.1). Both antecubital fossae have had several blood draws.

Solution

A right internal jugular vein (28 cm) split ash was placed in the OR under local anesthesia and minimal sedation, using the micropuncture technique guided by ultrasound (Site Rite®) and fluoroscopy. ICU Nurses were educated by the surgeon about cephalic veins as "lifelines" in dialysis patients.

Comments

This I.V. access abuse in hospitals is commonplace and is a major factor in preventing successful placement of native vein fistulae. The author has had little success educating medical staff and phlebotomists. However, the educated patient (and family) is the best preventive measure to this vein destruction.

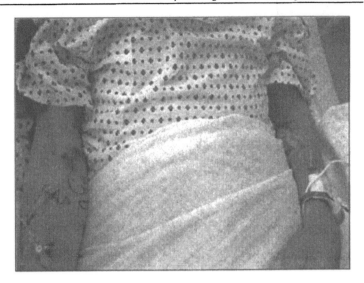

Fig. A.11.1.

Case Scenario #12: More Vein Abuse

This elderly woman had surgery scheduled for a right IJ cuffed catheter and a left forearm AV graft. The arm was marked the night before surgery with a blue line outlining the graft tract as part of the patient information for consent. Also, by marking OR sites and incisions the safety of correct side selection is increased. However, the anesthesiologist placed this I.V. line in the left antecubital fossa in the pre op holding area.

Solution
Only the right IJ catheter was placed. The I.V. was removed after surgery. Patient was returned one week later for graft placement.

Comment
Unbelievable!

Case Scenario #13: Cephalic Vein I.V.

Another common and sad example of cephalic vein I.V. placement on a medical floor in a patient waiting for access placement next morning (Figs. A.13.1, A.13.2). The guilty looking nurse in the picture (Fig. A.13.3) did not insert the I.V. She just volunteered as an actress for this picture.

At first site the reader may think of a hand subjected to partial amputation from peripheral ischemia. However, this is a chainsaw traumatic amputation many years ago. A successful primary AV fistula was placed at the wrist and is currently being used (CV).

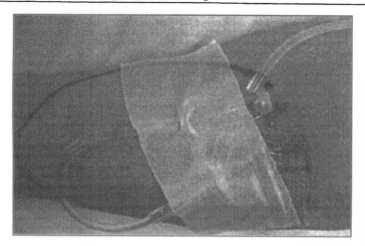

Fig. A.12.1.

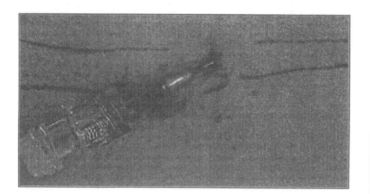

Fig. A.13.1.

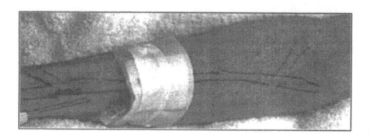

Fig. A.13.2.

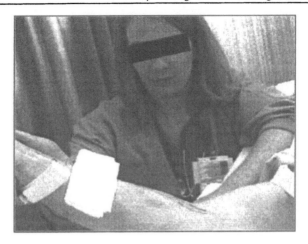

Fig. A.13.3.

Case Scenario #14: Traumatic Amputation

Comments

In this case the impaired arm was used; this principle may not always apply. The author has less favorable experience using a paralyzed arm, i.e., after a stroke, because of contraction, atrophy causing swelling after access placement (lack of muscle pump for edema prevention).

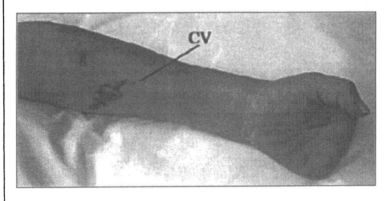

Fig. A.14.1.

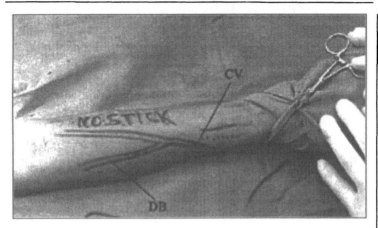

AI

Fig. A.15.1.

Case Scenario #15: Cephalic Vein Branching

Young female with a right forearm cephalic vein (CV) with early branching, a fairly common anatomy (Fig. A.15.1). The more dorsal branch (DB) continues into the upper arm CV and volar branch will become the median antecubital vein that will lead into the basilic vein at the medial aspect of the distal upper arm. Unless one of these branches is clearly significantly larger the author does not ligate the smaller at this initial surgery. Later, the less developed branch may be ligated to increase flow and size of the best developed vein branch.

Case Scenario #16: Clotted Primary AV Fistula with Stenosis

Debilitated, diabetic 50 year old rural man with signs from multiple central vein dialysis catheter placements (Fig. A.16.1). Left subclavian (!!) vein catheter of 2 months is his only dialysis access, which currently is malfunctioning (blue port does not pull). There is a left wrist old radiocephalic fistula (not shown) clotted at the anastomosis, but vein is still open (Fig. A.16.2).

Pre-Op Evaluation
Duplex Doppler shows occluded L IJ, open R IJ, open L SCV with current catheter; (we did not perform venogram of central veins. In an access center this would be done at time of surgery or radiology intervention. The cephalic vein has a tight area at (CV) with diameter of 2mm. (Duplex Doppler examination with a tourniquet on upper arm) (Fig. A.16.2).

Al

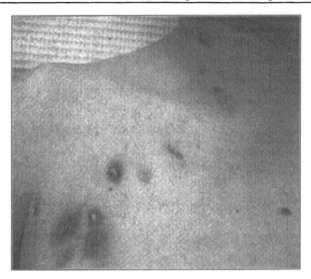

Fig. A.16.1.

Options

1. Place RIJ split ash catheter. Remove L SCV catheter. Place L forearm PTFE graft.
2. Place L forearm PTFE graft. Keep L SCV catheter for ~ 2 weeks until graft used.
3. Primary AVF R forearm. R IJ split ash.
4. Change L SCV catheter over guidewire.

Solution

1. At surgery a L PTFE loop graft was placed to fairly large superficial antecubital veins (no images).
2. R IJ attempted catheter fails, guidewires stopped at or below clavicula; location supports total occlusion of SCV RIJ junction (and stenotic site from subclavian vein catheters). Procedure aborted.
3. Left SCV manipulated. Poor "pull" of venous line, injected without VES. Red, arterial port works nicely (20cc/2 sec). Sutures to skin removed. Outer cuff at exit site removed, this is an Opti Flow® (Bard) catheter. Communicated to nephrologist and dialysis unit to try to get by with this catheter for 2 weeks, then use PTFE and remove catheter.

Comments

In this case the antecubital veins were chosen because of the mid forearm cephalic vein stenosis. The long term graft survival of these PTFE conversions to already dilated veins may be in the 90% range (Appendix V, Fig. 5A). The pre op "vein mapping" with duplex Doppler is critical in cases like this in order to choose the optimal anastomosis site.

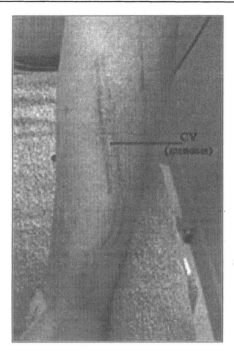

Fig. A.16.2.

Case Scenario #17: Upper Arm AVF with Central Stenosis

The patient is 60 year old with a right upper arm brachio-cephalic fistula (arrow) (Fig. A.17.1). There is a slightly dilated vein with a strong, hard pulse suggesting a more central stenosis. There is a pulse (flow) going down the forearm cephalic vein (at X, Fig. A.17.1).

Options
1. Interventional radiology for fistulagram and possible balloon angioplasty.
2. Create a new fistula L upper arm where there is a quite good cephalic vein palpable from the antecubital fossa to the shoulder (Fig. A.17.1).

Solution
Patient was referred for fistulagram, showing two outflow stenoses, one mid cephalic upper arm (arrow, Fig. A.17.2) and at the cephalic-subclavian level (arrows, Fig. A.17.3), both of which were subjected to balloon angioplasty (Figs. A.17.4, A.17.5). The effect of these radiologic are likely to be temporary.

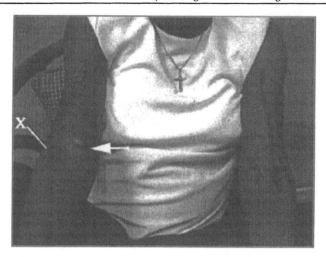

Fig. A.17.1.

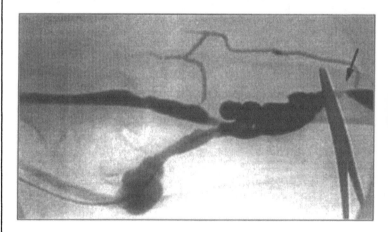

Fig. A.17.2.

AI

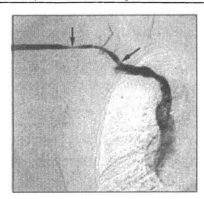

Fig. A.17.3.

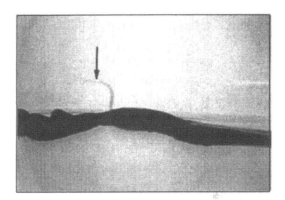

Fig. A.17.4.

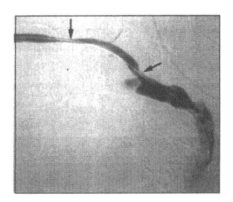

Fig. A.17.5.

Case Scenario #18: Intering PTFE Interposition Graft

A long venous graft/outflow stenosis may be technically awkward to patch (Fig. A.18.1). This area had previously been subjected to patch PTFE angioplasty (arrow, Fig. A.18.1). Also, the patched area is often badly scarred, making the "diseased" segment even longer. The new Intering graft provides an excellent material for interposition between old graft and a wide open vein across the antecubital fossa (Figs. A.18.3, A.18.4).

Solution

The Intering graft (W.L. Gore & Associates Inc, Medical Products, Flagstaff, AZ 86003. Ph: 800-437-8181. www.goremedical.com) is usable when the antecubital fossa

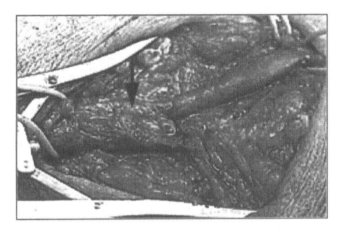

Fig. A.18.1.

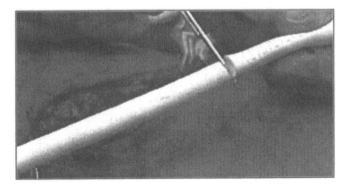

Fig. A.18.2.

is to be crossed. Unlike other "ringed" grafts on the market, the enforcement rings in this graft consist of condensed PTFE. The grafts currently are available in various lengths with 5 or 10 cm ringed sections with normal (standard) stretch PTFE of 5-20 cm on each side. The section without blue marking line represents the ringed portion (Fig. A.18.2). The graft can be cut and sutured through the rings (Fig. A.18.3).

Comments

The author has used the Intering in 9 cases. One has thrombosed. At surgery the rings now appear as bright rings (Fig. A.18.4).

Fig. A.18.3.

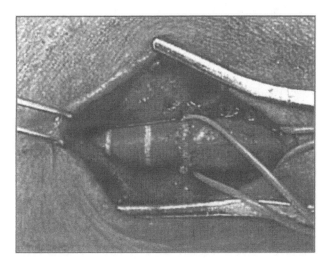

Fig. A.18.4.

Case Scenario #19: Outflow Stenosis

    Elderly lady with decreasing dialysis efficacy. The 3 cm long hard outflow stenosis can be palpated (Fig. A.19.1), and seen on Duplex Doppler (between arrows). At surgery the stenosis is mainly in the median antecubital vein (MAV) leading up to the basilic vein (BV). The BV is in moderate spasm from surgery (Fig. A.19.2).

Solution

    A 4 cm segment of the 6 mm Intering graft was placed across the antecubital fossa (Fig. A.19.3).

Comments

    This patient also demonstrates a bad but common dialysis unit habit, i.e., repeat needle punctures at the same small area (arrows, Fig. A.19.4), often contributing to fibrosis, graft stenosis and pseudoaneurysm formation.

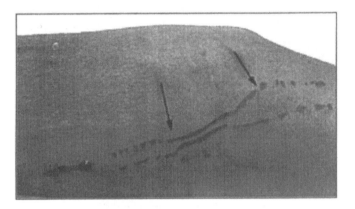

Fig. A.19.1.

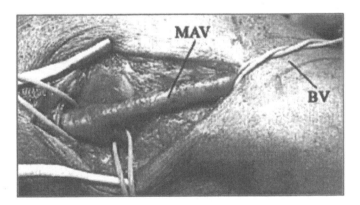

Fig. A.19.2.

AI

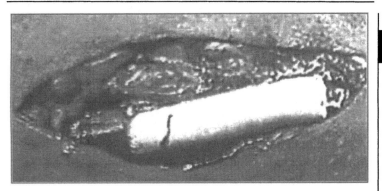

Fig. A.19.3.

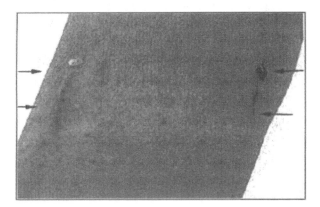

Fig. A.19.4.

Case Scenario #20: PTFE Patch Angioplasty

Classic case with AV graft thrombosis preceded by increasing venous dialysis machine pressures last several months. Preoperatively, two outflow veins can be felt patent (i.e., median antecubital vein (MAV) and the branch leading to the upper arm cephalic vein (CV)) (Fig. A.20.1). At surgery there is also a vena communicates that can be seen inside the opened graft (arrow, Fig. A.20.2). An Accuseal PTFE patch was sewn onto the graft-vein after successful declotting. Note the large outflow veins (arrows, Fig. A.20.3).

Comments

In this case a patch was chosen to preserve the third (v. communicates) outflow vein. In other situations the Intering graft interposition may be suitable (see Cases 18-19).

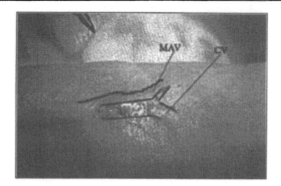

Fig. A.20.1.

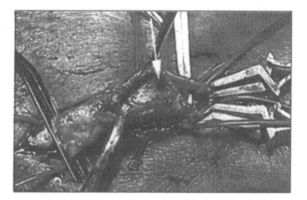

Fig. A.20.2.

Fig. A.20.3.

Case Scenario #21: Accuseal Patch Angioplasty

AI

Thrombosed PTFE AV graft. At surgery there are, in addition to a large antecubital vein, two additional outflow veins, one of which is the vena anastomotica (VA) (Fig. A.21.1) and one vein going distally (V) (Fig. A.21.1). Both of these veins were easily flushed with saline without resistance. To preserve these venous outflows a patch angioplasty is the proper procedure (Fig. A.21.2-3), rather than an interposition graft.

Comments

Each case is different and must be judged on its own merit; factors involved include outflow vein anatomy, length of stenosed segment, previous patch angioplasty and surgeons' technique.

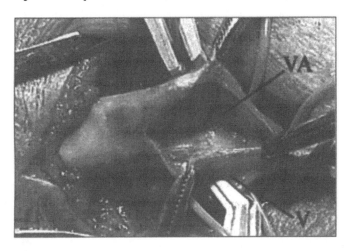

Fig. A.21.1.

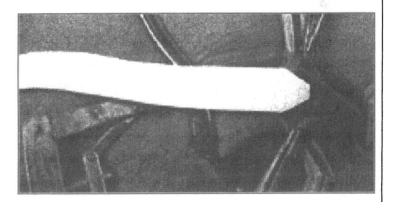

Fig. A.21.2.

AI

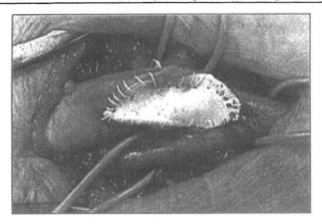

Fig. A.21.3.

Case Scenario #22: Outflow Vein Occlusion

Middle aged woman with lupus nephropathy. The PTFE graft loop has been anastomosed to the basilic vein (BV) mid-forearm (Fig. A.22.1). There is an almost complete obstruction of the basilic vein proximally with large venous collateral plexus going into distal forearm and hand (Fig. A.22.1), more dramatically visualized on a fistulagram (Fig. A.22.2). Duplex Doppler reveals no usable veins in the antecubital fossa. A sizable basilic vein is first reconstituted at mid upper arm at arrow (Fig. A.22.3).

Solution

At surgery a PTFE Intering graft was placed across the antecubital fossa between the old graft and upper arm basilic vein (BV). The basilic vein distally was ligated relieving pressure/pain symptoms. The 5 cm Intering section (IR) was placed at the elbow joint level (Fig. A.22.4), as indicated by the skin markings.

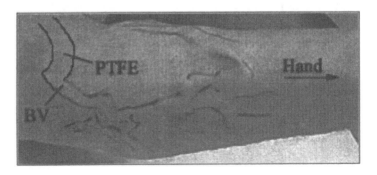

Fig. A.22.1.

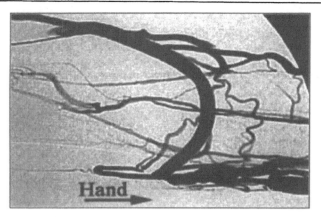

Fig. A.22.2.

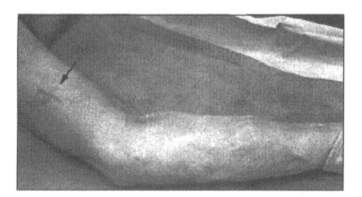

Fig. A.22.3.

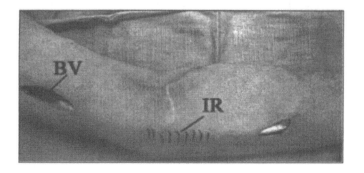

Fig. A.22.4.

Comments

This represents another use for the Intering graft when passing the elbow joint. The Intergraft is currently available as 5 or 10 cm segments with 20 and 5 cm stretch Gore Tex graft on either side respectively. The ring segment can be cut and sutured if needed. The rings consist of condensed PTFE (see Fig. A.18.2-A.18.4).

Case Scenario #23: Exposed Sutures

This 17 year old girl comes with this clotted left forearm PTFE graft 4 weeks after placement (Fig. A.23.1). Antecubital fossa detailed in Fig. A.23.2. Simple physical exam shows an adequate cephalic vein right forearm. At surgery a right IJ dual lumen catheter and a right forearm primary AV fistula placed. The AV fistula developed nicely. She had a living donor kidney transplant before the AV fistula was used. The IJ catheter was kept through post transplant day 5 for anti thymocyte globulin infusion. The AV fistula is still in place, working.

Comments

This case demonstrates a combination of poor judgment or lack thereof as well as poor surgical technique on the surgeon's part.

Case Scenario #24: Cutaneous Fistula

Infected PTFE right upper arm graft was removed 6 weeks ago. Patient presents with a cutaneous fistula (CF) (Fig. A.24.1) over the arterial anastomosis site where a small remnant of the graft was oversewn. At surgery the brachial artery (BA) was exposed and controlled above and below the infected PTFE (Fig. A.24.2). The infected piece of PTFE was excised from the artery (Fig. A.24.3), leaving a defect in the artery. Direct closure would have severely stenosed the artery. Therefore, the artery was divided at 45° angle; each end of the artery rotated 90° and reanastomosed with 7-0 Prolene®, BV-1 needle (Fig. A.24.4).

Comments

It is common practice (including the author's) to leave a small remnant ‑ 0.5-1.0 cm of the PTFE oversewn at the arterial anastomosis. In the majority of cases this area heals with no further problems. Should problems arise (as in this case), it is safer to return in a relatively infection-free area and reconstruct the artery with vein patch, resection or direct closure pending local anatomy (see Chapter 4, Fig. 4.28).

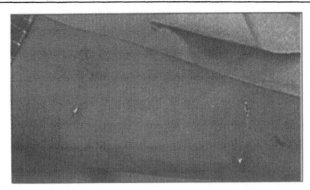

Fig. A.23.1.

Fig. A.23.2.

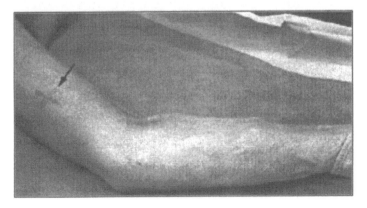

Fig. A.23.3.

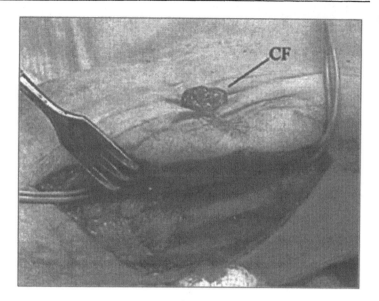

Fig. A.24.1.

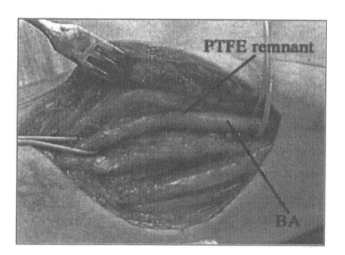

Fig. A.24.2.

AI

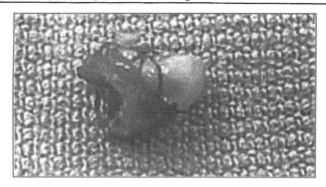

Fig. A.24.3.

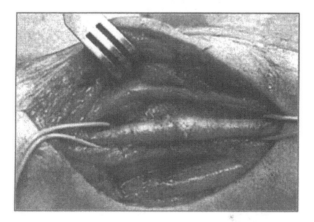

Fig. A.24.4.

Case Scenario #25: Infected Pseudoaneurysm

Infected, pulsating pseudoaneurysm at brachial artery PTFE anastomosis above the antecubital fossa (Fig. A.25.1). The graft runs lateral up on the upper arm with the venous anastomosis to proximal basilic/brachial vein. This problem needs to be addressed urgently.

Solution

At surgery, the brachial artery (BA) was controlled above and below the abscess (Fig. A.25.2) (Median nerve (MN) and brachial veins (BV) also shown). The affected arterial defect was resected and the artery reanastomosed (similar to previous Case #24) (Gore Tex® suture, CV-6, TT-9) (Fig. A.25.3). The entire PTFE graft was removed; the tract drained with 1/2" Penrose drains for 48 hours.

Comments

During surgery the author prefers to have a sterile inflatable tourniquet ready in case of sudden rupture of the friable infected pseudoaneurysm. Cases like this are not very common, and often represent a combination of patient and doctor delay.

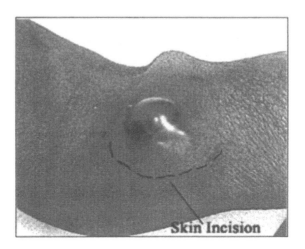

Fig. A.25.1.

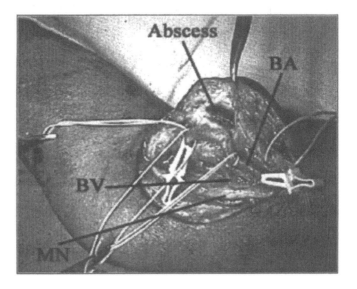

Fig. A.25.2.

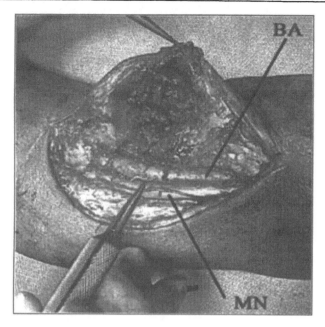

Fig. A.25.3.

Case Scenario #26: AV Graft Anastomosis Seroma

Sixty year old female one year after PTFE placement with this expanding nontender, non -pulsating, bulging mass (Fig. A.26.1). Arterial (pull) and venous (return) graft sides are marked. At surgery, encapsulated gelatinous material is found (Fig. A.26.2), typical for seroma formation. The capsule was completely excised

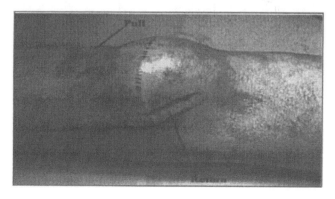

Fig. A.26.1.

(Fig. A.26.3-4). The graft arterial (A) and venous (V) limbs are exposed at the bottom of the cavity (Fig. A.26.5). The cavity was drained with a small JP drain for 48 hours. No recurrence of this seroma occurred at nine days post operatively (Fig. A.26.6).

Fig. A.26.2.

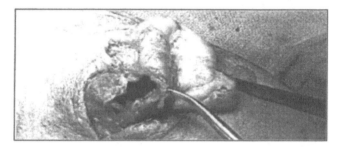

Fig. A.26.3.

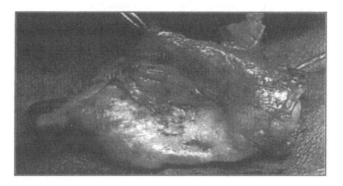

Fig. A.26.4.

AI

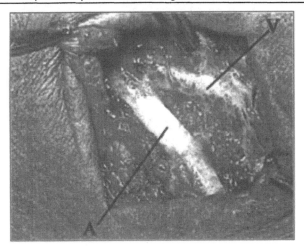

Fig. A.26.5.

Fig. A.26.6.

Case Scenario #27: Seroma

This is a 72 year old male with another tense (Fig. A.27.1) seroma that failed excision and drainage (Fig. A.27.1). He relapsed and another gelatinous mass was removed 3 weeks later; at this time the floor in the cavity including the PTFE graft was extensively cleaned from all gelatinous material. The human thrombin spray (Tisseel®, Baxter 1627 Lake Look Road LC-IV, Deerfield, IL 60015. Phone: 800-423-2090; www.tissuesealing.com) was applied and a small suction drain placed for 48 hours. There was no recurrence at 3 weeks postop (Fig. A.27.2). At the time of this publication going to the printer (12 weeks post op) still no recurrence.

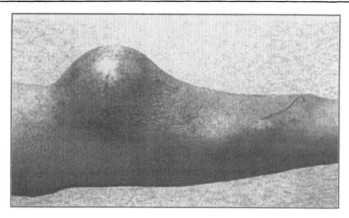

Fig. A.27.1.

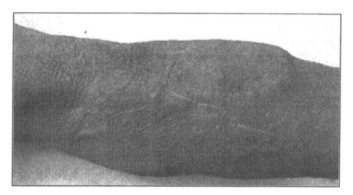

Fig. A.27.2.

Case Scenario #28: Chronic Steal

This patient is a 63 year old with type II diabetes for 20 years and a left upper arm brachiocephalic fistula of 2.5 months (Fig. A.28.1) (9/11/01). This fistula is working nicely. It has 2 aneurysms with a short segment of normal fistula vein in between (arrow) of about 3-4 cm. He now comes with dry distal ulcerations on all but digit V (Fig. A.28.2). There is numbness but no pain. Both hands are cold, more so on the left. There are early dry ulcers on digit III right side. Duplex Doppler estimated flow is 2.2 L/min. Both RA and UA are open but narrowed and severely calcified. Left sided finger pressures are about 70 mmHg. Right hand finger pressures are about 190 mmHg.

Options

1. Ligate the left arm fistula and place an IJ dual lumen, tunneled catheter as the only access.
2. Band the fistula between the aneurysms.

Solution

Patient scheduled for banding surgery; however, nephrologist wants 3 weeks anti-platelet trial with Plavix®. At five weeks later on Plavix® 75 mg daily (10/16/01), all ulcers are smaller (Fig. A.28.3), but there is a new one on the index finger; still there is no pain. Patient to return in another four weeks.

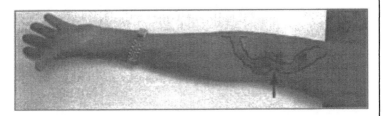

Fig. A.28.1.

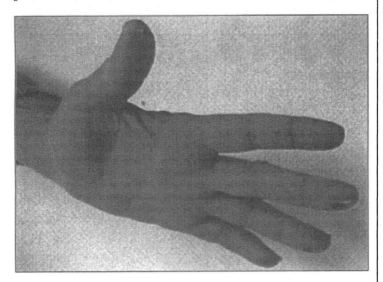

Fig. A.28.2.

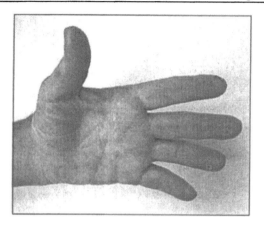

Fig. A.28.3.

Case Scenario #29: Shiny Pseudoaneurysm

This type of quite small, paper thin aneurysm will eventually develop necrosis and bleed (Fig. A.29.1, A.29.2). Occasionally the skin is so thin that the turbulent blood (red cell) flow can be seen. At surgery, proximal and distal control of the graft must be obtained. Alternatively, if there is only one or two small aneurysms a stitch in the graft may suffice in a bloodless arm obtained with an upper arm tourniquet.

Comments

Aneurysms come in all forms and shapes; do not operate just because there is an aneurysm, or because the referring doctor and the patient are concerned. Surgery is indicated when infection is present, risk of rupture through shiny skin or under a scab. Other indications include bothersome aneurysm because of size and location, and when associated with other complications such as stenosis.

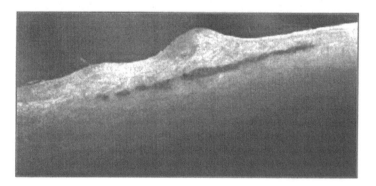

Fig. A.29.1.

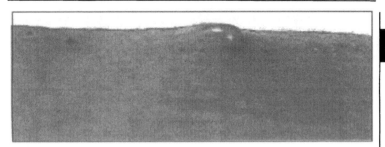

Fig. A.29.2.

Case Scenario #30: Multiple Shiny Aneurysms

Patient with scleroderma. Her upper arm entire PTFE graft is covered with small, paper-thin pseudoaneurysms (Fig. A.30.1). Eventually, these are likely to necrose (and be inflamed) causing significant bleeding. In some of these turbulent blood flow can be seen with the bare eye. After much discussion and workup (duplex Doppler, finger pressures) the patient had a forearm AV graft placed in the same extremity (Fig. A.30.1). She developed significant hand ischemia. The upper arm graft was temporarily banded (at B, Fig. A.30.2) but used for dialysis another 3

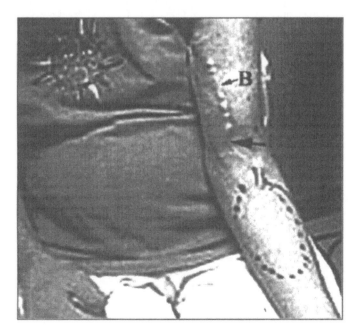

Fig. A.30.1.

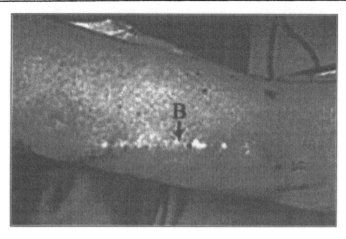

Fig. A.30.2.

weeks until the forearm graft was ready. Then the upper arm graft was ligated close to the arterial anastomosis (clipped with large hemoclip) at arrow (Fig. A.30.1). No further hand ischemia occurred.

Case Scenario #31: Multiple Pseudoaneurysms

In cases of multiple small (paper thin) aneurysms a longer (Fig. A.31.1-2) segment of the graft with the skin may be excised. Inside the graft large defects demonstrate the often extensive destruction of the graft (Fig. A.31.3) with multiple communicating channels.

Case Scenario #32: Forearm PTFE Pseudoaneurysms

Patient with an expanding (<4 weeks) dialysis needle induced pseudoaneurysm (Fig. A.32.1). There is some pain but the protruding lump was bothersome and indicated its resection.

At surgery the aneurysm was completely mobilized and the PTFE graft controlled on each side (Fig. A.32.2). Before the second anastomosis the aneurysm and outflow vein are flushed with heparinized (10u/ml) saline (Fig. A.32.3).

Case Scenario #33: Upper Arm PTFE AV Graft

This case illustrates several aspects of upper arm access. First, the major section of the graft should be on the lateral or outermost aspect of the arm where the skin is firmer and less mobile than in the medial aspect (Fig. A.33.1-2); this is most obvious in females and after significant weight loss. Also, the lateral aspect is less painful for needle sticks.

Fig. A.31.1.

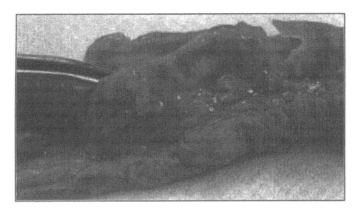

Fig. A.31.2.

Fig. A.31.3.

AI

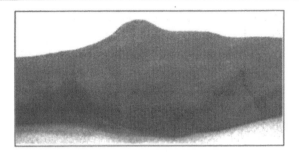

Fig. A.32.1.

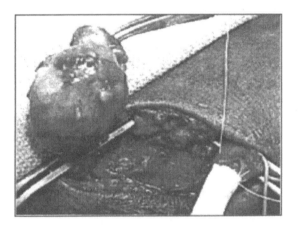

Fig. A.32.2.

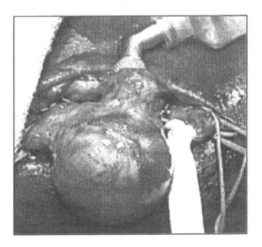

Fig. A.32.3.

    Second, the selection of anastomosis sites is greatly facilitated through preoperative Doppler "vein mapping", where the surgeon is present to mark the exact site (arrow).

    Third, select the most distal vein anastomosis site to facilitate future revision and extension up the vein.

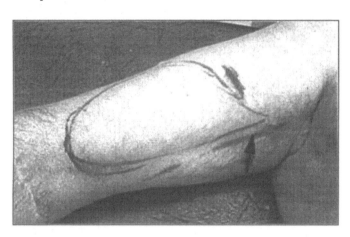

Fig. A.33.1.

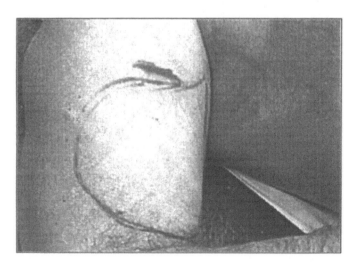

Fig. A.33.2.

Case Scenario #34: No Dialysis Access

A1   This 51 year old overweight woman has no current working access. She has a large hemangiomic nevus, excluding the left arm for access (Fig. A.34.1). At time of left IJ percutaneous catheter placement 10 days ago patient became hoarse. There is a failed right forearm AV graft for 6 months. She has had several right IJ catheters. Pre op duplex Doppler shows the cephalic vein patent along upper arm to the subclavian vein (visible in Fig. A.34.2). Right IJ is occluded midneck but open at the collar bone level.

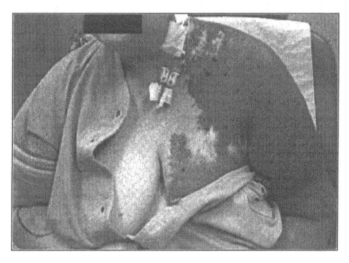

Fig. A.34.1.

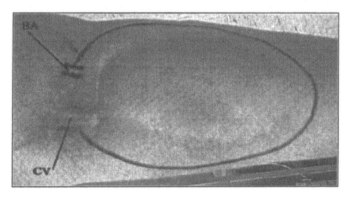

Fig. A.34.2.

Solution

At surgery a PTFE AV graft was placed around the old graft right forearm (Fig. A.34.2). A low stick under direction with Site Rite using micropuncture technique was used to place a 28 cm right IJ split ash catheter. The left percutaneous IJ catheter was removed.

Comments

Patient hoarseness resolved slowly within 7-10 days. This is the only time the author has come across this complication. Aberrant recurrent laryngeal nerve? Hematoma related to nevus? Or a "bad" stick?

Case Scenario #35: A Case of System Problems

Many case scenarios illustrate the importance of history taking, physical exam and the use of duplex Doppler in coming to a workable solution. During the exam the author uses a marking pen to outline old grafts, veins and arteries (Fig.A.35.1). This arm belongs to a slightly debilitated and obese 38 year old lady in whom I placed a primary AV fistula (AV) at the wrist 11 months ago. The dialysis unit nurse told the patient that the fistula has not matured and that she needs a graft; the patient, because of transportation problems, went to a closer local hospital and the surgeon placed an AV loop graft (PTFE) that "never worked" in the patient's own words. There is a large palpable basilic vein (BV) and a large cephalic vein (CV) at the antecubital fossa. Duplex Doppler confirms these findings; in addition the diameter of the cephalic vein 3 cm from the anastomosis is 8 mm, with a flow rate of about 1200 cc/min. Also, the cephalic vein divides beneath the clotted PTFE graft with a large branch likely connecting to the BV at the antecubital fossa.

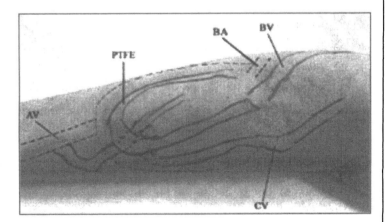

Fig. A.35.1.

Solution

This is what I did: a new PTFE (Gore Tex® stretch 7 mm tapered) was placed from the brachial artery (BA) to the cephalic vein (CV) (dotted line). The primary AV fistula was ligated at the wrist. Patient did not show for follow up. Phone call to dialysis unit indicates graft is being used successfully, four months after surgery.

Comments

This case demonstrates the system multi-factorial elements leading up to this case scenario. The patient (obese and debilitated) is perhaps the main factor. It could be argued that she then needed greater attention and firm direction from dialysis and nursing home medical personnel. Placing a PTFE on top of a large AV fistula (although deep) is hard to defend.

Case Scenario #36: Superior Vena Cava Occlusion

34 year old male with history and scars from many catheters and graft placements (Fig. A.36.1). Dilated subcutaneous veins on upper torso suggest superior vena cava occlusion, confirmed with central venogram (Fig. A.36.2) where there is a 9 cm absence of SVC. Currently he is kept alive with a right femoral vein cuffed dual lumen catheter. He also has had a left groin PTFE graft (Fig. A.36.3).

Solution

Duplex Doppler shows open femoral vein and superficial femoral artery above old graft left side (Fig. A.36.3). Suggest placement of PTFE graft at this site. A left groin stretch loop graft was placed between the femoral artery and the saphenous vein as it entered the femoral vein (Fig. A.36.3). The graft was successfully used three weeks post surgery

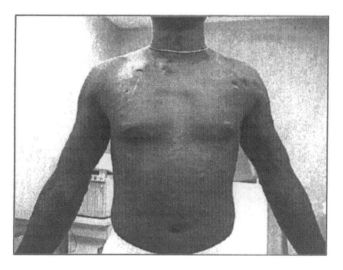

Fig. A.36.1.

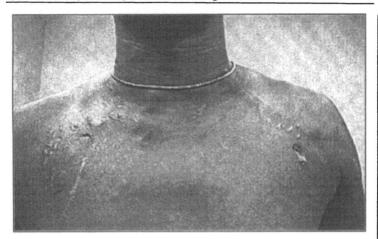

Fig. A.36.2.

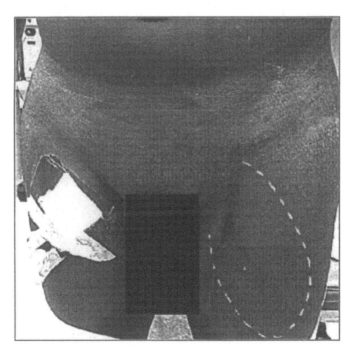

Fig. A.36.3.

Case Scenario #37: Upper Arm Pseudoaneurysm

Patient is a 52 year old diabetic male with an expanding pseudoaneurysm of about 1 year in a PTFE graft of 4 years (Fig. A.37.1). Just the size and the inconvenience of its presence indicate removal (Fig. A.37.2). Preoperative duplex Doppler shows wide open arterial anastomosis (A) (Fig. A.37.1). However, there is a very tight (pinhole) stenosis at the venous anastomosis site (V) (Fig. A.37.3).

Solution

The plan was to resect the aneurysm, revise the venous outflow stenosis and use the remaining segment for dialysis (between arrows) (Fig. A.37.1).

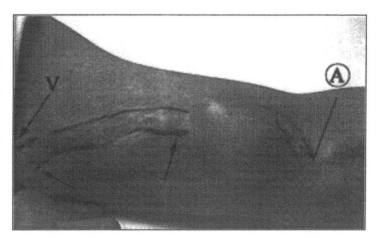

Fig. A.37.1.

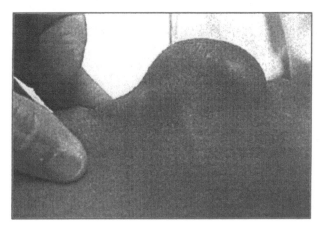

Fig. A.37.2.

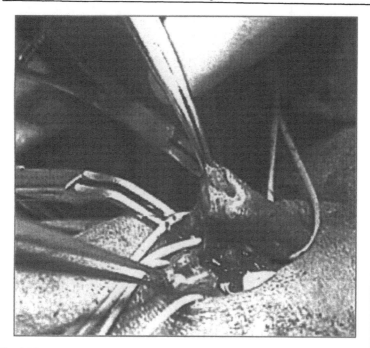

Fig. A.37.3.

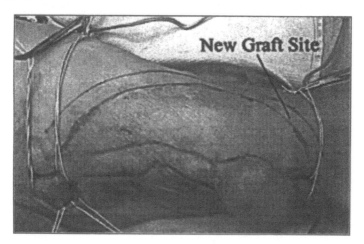

Fig. A.37.4.

At surgery the venous outflow stenosis was 1 cm long into a very scarred, fibrotic graft (Fig. A.37.3). However, the brachial vein at this site is huge ~ 15-20 mm (Fig. A.37.4-5). The decision was made to place a new graft around the old graft (Fig.

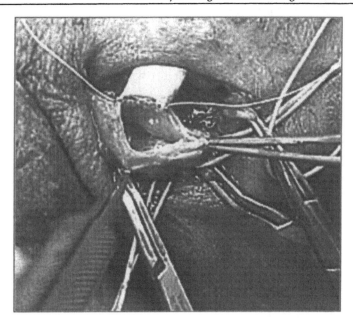

Fig. A.37.5.

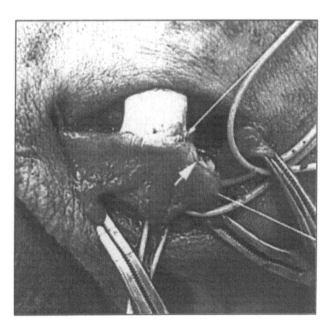

Fig. A.37.6.

A.37.4). The aneurysm (collapsed) and the old graft was left in place. The ends of the grafts were not oversewn. In order to anastomose the new graft into virgin venous tissue the venotomy was extended away from the old scar (Fig. A.37.5), leaving a large opening in the vein. This abundant venotomy (arrow) was closed by continuing one of the distal corner sutures (Fig. A.37.6).

A split ash right IJ catheter was placed as well.

Case Scenario #38: Distal Ischemia in Diabetic Patients

American Indian woman with diabetes and metastatic calcification deposits. In addition to bilateral hand amputations she has both legs amputated above the knee. The last two years she has been dialyzed through catheters, currently one in left subclavian vein(!!). Surprisingly, her R SCV is open. There is a previously failed L forearm graft prior to the hand amputation.

Duplex Doppler shows severe brachial artery wall calcifications (Fig. A.38.1), but adequate lumen (Fig. A.38.2).

Solution

R Upper arm PTFE loop. Duplex Doppler used to map basilic/brachial veins for optimal anastomosis site (arrow) (Fig. A.38.2). By choosing a high proximal brachial artery anastomosis site, distal ischemia is diminished. This patient did not develop further signs of ischemia (Also see Case scenario #33).

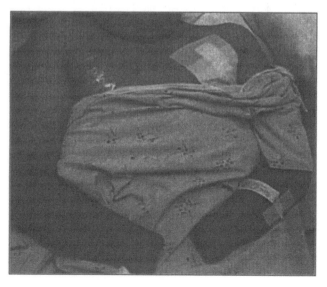

Fig. A.38.1.

AI

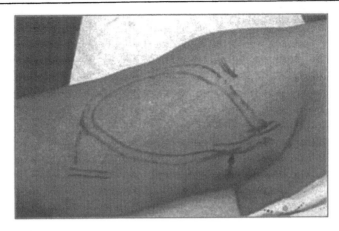

Fig. A.38.2.

Case Scenario #39: Loose Cuffed Tunneled Catheters

This 60 year old diabetic patient comes for elective removal of cuffed dialysis (OptiFlow®) catheter (2 cuffs) (Fig. A.39.1). The catheter slides out fairly easily. The author uniformly places a stitch around the tract to prevent backbleeding (Fig. A.39.1-2). The patient is also advised to hold pressure for about 1 hour, and avoid strenuous exercise, and be upright for most of the next 12 hours.

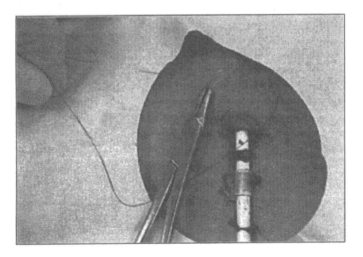

Fig. A.39.1.

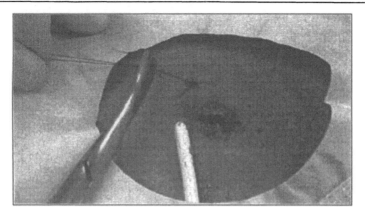

Fig. A.39.2.

Comments

Cuffed catheters are often infected (staph. epidermidis) with little or no systemic reaction. The bacteria in these cases are staph epidermidis species. No antibiotics are needed in these cases, only removal of the catheter.

Case Scenario #40: Subclavian Vein Catheter

This gentleman was referred to me with a malfunctioning left subclavian (!!) vein Tesio catheter as the only access, placed 4 months ago. He has had 7 PTFE grafts in upper extremities (Fig. A.40.1). Preoperative Duplex Doppler (with the author present) shows right IJ open, right SCV open. Also, there is a large brachial-basilic vein upper arm (BV).

Solution

1. A right IJ 28 cm split ash catheter cuffed tunneled catheter was placed, using Site Rite directed micro puncture technique (Chapter 5). .
2. A right upper arm PTFE stretch graft was placed (Fig. A.40.2). The main segment of the graft should run on the anterior/lateral aspect of the upper arm. (see case #33).
3. The left SCV Tesio catheters were removed.

Comments

Do not (ever) use subclavian veins for dialysis access (See Chapter 5, Table 5.3).

Case Scenario #41: Subclavian Vein Occlusion

Patient with right subclavian vein occlusion (Fig. A.41.1) evidenced by arm swelling and distended neck and chest wall veins. Radiology has failed balloon angioplasty. Note the protruding external jugular vein (arrow, Fig. A.41.1) and marked chronic edema of right hand (Fig. A.41.2). A PTFE graft had been placed six months ago in

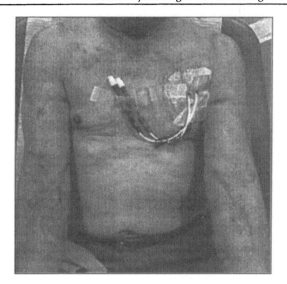

Fig. A.40.1.

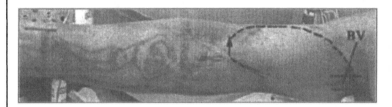

Fig. A.40.2.

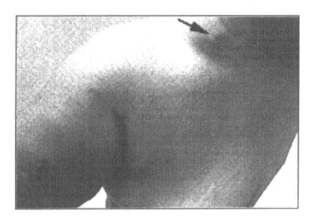

Fig. A.41.1.

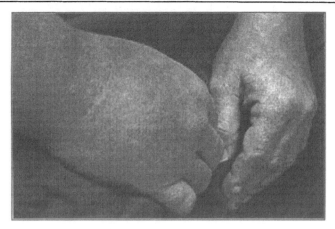

Fig. A.41.2.

her left arm, but patient did not show for surgery and graft ligation of this right upper arm graft ("Had no ride"). After ligation, swelling went away, but marked discoloration of her hand remained.

Case Scenario #42: Kinked Split Ash Catheter

This small kink "bubble" on a split ash catheter was discovered on post insertion X-ray (Fig. A.42.1-2). There were no flow problems in the OR nor on dialysis next day. There are no follow up X-rays. The author has only seen this case with this type kink only involving one port.

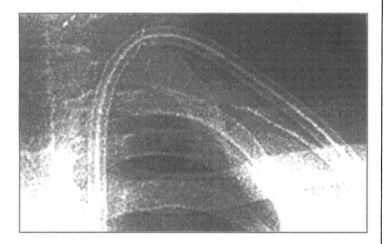

Fig. A.42.1.

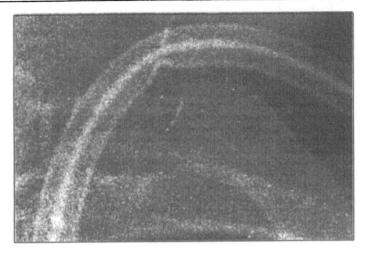

Fig. A.42.2.

Case Scenario #43: Subclavian Catheter

In the future this type picture will only be viewed in the medical historical library; this one is from October, 2000. Not only is this catheter in the wrong vein, also it is not properly primed (filled with blood) (Fig. A.43.1), and thrombus (Fig. A.43.2). When removed, these stiff catheters outline the venous anatomy (Fig. A.43.2). Also see Chapter 5, Tables 5.1, 5.2 and 5.3 for current indications for catheters.

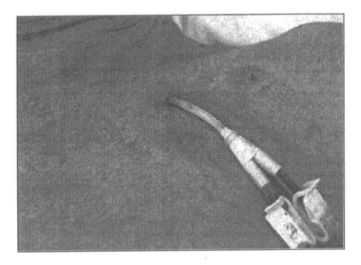

Fig. A.43.1.

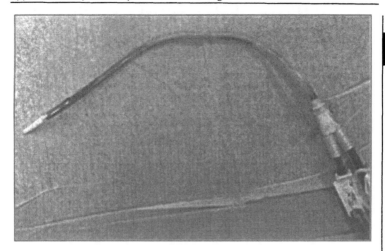

Fig. A.43.2.

Case Scenario #44: Poor Catheter Care

This somewhat debilitated patient came for placement of permanent access. On removing the dressing covering the Tesio catheters this is what it looks like (Fig. A.44.1). No signs of infection, and the catheter has worked just fine for now almost one year.

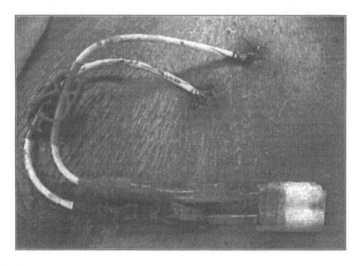

Fig. A.44.1.

Al

Comments

The author thinks it makes common sense for the dialysis unit (may it be the nephrologist, RN, DON, technician) to change/check the catheter dressing, clean the area, including the catheter; sutures should be removed 10-14 days after catheter placement.

Case Scenario #45: Malplaced Dual Lumen Catheter

These two elderly ladies share a few problems in addition to advanced age (Fig. A.45.1-2). Both have left subclavian vein cuffed, tunneled catheters that are nonfunctioning (lady in Fig. A.45.1 also has a gastrostomy feeding tube). Catheter

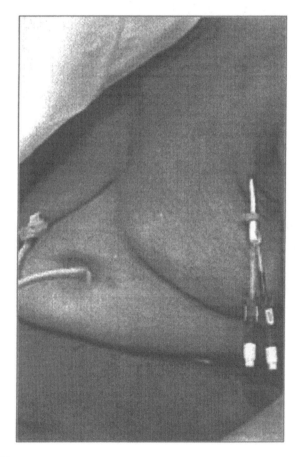

Fig. A.45.1.

in Figure A.45.1 is a split ash (poorly primed, blood in the red port) (same dialysis unit as cases 43 and 44) and catheter in Figure A.45.2 is an OptiFlow sutured to the wing. The long scar leading up to the exit site suggests some unusual surgical technique at time of placement (Fig. A.45.2). The main problem with these examples is low and poorly positioned exit sites. This suggests inability to access the central vein or in this case being occluded due to sucking against the lateral SVC wall (Fig. A.45.3).

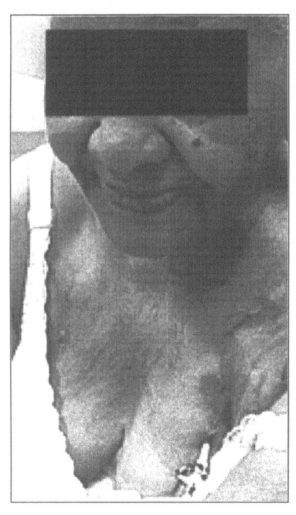

Fig. A.45.2.

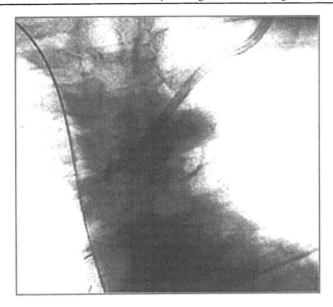

Fig. A.45.3.

Comments

Much morbidity and suffering would be prevented by paying attention to a few basic principles, and details of the steps for the specific procedure to be performed (skills and knowledge), or is it a matter of attitude?

Consider each catheter placement a life saving procedure. Use micropuncture technique, portable ultrasound and fluoroscopy. Also, mark the catheter tract and exit site. The skin moves considerably from supine (head down) to standing, especially in females and obese individuals (see Chapter 5 for catheter placement techniques).

Case Scenario #46: Catheter Infection

Patient, a 47 year old female, arrives in the office for permanent access placement. The only access is a right IJ tunneled (2 cuff) OptiFlow catheter. She has a temperature of 102 F, and has chills and sweatening. On exam there are no forearm cephalic veins, but a possible L upper arm cephalic vein (to be worked up later!) (Fig. A.46.1). She had not dialyzed in five days. The catheter tract is red, somewhat tender (arrow) (Fig. A.46.2). The outer cuff is exposed.

Treatment

Patient was urgently taken to dialysis treatment, given 1g vancomycin IVPB at completion. Then the catheter was removed; did not require any surgical incision. Infected catheters can usually easily be pulled. One stitch was placed around the tract at x (Fig. A.46.2) to prevent backbleeding. Blood cultures grew *Staph aureus*.

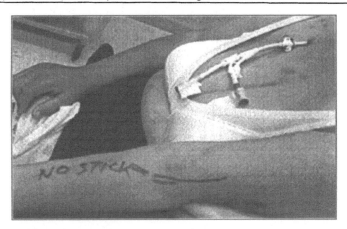

Fig. A.46.1.

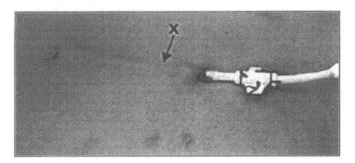

Fig. A.46.2.

After 48 hours of no fever, clinically improved, wbc = 6.7, a left IJ split ash catheter was placed. A left arm AV access (Fig. A.46.1) is scheduled in 10-14 days.

Comments

The author strongly advises against suturing the wings to the skin. Only one suture at exit site is needed, tied around catheter and into the groove remaining when the wing is removed, i.e., split ash. To all manufacturers of catheters: Remove the wings? To all surgeons: Do not suture the wing to the skin.

Case Scenario #47: Subclavian Vein Catheter

Patient (69 year old) with a left subclavian (!!) vein catheter as only access (Fig. A.47.1). There is an old clotted left forearm AV graft. There are scars from previous chest/neck catheters (arrows) (Fig. A.47.2). Duplex shows right IJ open, right subclavian vein open.

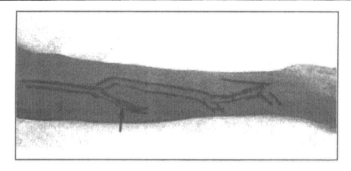

Fig. A.47.1.

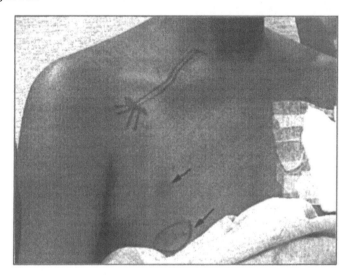

Fig. A.47.2.

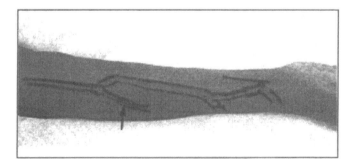

Fig. A.47.3.

Solution

1. A right forearm radio-cephalic primary AV fistula was created. The proximal side branch was left intact (arrow) (Fig. A.47.3).
2. A right subclavian vein cuffed, tunneled catheter placed using the micropuncture technique, portable ultrasound and fluoroscopy (subcutaneous tract drawn in Fig. A.47.2).
3. The left IJ cuffed tunneled catheter was removed.

Comments

A most burning question remains; why wasn't her right cephalic vein used in the first place; and how did it survive I.V.'s and blood draws for so long?

Case Scenario #48: The Elderly

This 94 year old mentally alert man had a right infected IJ catheter removed. After one week (3 sessions) of femoral vein catheter dialysis a left IJ was placed, now working well two weeks later. This visit is for preoperative permanent access placement. His main concern was not to be able to have another catheter should this one fail. After discussion with him and family we decide to just keep the current catheter. Ten days later he suffered a stroke.

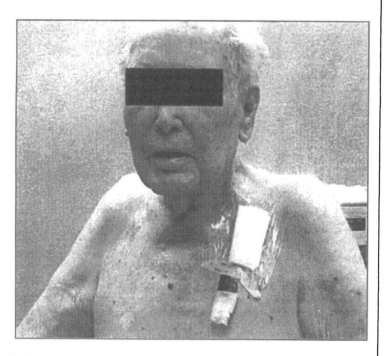

Fig. A.48.1.

Comments

The management of the elderly with ESRD is challenging and requires much common sense. Factors to be considered include: short survival, severe comorbidity, mental capacity, fragile skin, social and family support. The author initially relies more on temporary cuffed tunneled catheters.

Case Scenario #49: Osteomyelitis of Clavicle-Sternal Joint

Patient is a 29 year old diabetic on dialysis for 6 months with a percutaneous right IJ catheter as only access. She has a hard lump over the clavicle-sternal joint since ~ 4-6 months after a SCV stick for I.V. access (Fig. A.49.1). She has been on long term I.V. antibiotics (mostly in ICU). She comes for permanent access (will need a PTFE graft). After discussion with referring doctor the left IJ was exchanged with a split ash over a single guidewire technique (Chapter 5, Fig. 5.2). Permanent access will be placed when infection free (negative blood cultures and normal white cell count) and stable off antibiotics for >4 weeks.

Comments

A native AV fistula would be most desirable in this setting. Patient has no acceptable peripheral veins at this time.

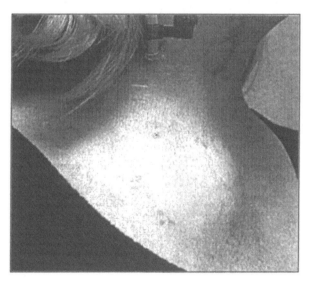

Fig. A.49.1.

Case Scenario #50: Malplaced PD Catheter

Young girl with poor drainage of PD catheter, the pigtail was lodged in the right upper quadrant (Fig. A.50.1).

Solution

The old catheter was removed and a new inserted at the periumbilical transrectus (arrow) (see chapter 6 for surgical technique).

Comments

The below umbilicus midline incision for PD catheter insertion is suboptimal. It also tends to misdirect the pigtail catheter into the right or left side of the abdomen, to be engulfed by the omentum. (Chapter 6).

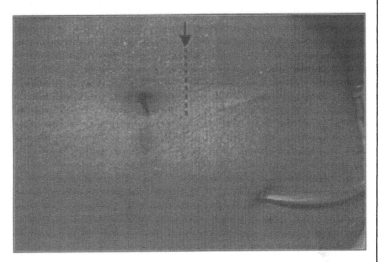

Fig. A.50.1.

Prescription Drug Administration in ESRD and Dialysis: A Preliminary Guide

Mark Durso

This brief appendix is a guide to altered drug effects in renal failure patients. It is not meant to be a prescription reference. For specific drug choices and dosing, refer to your institution's clinical pharmacist.

Prescription Drug Administration in ESRD and Dialysis

It is important to understand the physiologic and biochemical effects of drugs in patients with renal disease. A decrease in glomerular filtration rate (GFR) associated with ESRD impacts the metabolism and clearance of these drugs. To further complicate matters, levels of many drugs are affected by the processes of CAPD or hemodialysis. This chapter will briefly summarize some of the strategies used in determining appropriate dosage and drug monitoring, as well as specific adjustments and considerations of many commonly used drugs in a table.

Regardless of renal function status, drug concentration is a function of several factors. Bioavailability is the rate and amount of a dose to enter systemic venous circulation to yield the maximum drug concentration in the shortest time. While drugs administered intravenously retain an essentially normal bioavailability in ESRD patients, it is orally administered medications that can be affected at multiple levels in ESRD patients.

Drug absorption across gastrointestinal membranes may be reduced or eliminated in ESRD. Uremic gastrointestinal symptoms, including nausea and vomiting, may prevent sufficient drug breakdown in the stomach. Enzymatic changes in uremia, including those in intestinal epithelia and salivary urea may metabolize drug compounds too early, interfere with acid hydrolysis or chelate active compounds.

Gastroparesis lengthens the time a drug remains in the stomach, which delays bioavailability, while diarrhea shortens the intestinal contact time, reducing or eliminating absorption by the small bowel.

Drugs metabolized or changed by the liver may be impacted by these gastrointestinal factors, by diminished compound binding to the plasma protein or by decreased transformation of the drug on arrival to the liver, resulting in larger amounts of nonmetabolized drug being cleared by the portal system, effectively diminishing potential bioavailability.

Drug distribution may be highly variable in ESRD patients. Distribution is the amount of dispersal of drug over a given time, and is a factor of drug amount, plasma concentration and solubility factors. Protein-bound drugs are water soluble, and demonstrate a narrow distribution in the extracellular fluid space. This can be affected by patient fluid status (edema vs dehydration), ascites and muscle wasting. Decreased binding of these drugs can occur in ESRD with low albumin

Access for Dialysis: Surgical and Radiologic Procedures, 2nd ed., edited by Ingemar J.A. Davidson. ©2002 Landes Bioscience.

concentrations, resulting in higher amounts of unbound circulating compound, which may be quickly eliminated or accumulate to toxic levels.

The overall drug metabolism and elimination is affected by ESRD. Drug end products may be exclusively cleared by the kidney, representing a potentially toxic metabolite build up with a regularly administered drug.

Calculations for Drug Dosing and Removal by Dialysis

Calculating creatinine clearance is the first step in safe and effective drug dosing in patients with renal failure. There are many formulae for estimating creatinine clearance, including the Cockcroft and Gault formula for patients with a steady state serum creatinine level:

$$\text{Creatinine Clearance (CrCl)} = \frac{(140\text{-age}) \times \text{IBW (kg)}}{72 \times \text{serum creatinine (mg/dl)}}$$

Loading Dose (same for ESRD/nonESRD patients)

This calculation is most useful for drugs with long half lives.

$$LD = Vd \times IBW \times Cp$$

where LD is loading dose; Vd is volume of distribution in litres; and IBW is ideal body weight in Kg, defined for men as 50Kg + 2.3Kg per inch for each inch above 5ft; and for women as 45.5Kg + 2.3Kg per inch for each inch above 5ft. Cp is desired plasma concentration of the drug.

Drug Removal by Dialysis

Most drug removal during dialysis occurs by diffusion from plasma across the dialysis membrane along a concentration gradient. Most drugs > 500D, those >90% protein bound, or those with large tissue distribution volumes do not cross this membrane.

The amount of drug removed during dialysis is increased by the use of porous membranes (i.e., CVVHD, CAVHD), large surface area dialyzers or double dialyzers, an increased blood flow rate, increased dialysate flow rate, or an increased treatment duration.

Estimating Hemodialysis Drug Clearance

$$CL_{HD} = CL_{UREA} \times (60/\text{molecular weight of drug})$$

where CL_{HD} is the hemodialysis drug clearance; CL_{UREA} is the dialyzer urea clearance (usually 150 ml/min) and (60/molecular weight of drug) is a standard proportion.

Estimating Peritoneal Dialysis Drug Clearance

The peritoneal membrane is more porous than standard hemodialysis membranes, but is still limited in effectiveness in clearing many drugs. The general rule of thumb is if a drug is not cleared by hemodialysis, it will not be cleared by peritoneal dialysis. Peritoneal dialysis is effective in clearing drugs that are not extensively protein bound, those distributed in extracellular fluid, or with small molecular weights. The rate and extent of small molecular weight drug removal depends on the volume of peritoneal dialysate exchanged, as there is a large concentration gradient between the dialysate and blood. Therefore, concentration is greatly increased in drugs added to the peritoneal dialysate.

Continuous renal replacement therapy, including CAVH, CAVHD and CVVHD involves the use of a large bore membrane with a lower, continuous blood flow (15-20 ml/min in CAVHD and CVVHD and 10-30 ml/min in CAVH). Drugs > 80% protein bound are not removed, nor are drugs with a volume of distribution >0.71/Kg.

All

All

Drug	Brand Name	Usual Dose	% Usual Dose for GFR > 50 ml/min	% Usual Dose for GFR 10-50 ml/min	% Usual Dose for GFR < 10 ml/min
Immunosuppression					
Anti T Cell (poly)					
ATG (equine)	Atgam	15 mg/kg/d over 4h	100	100	100
Anti T Cell (mono)					
Muromonab CD3	OKT3	5 mg/d	100	100	100
Nuc Synthesis Inhib					
Azathioprine	Imuran	1-3 mg/kg/d	100	75	50
Mycophenolate Mofetil	CellCept	1 g bid	100	100	NR
Cytokine Tran Inhib					
Cyclosporine A	Sandimmune	5 mg/kg bid	100	100	100
Cyclosporine emuls	Neoral	5 mg/kg bid	100	100	100
Tacrolimus (FK506)	Prograf	0.15-0.30 mg/kg bid	100		
Steroids					
Methylprednisolone	Solumedrol	2 mg/kg/d tapered	100	100	100
Prednisone	Deltasone	15 mg/d tapered	100	100	100

Drug	Brand Name	Usual Dose	% Usual Dose for GFR > 50 ml/min	% Usual Dose for GFR 10-50 ml/min	% Usual Dose for GFR < 10 ml/min
Antivirals					
Acyclovir	Zovirax	5 mg/kg q8h	5 mg/kg q8h	5 mg/kg q12-24h	2.5 mg/kg q24h
Ganciclovir	Cytovene	5 mg/kg q12h	q12h	q24-48h	q48-96h
CMV hyperimm glob	Cytogam	150 mg/kg tapered	28 mg/kg	15 mg/kg	6 mg/kg
Antibiotics					
TMP-SMX					
160 mg-800 mg	Bactrim DS	160/800 q12	160/800 q12	160/800 q18	160/800 q24
PCN + β Lactams					
Amoxicillin/clavulanate	Augmentin	250-500mg q8	q8h	q8-12h	q24h
Ticarcillin/clavulanate	Timentin	3.1g q4h	100	2g q4-8h	2g q12h
Ampicillin/sulbactam	Unasyn	1.5-3g q6	q6	Q8-12	q24
Quinolones					
Levofloxacin	Levaquin	500mg q24	100	500mg load 250mg q24-48	500mg load 250mg q48
Ciprofloxacin	Cipro	250-500mg q12 po	100	50-75	50
Cephalosporins					
Cefazolin	Ancef	1-2g q8	q8	q12	q24-48
Ceftriaxone	Rocephin	250mg-2g q12	100	100	100

All

Drug	Brand Name	Usual Dose	% Usual Dose for GFR > 50 ml/min	% Usual Dose for GFR 10-50 ml/min	% Usual Dose for GFR < 10 ml/min
Aminoglycosides					
Amikacin	Amikin	7.5mg/kg q12	CONSIDER DOSING BASED ON DRUG LEVELS AND PHARMACO KINETICS		
Tobramycin	Nebcin	1.7 mg/kg q8	CONSIDER DOSING BASED ON DRUG LEVELS AND PHARMACO KINETICS		
Gentamicin	Garamycin	1.7 mg/kg q8	CONSIDER DOSING BASED ON DRUG LEVELS AND PHARMACO KINETICS		
Other ABX					
Vancomycin	Vancocin	1g q12 check levels	1g q12-24 check levels	1g q24-96 check levels	1g q4-7days check levels
Antifungals					
Fluconazole	Diflucan	200-400mg q24	100	50	50
Ketoconazole	Nizoral	200mg q24	100	100	100
Amphotericin B	Fungizone	0.4-1.0mg/kg q24	0.4-1.0mg/kg q24	0.4-1.0mg/kg q24	0.4-1.0mg/kg q48
CV Drugs					
Ca Chan B					
Diltiazem	Cardizem	30-90mg q6-8	100	100	100
Nifedipine	Procardia	10-20mg q6-8	100	100	100
Verapamil	Calan	80mg q8	100	100	100
Amlodipine	Norvasc	5mg q24	100	100	100

All

Drug	Brand Name	Usual Dose	% Usual Dose for GFR > 50 ml/min	% Usual Dose for GFR 10-50 ml/min	% Usual Dose for GFR < 10 ml/min
α Adren R-B					
Doxazosin	Cardura	1-16mg q24	100	100	100
Central α-Adren A					
Clonidine	Catapres	0.1-0.6mg bid	100	100	100
β Blockers					
Metoprolol	Lopressor	50-100mg bid	100	100	100
Atenolol	Tenormin	50-100mg q24	100% q24	50% q48	30-50% q96
Labetalol (α + β)	Normodyne	200-600mg bid	100	100	100
Other CV					
Digoxin	Lanoxin	1-1.5mg load 0.25-0.5mg q24	100% q24	25-75% q36	10-25% q48
ACE inhibitors					
Enalapril	Vasotec	5-10mg q12	100	75-100	50
Captopril	Capoten	25mg q8	100	75	50
Angiotensin II R-A					
Losartan	Cozaar	50mg q12	100	100	100

All

Drug	Brand Name	Usual Dose	% Usual Dose for GFR > 50 ml/min	% Usual Dose for GFR 10-50 ml/min	% Usual Dose for GFR < 10 ml/min
Statins (Anti CHO)					
Lovastatin	Mevacor	20-80mg q24	100	100	100
Pravastatin	Pravachol	10-40mg q24	100	100	100
Fluvastatin	Lescol	2-10mg q24	100	100	100
Simvastatin	Zocor	5-40mg q24	100	100	100
Atorvastatin	Lipitor	10-80mg q24	100	100	100
Antiplatelet					
Aspirin		650mg q4	q4	q4-6	avoid
Clopidogrel	Plavix	75mg QD	100	100	100
Diabetic Agents					
Insulin NPH Regular Regular/NPH mix Lente	Humulin N Humulin R Humulin 70/30; 50/50 Humulin L	variable	100	75	50
Glyburide	Micronase	1.25-20mg q24	no data	avoid	avoid
Glipizide	Glucotrol	2.5-15mg q24	100	50	50
Metformin	Glucophage	500-850mg bid	50	25	avoid

All

Drug	Brand Name	Usual Dose	% Usual Dose for GFR > 50 ml/min	% Usual Dose for GFR 10-50 ml/min	% Usual Dose for GFR < 10 ml/min
GI Drugs					
H₂ R-A					
Famotidine	Pepcid	20-40mg qhs	50	25	10
Ranitidine	Zantac	150-300mg qhs	75	50	25
Cimetidine	Tagamet	400mg bid or 400-800mg qhs	100	50	25
Proton Pump Inhib					
Omeprazole	Prilosec	20-60mg q24	100	100	100
Lansoprazole		15-60mg q24	100	100	100
Gastric Empty					
Metoclopramide	Reglan	10-15mg qid	100	75	50
Diuretics					
Furosemide	Lasix	40-80mg bid	100	100	100
Bumetanide	Bumex	1-2mg q8-12h	100	100	100
CNS/Endocrine					
Levothyroxine	Synthroid	1.0-1.6mcg/kg/day	100	100	100
Phenytoin	Dilantin	1000 mg load 300-400mg q24	100-measure free levels	100-measure free levels	100-measure free levels

All

Drug	Brand Name	Usual Dose	% Usual Dose for GFR > 50 ml/min	% Usual Dose for GFR 10-50 ml/min	% Usual Dose for GFR < 10 ml/min
Vascular Drugs					
Warfarin	Coumadin	2-10mg q24	100	100	100
Pentoxifylline	Trental	400mg tid	q8-12	q12-24	q24
LMW heparin	Lovenox	1-1.5mg/kg qd	100	Adjustment needed	for GFR <30ml/min
Pain Drugs					
Codeine		30-60mg q4-6	100	75	50
Hydrocodone		5-10mg q4-6	100 may be needed	Dose reduction may be needed	Dose reduction
Propoxyphene	Darvocet	65mg q6-8	100	100	avoid
Oxycodone		5mg q6	100	100	use with caution
Meperidine	Demerol	50-100mg q3-4	100	avoid	avoid
Morphine		2-10mg q4	100	75	50

NR = not recommended

All

Drug	Dose Adjustment for HD	Drug Cleared by CAPD	Drug Cleared by CRRT
Immunosuppression			
Anti T cell (poly)			
ATG (equine)	dose after hd	n/a	n/a
Anti T Cell (mono)			
Muromonab CD3	no	no	no
Nuc Synthesis Inhib			
Azathioprine	yes	unk	use GFR 10-50
Mycophenolate Mofetil	no	no	unk
Cytokine Tran Inhib			
Cyclosporine A	no	no	100%
Cyclosporine emuls	no	no	100%
Tacrolimus (FK506)	no	no	unk
Steroids			
Methylprednisolone	yes	unk	GFR 10-50
Prednisone	no	unk	GFR 10-50
Antivirals			
Acyclovir	dose after hd	use GFR<10	3.5 mg/kd/d
Ganciclovir	dose after hd	use GFR<10	2.5 mg/kd/d
CMV hyperimm glob	dose after dialysis	use GFR<10	use GFR 10-50
Antibiotics			
TMP-SMX 160 mg-800 mg	dose after hd	160/800 q24	160/800 q18
PCN + β Lactans			
Amoxicillan/clavulanat	dose after dialysis	250 mg q12	NA
Ticarcillin/clavulanate	3.1g after dialysis	use GFR<10	use GFR 10-50
Ampicillin/sulbactam	dose after hd	1.5-3g q24	2.25g q12
Quinolones			
Levofloxacin	dose gfr<10	dose GFR<10	dose GFR 10-50
Ciprofloxacin	250mg q12	250mg q8	200mg iv q12
Cephalosporins			
Cefazolin	0.5-1.0g after hd	0.5g q12	use GFR 10-50
Ceftriaxone	dose after hd	750mg q12	use GFR 10-50

Drug	Dose Adjustment for HD	Drug Cleared by CAPD	Drug Cleared by CRRT
Aminoglycosides			
Amikacin	Consider dosing based on drug levels and pharmaco kinetics		
Tobramycin	Consider dosing based on drug levels and pharmaco kinetics		
Gentamicin	Consider dosing based on drug levels and pharmaco kinetics		
Other ABX			
Vancomycin levels	use gfr<10	use GFR<10	500 q24-48 check
Antifungals			
Fluconazole	200mg after hd	use GFR<10	use GFR 10-50
Ketoconazole	no	no	no
Amphotericin B	no	use GFR<10	use GFR 10-50
CV Drugs			
Ca Chan B			
Diltiazem	no	no	use GFR 10-50
Nifedipine	no	no	use GFR 10-50
Verapamil	no	no	use GFR 10-50
Amlodipine	no	no	use GFR 10-50
α Adren R-B			
Doxazosin	no	no	use GFR 10-50
Central α-Adren A			
Clonidine	no	no	use GFR 10-50
β Blockers			
Metoprolol	50mg	no	use GFR 10-50
Atenolol	25-50mg	no	use GFR 10-50
Labetalol (α+β)	no	no	use GFR 10-50
Other CV			
Digoxin	no	no	use GFR 10-50
ACE Inhibitors			
Enalapril	20-25	no	use GFR 10-50
Captopril	25-30	no	use GFR 10-50
Angiotensin II R-A			
Losartan	no data	no data	use GFR 10-50
Statins (Anti CHO)			
Lovastatin	no data	no data	use GFR 10-50
Pravastatin	no data	no data	use GFR 10-50
Fluvastatin	no data	no data	use GFR 10-50
Simvastatin	no data	no data	use GFR 10-50
Atorvastatin	no data	no data	use GFR 10-50

Drug	Dose Adjustment for HD	Drug Cleared by CAPD	Drug Cleared by CRRT
Antiplatelet			
Aspirin	after hd	no	use GFR 10-50
Clopidogrel	no data	no data	no data
Diabetic Agents			
Insulin			
NPH	no	no	dose GFR 10-50
Regular			
Regular/NPH mix			
Lente			
Glyburide	no	no	avoid
Glipizide	no data	no data	avoid
Metformin	no data	no data	avoid
GI Drugs			
H$_2$ R-A			
Famotidine	no	no	dose GFR 10-50
Ranitidine	1/2 dose	no	dose GFR 10-50
Cimetidine	no	no	dose GFR 10-50
Proton Pump Inhib			
Omeprazole	no data	no data	no data
Lansoprazole	no data	no data	no data
Gastric Empty			
Metoclopramide	no	no data	dose GFR 10-50
Diuretics			
Furosemide	no	no	no
Bumetanide	no	no	no
CNS/Endocrine			
Levothyroxine	no	no	no data
Phenytoin	no	no	no
Vascular Drugs			
Warfarin	no	no	no
Pentoxifylline	no data	no data	no data
LMW heparin	contraindicated	contraindicated	contraindicated
Pain Drugs			
Codeine	no data	no data	use GFR 10-50
Hydrocodone	no data	no data	no data
Propoxyphene	avoid	avoid	n/a
Oxycodone	no data	no data	no data
Meperidine	avoid	avoid	avoid
Morphine	no data	no data	use GFR 10-50

Access for Dialysis:
A Preoperative Guide for Patients and Their Families

Ingemar J.A. Davidson, Wilson V. Garrett, Angela Kuhnel, Carolyn E. Munschauer and Gregory J. Pearl

This written summary is given to the patient to read and take home. A signed copy is kept in the patient's office record.

Background

You have been diagnosed with kidney (renal) failure. In medical textbooks this diagnosis is often labeled as "end stage renal disease (ESRD)". Most likely you were referred to me, the surgeon, by an internal medicine doctor specializing in kidney disease (nephrologist). You may have known about your failing kidneys for a long time, or it may have come as a surprise just recently. The most common causes of kidney failure are: diabetes, high blood pressure, infection and polycystic kidney disease.

Your kidneys help remove waste and toxins from your body, and also help make healthy red blood cells. If your kidneys stop working, these functions can be replaced by dialysis. If you do not have dialysis, the toxins will stay in your body and make you very sick.

Kidney function is measured from your blood as the concentration of creatinine (normal serum creatinine is 0.5 - 1.2 mg/dl). A better way is a test called creatinine clearance, or how much blood your kidney can clean per minute. The normal value is 100 ml/min or greater. When this number drops to about 10 ml/min (which corresponds to 10% of your normal kidney function),you will need dialysis to live. The signs of kidney failure which show you may be close to needing dialysis are: nausea, headache, anemia, swelling (edema) and high blood pressure. Most patients also feel exhausted or tired. Not every patient has all these symptoms, but usually have at least one or two.

Your case is unique. Even though it may sound like all patients with kidney failure needing dialysis are similar, this is far from the truth. Because there are many different causes of renal failure, the treatment and the management needed is often very different for different patients. Today we will be discussing your unique diagnosis and treatment.

The Treatment of Kidney Failure

In general, there are several options that can treat renal failure, depending on your special circumstance. Below are outlined the various treatments and I will explain to you where you best fit in.

Access for Dialysis: Surgical and Radiologic Procedures, 2nd ed.,
edited by Ingemar J.A. Davidson. ©2002 Landes Bioscience.

Kidney Transplantation

Kidney transplantation from a relative, a friend or from a cadaver organ donor (from a dead person) is the only definitive treatment of kidney failure. This can be done before you need dialysis or at any time thereafter. Because there are not enough kidneys available for everybody who needs one, only a fraction of all kidney failure patients on dialysis get transplanted. You will need special testing to determine if you are healthy enough for the transplant. Many patients are afraid they do not have enough money for a transplant, but Medicare or your insurance usually will pay for it. If you want more information on kidney transplant, ask me or your kidney doctor.

Dialysis

There are two types of dialysis: 1. hemodialysis and 2. peritoneal dialysis. Hopefully, before you see your surgeon, you have discussed your best options with your nephrologist. The surgical team can also advise you on which type may be best for you, but we concentrate on the technical aspect of your surgery necessary for your dialysis to take place.

Dialysis is sometimes called an artificial kidney, because a machine or filter takes the place of your kidneys. Before you can start dialysis you need surgery to create a way for dialysis to happen. This is called dialysis access, and will be your "lifeline" for the rest of your life. That is why you have come to see me.

Hemodialysis

Hemodialysis means that your blood is pumped through a filter to clean waste products from your blood. Waste products build up from the things you eat, drink and medication you take. The dialysis machine moves your blood quickly through the filter and returns it to your body after it is clean. This requires a tube to remove the blood from your body and then return the blood to your system after cleansing. Your veins are not big enough to move blood fast enough. Vascular access surgery creates a site for insertion of the tube that removes and returns your blood. Depending on your special circumstances, there are three different ways this can be achieved. First, in urgent situations a catheter can be inserted in a vein in your neck or groin. Second, most ideally, 6-12 months before you need dialysis, a vein and artery in your arm are surgically connected together, allowing the vein to grow. During dialysis, two needles will be placed in your arm; one needle removes blood, where it flows through the dialysis machine and is returned through the second needle back into your body. This is called a fistula. Third, if you have no suitable vein of your own, if your veins are too small or scarred to grow big, an artificial blood vessel the size of a plastic straw is surgically placed under your skin in the arm and connected to your artery and a vein. This allows for needles to be placed in the same way as a fistula, allowing blood to go from you to the machine and back. This artificial blood vessel (graft) requires 2-4 weeks before it can be used. If you need urgent dialysis, both a catheter in your neck and another surgery (fistula or graft) may be required at the same time.

Most dialysis surgery can be done as an outpatient surgery. You will go to the day surgery unit at the hospital. Surgery nurses will get you ready for the operation. An anesthesiologist will most likely inject anesthesia into your armpit to completely put your arm to sleep. You will also be getting IV drugs to calm you down. Occasionally, general anesthesia is needed.

After surgery, you will return to the day surgery unit. You will be given writen post surgery instructions. You may need a pain killer for 2-3 days. Most importantly, you must elevate your arm and exercise by making fists around a soft ball. Also, in cases of graft placement, an ice pack may be given to be placed over the surgical area for the next 6-12 hours. If you have questions or problems after your surgery, you will be given my office phone number to call. As your access surgeon, you can see me whenever you have access problems or concerns, before or after you start dialysis. You can call me yourself, or ask your dialysis nurses to call and explain the problem. You have a choice of the doctors who care for you before and after you start dialysis.

Your kidney doctor (nephrologist) may choose for you to start dialysis in the hospital, where you will be closely monitored for the first 2-3 days of dialysis. After that, you will go to an outpatient dialysis center 3 days a week.

Peritoneal Dialysis

The second major type of dialysis treatment is called peritoneal dialysis. In this case, a plastic tube is surgically placed inside your abdominal cavity and exited through your abdominal wall. Part of the tube is outside your body. Once the tube heals, nurses teach you how to put fluid in and take it out of your abdomen through the tube (exchange). The tissue covering your internal organs acts like a dialysis filter and helps pull waste products into the fluid. When you remove the fluid you remove the waste. Some patients can do their exchanges at night by using a special machine. Some patients need to do their exchanges during the day. You have to learn some simple technical procedures from the nurses before you start, because you are doing your own dialysis. Many patients like peritoneal dialysis because they have more freedom and do not have to go to the dialysis unit. This type of dialysis requires some basic technical abilities on your part, but offers more freedom and avoids needle sticks and long hours by the dialysis machine in the dialysis unit. However, it requires much greater responsibility by you, the patient to do the exchanges as they are ordered by your nephrologist.

The more you know about your condition and treatment options, the better decisions you can make. You are encouraged to ask questions about your specific kidney disease, and surgical and dialysis issues. You are likely to do better if you have a more complete understanding and participate in the decision making. We encourage to involve family members as well.

I have read this guide and discussed it with my surgeon. I have been told I can ask questions about my condition and my care. I understand the answers my surgeon has given me. I know that this surgery is necessary for starting dialysis. I understand that this operation, like all surgery, has risks. My surgeon has reviewed these risks with me.

Dallas, _____ _____ _____
　　　　　month　　day　　　year

_____　　_____
patient name signature　　　　　　　　　surgeon name signature (or alternate)

_____　　_____
patient name printed　　　　　　　　　　surgeon name printed (or alternate)

Postoperative Instructions for Access Procedures

Ingemar J.A. Davidson, Angela Kuhnel, Carolyn E. Munschauer, Gregory J. Pearl and Bertram L. Smith

Example of the author's written instructions to patients and family members leaving the day surgery unit

Currently in use at Baylor University Medical Center and Medical City Hospital, Dallas, Texas

- Primary (Native) AV Fistula
- PTFE AV Graft
- AV Graft Declotting and Revision
- Peritoneal Catheter
- Dual Lumen Cuffed Tunneled Catheter
- Removal of Dialysis Catheter
- Sample Vascular Access Patient Exam Forms
- Sample t-PA Catheter Clearance Protocols

Access for Dialysis: Surgical and Radiologic Procedures, 2nd ed., edited by Ingemar J.A. Davidson. ©2002 Landes Bioscience.

Example of the author's written instructions to the patients and family members leaving the Day Surgery Unit.

Postoperative Instruction

Primary (Native) AV Fistula Placement for Hemodialysis

1. During surgery, your vein and an artery were connected in order to create a useable vein for your future dialysis treatment.
2. Some pain and swelling is normal. To decrease swelling and pain, keep your arm elevated whenever possible on pillows above your heart for 24 to 48 hours. **Do not** put heating pads on your arm. This will make it feel worse. You may use your arm and hand without endangering your surgery.
3. Your arm may bruise slightly. There might be some blood oozing through the dressing. You will feel a buzz or thrill through the bandage, which will increase the next several days. This is normal.
4. You will be given a prescription for pain medication. This should provide adequate relief, although some discomfort is to be expected for the next several days. Please call if your pain becomes severe or persistent. If needed, take pain medication prescribed for you. It is not advisable to take prescription pain medication for more than two days after surgery. Pain medicine may cause constipation. We recommend taking a mild laxative such as milk of magnesia or Dulcolax to prevent constipation.
5. There are usually no stitches on the outside that need to be removed. The small steri-strip tapes may stay on for 7 to 10 days. A loose circular dressing may be changed and stay on for several days to protect your surgical area.
6. If your wound gets wet do not rub it dry. Gently pat with a clean towel.
7. If we are unable to contact you, call our office tomorrow (000-000-0000) to report your progress, or if you have problems or questions.
8. Make an appointment to see the surgeon in about 14 days (000-000-0000).
9. Notify the surgeon if:
   - You have numbness, tingling or pain in your hand
   - You are unable to move your hand
   - Your arm is red, painful and swelling get worse
   - Your wound continues to ooze blood
   - You notice drainage
   - You have a fever above 100⁰ F
10. Resume your preoperative medications following surgery. Normal activities such as work, driving, housework and sex can be resumed after surgery at your discretion, using common sense.

Signed_____ Patient
_____ RN

AIV

Example of the author's written instructions to the patients and family members leaving the Day Surgery Unit.

Postoperative Instruction

AV Graft Placement for Hemodialysis

1. During surgery, an artificial blood vessel tube was connected between an artery and a vein in your arm.
2. Some pain and swelling is normal. To decrease pain and swelling, keep your arm elevated whenever possible on pillows above your heart for 24 to 48 hours. **Do not** put heating pads on your arm. This will make it feel worse. An ice pack outside the gauze dressing for 6-12 hours will markedly decrease swelling and pain.
3. Your arm may bruise. It will feel warm. Some blood might ooze from the wound.
4. You will be given a prescription for pain medication. This should provide adequate relief, although some discomfort is to be expected for the next several days. Please call if your pain becomes severe or persistent. If needed, take pain medication prescribed for you. It is not advisable to take prescription pain medication for more than two days after surgery. Pain medicine may cause constipation. We recommend taking a mild laxative such as milk of magnesia or Dulcolax to prevent constipation.
5. There are usually no stitches on the outside that need to be removed. The small steri-strip tapes need to stay on for 7 to 10 days.
6. Take the bandage off tomorrow, or at next dialysis session.
7. Do not get your wound wet for 4 days after surgery. When you do get it wet, do not rub it dry. Gently pat it dry with a towel.
8. If we are unable to contact you, call our office tomorrow (000-000-0000) to report your progress, or if you have problems or questions.
9. Make an appointment to see the surgeon in 10-14 days (000-000-0000).
10. Notify the surgeon if:
    - You have numbness, tingling or pain in your hand
    - You are unable to move your hand
    - Your arm is red, or pain and swelling get worse
    - Your wound continues to ooze blood
    - You notice drainage
    - You have a fever above 100° F
10. Resume your preoperative medications following surgery. Normal activities such as work, driving, housework and sex can be resumed after surgery at your discretion, using common sense.

Signed_____ Patient
         _____ RN

Example of the author's written instructions to the patients and family members leaving the Day Surgery Unit.

Postoperative Instruction

AV Graft Thrombectomy or Revision

1. During surgery, blood clots were removed from your dialysis graft. Also, problems with your graft may have been surgically corrected.
2. Unless we tell you, your AV graft can be used right away for dialysis treatment.
3. Some pain and swelling is normal. Your arm may bruise. To decrease pain and swelling, keep your arm elevated on pillows above your heart for 24 to 48 hours. **Do not** put heating pads on your arm. This will make it feel worse.
4. You will be given a prescription for pain medication. Pain medication is usually only needed for 24-48 hours after surgery. This should provide adequate relief, although some discomfort is to be expected for the next several days. Please call if your pain becomes severe or persistent. If needed, take pain medication prescribed for you. It is not advisable to take prescription pain medication for more than two days after surgery. Pain medicine may cause constipation. We recommend taking a mild laxative such as milk of magnesia or Dulcolax to prevent constipation.
5. There are usually no stitches on the outside that need to be removed. The paper tapes covering the incision may be removed after 7-10 days.
6. Take the bandage off tomorrow.
7. Signs of infection are redness or swelling, pain that gets worse or fever above 100° F. Call our office if you have any of these problems.
8. If we are unable to contact you, call our office tomorrow (000-000-0000) to report your progress, or if you have problems or questions. You may need a follow up visit with the surgeons office in 5-7 days.
9. Notify the surgeon if:
   - You have numbness, tingling or pain in your hand
   - You are unable to move your hand
   - Your arm is red, or pain and swelling get worse
   - Your wound continues to ooze blood
   - You notice drainage
   - You have a fever above 100˚ F
10. Resume your preoperative medications following surgery. Normal activities such as work, driving, housework and sex can be resumed after surgery at your discretion, using common sense.

Signed_____ Patient
        _____ RN

AIV

Example of the author's written instructions to the patients and family members leaving the Day Surgery Unit.

Postoperative Instruction

Peritoneal Dialysis Catheter (Tenckhoff)

1. During surgery, a catheter was placed in your abdominal cavity for dialysis.
2. Some pain is normal. Do not lift anything that weighs more than 5 lbs for the next 2 weeks. Do not **strain or tighten** your abdominal muscles.
3. You will be given a prescription for pain medication. Pain medication is usually only needed for 2 or 3 days after surgery. This should provide adequate relief, although some discomfort is to be expected for the next several days. Please call if your pain becomes severe or persistent. If needed, take pain medication prescribed for you. It is not advisable to take prescription pain medication for more than two days after surgery. Pain medicine may cause constipation. We recommend taking a mild laxative such as milk of magnesia or Dulcolax to prevent constipation.
4. There are no stitches on the outside that need to be removed.
5. Leave the bandage on until you go to the dialysis unit for training.
6. Signs of infection are redness or swelling, pain that gets worse or fever above 100° F. Call our office if you have any of these problems.
7. If we are unable to contact you, call our office tomorrow (000-000-0000) to report your progress, or if you have problems or questions.
8. If you do not have an appointment for peritoneal dialysis training, call your dialysis center or kidney doctor (nephrologist) tomorrow.
10. Resume your preoperative medications following surgery. Normal activities such as work, driving, housework and sex can be resumed after surgery at your discretion, using common sense.

Signed_____ Patient
_____ RN

Example of the author's written instructions to the patients and family members leaving the Day Surgery Unit.

Postoperative Instruction

Dual Lumen Catheter Insertion

1. During surgery, a catheter was placed in a vein in your neck or chest for dialysis.
2. Some pain is expected. Your chest or neck may bruise.
3. You will be given a prescription for pain medication. Pain medication is usually only needed for 24-48 hours after surgery. This should provide adequate relief, although some discomfort is to be expected for the next several days. Please call if your pain becomes severe or persistent. If needed, take pain medication prescribed for you. It is not advisable to take prescription pain medication for more than two days after surgery. Pain medicine may cause constipation. We recommend taking a mild laxative such as milk of magnesia or Dulcolax to prevent constipation.
4. There is one small incision on your neck covered with a small steri-strip tape. Leave this tape on for about 7-10 days or until it falls off. At the catheter exit site on your upper chest there is a nylon stitch through the skin also tied to the catheter. This stitch should come out in about 14 days. You may have this done in my office or by your dialysis unit.
5. Do not remove the bandage. Do not push or pull on the catheter. Only the nurses at dialysis should remove the bandage.
6. Signs of infection are redness or swelling, pain that gets worse or fever above 100° F. Call our office if you have any of these signs of infection.
7. If we are unable to contact you, call our office tomorrow (000-000-0000) to report your progress, or if you have problems or questions.
8. Resume your preoperative medications following surgery. Normal activities such as work, driving, housework and sex can be resumed after surgery at your discretion, using common sense.

Signed_____ Patient
　　　　_____ RN

AIV

Example of the author's written instructions to the patients and family members leaving the Day Surgery Unit.

Postoperative Instruction

Dual Lumen Catheter Removal

1. During surgery, the dialysis catheter in your neck or chest was removed.
2. There are usually no stitches on the outside that need to be removed.
3. Keep the bandage on until your next dialysis treatment. If you do not go to dialysis, take the bandage off in 2 days.
4. If you notice bleeding through the bandage, apply gentle, firm pressure to the dressing and call our office. Do not remove the dressing to check the bleeding.
5. Signs of infection are redness or swelling, pain that gets worse or fever above 100° F. Call our office if you have any of these signs of fever or other catheter related problems that concern you.
6. If we are unable to contact you, call our office tomorrow (000-000-0000) to report your progress, or if you have problems or questions.

Signed_____ Patient
            _____ RN

Example of the authors physical exam and evaluation (E&M). The extent of evaluation will determine the level of ICD-9 coding (see Chapter 11 for details). The highlighted (bolded) areas are the most relevant for vascular access cases.

Vascular Access Physical Examination

Patient: _____ **Date:** _____
Location: _____
Surgeon: _____
Referring MD: _____
Reason for Visit:_____

AIV

1. **Constitutional:** any three of the following seven vital signs:
    BP Arm:_____; Sit: ____/____; Stand: ____/____; Supine: ____/____;
    Height:_____; **Weight:**_____; **Pulse:**_____ Regular:_____ Yes_____ No;
    Respiration:_____ ; **Temperature:**_____ Norm:_____ Yes_____ No;
    General appearance: _____

2. Eyes: Conjunctivae and lids: ____ Norm ____ Ab;
    Pupils and irises: Light: ____ Norm ____ Ab;
    Accommodation: ____ Norm ____ Ab;
    Ophthalmoscopic exam:_____ Yes_____ No; Disc:_____ Norm_____ Ab;
    Fundus:_____ Norm_____ Ab

3. Ears, Nose, Mouth and Throat External ears/nose: ___ Norm ___ Ab;
    Otoscopic exam: ___ Yes ___ No ___ Norm ___ Ab;
    Assessment of hearing:_____ Yes_____ No _____ Norm_____ Ab;
    Nasal mucosa, septum,turbinates:_____ Norm_____ Ab;
    Lips, teeth, gums: _____ Norm_____ Ab;
    Oropharynx Exam:_____ Yes_____ No Mucosa:_____ Norm_____ Ab
    Glands:_____ Norm_____ Ab; Palates:_____ Norm_____ Ab
    Tongue:_____ Norm_____ Ab; Tonsils/post pharynx:_____ Norm____ Ab

4. Neck: Overall appearance: __ Norm __ Ab Thyroid: ___ Norm ___ Ab

5. **Respiratory: Effort:** ____ Norm ____ Ab;
    Percussion of chest:_____ Norm_____ Ab; **Palpation
    of chest:**_____ Norm_____ Ab; **Auscultation of lungs:** ____ Norm ____ Ab

6. **Cardiovascular: Palpation of heart:** ___ Norm ___ Ab
    Auscultation of heart: ___ Norm ___ Ab;
    Carotid artery: L: ___ Norm ___ Ab; R: ___ Norm ___ Ab;
    Abdominal aorta: __ Norm ___ Ab; **Femoral artery: L:** __ Norm __ Ab;
    R: __ Norm __ Ab; **Ped pulse: L:** __ Norm __ Ab; **R:** __ Norm __ Ab;
    Extr: L: __ Norm __ Ab; **R:** __ Norm __ Ab

7. Chest (Breasts): Appearance: L: __ Norm ___ Ab; R: __ Norm __ Ab;
   Breast/ax: L: __ Norm __ Ab; R: __ Norm __ Ab
8. Gastrointestinal: Abdomen: ___ Norm ___ Ab;
   Liver/spleen: ___ Norm ___ Ab; Hernia: ___ Present Location: _____;
   Rectal exam: ___ Yes ___ No ___ Norm ___ Ab;
   Stool sample: ___ Yes ___ No
9. Genitourinary:
   MALE: Scrotum: ___ Norm ___ Ab; Penis: ___ Norm ___ Ab;
   DRE: ___ Yes ___ No ___ Norm ___ Ab;
   FEMALE: Pelvic exam: ___ Yes ___ No ___ Norm ___ Ab;
   Cultures: ___ Yes ___ No; Urethra: __ Norm ___ Ab;
   Bladder: ___ Norm ___ Ab; Cervix: ___ Norm ___ Ab;
   Uterus: __ Norm __ Ab; Adnexa/parametria: ___ Norm ___ Ab
10. Lymphatic: Palpation of lymph nodes in two or more areas:
    Cervical: ___ Norm ___ Ab; Axillary: ___ Norm ___ Ab;
    Inguinal: ___ Norm ___ Ab; Other: _____
11. **Musculoskeletal: Gait:** ___ Norm ___ Ab; **Digits: LUE:** __ Norm ___ Ab;
    RUE: ___ Norm ___ Ab; **LLE:** ___ Norm ___ Ab; **RLE:** ___ Norm ___ Ab
    Examination of joints, bones and muscles of one or more of the following
    six areas:

Region	Palpation	ROM	Stability	Strength
Head,Neck	___ Norm ___ Ab;	___ Norm ___ Ab;	___ Norm ___ Ab;	___ Norm ___ Ab
Spine, Ribs, Pelvis	___ Norm ___ Ab;	___ Norm ___ Ab;	___ Norm ___ Ab;	___ Norm ___ Ab
RUE	___ Norm ___ Ab;	___ Norm ___ Ab;	___ Norm ___ Ab;	___ Norm ___ Ab
LUE	___ Norm ___ Ab;	___ Norm ___ Ab;	___ Norm ___ Ab;	___ Norm ___ Ab
RLE	___ Norm ___ Ab;	___ Norm ___ Ab;	___ Norm ___ Ab;	___ Norm ___ Ab
LLE	___ Norm ___ Ab;	___ Norm ___ Ab;	___ Norm ___ Ab;	___ Norm ___ Ab

12. **Skin: General appearance:** ___ Norm ___ Ab;
    Palpation of skin and subq: ___ Norm ___ Ab
13. Neurologic: Cranial nerve exam: ___ Yes ___ No ___ Norm ___ Ab;
    Sensation: ___ Norm ___ Ab; DTR: LUE: ___ Norm ___ Ab;
    RUE: ___ Norm ___ Ab; LLE: Patellar: ___ Norm ___ Ab;
    Achilles: ___ Norm ___ Ab; RLE: Patellar: ___ Norm ___ Ab;
    Achilles: ___ Norm ___ Ab
14. **Psychiatric: Judgment/insight:** ___ Norm ___ Ab;
    Memory: ___ Norm ___ Ab; **Mood/affect:** ___ Norm ___ Ab;
    Mental status: Orientation to time/place/person: ___ Norm ___ Ab

Comments: _____
_____
_____
_____
_____
_____
_____
_____
_____
_____
_____

Signed: _____ **MD Date:** _____

Access Data Collection

ID: _____

Name Last: _____Name First: _____Age: _____

Sex: _____ OP date: _____

Diagnosis:_____

1. PTFE

arm _____ ua or fa _____ leg _____

prev ptfe _____ date _____ side _____ ua or fa _____ leg _____ side _____

prev ptfe _____ date _____ side _____ ua or fa _____ leg _____ side _____

_____

2. PRIMARY AVF

arm _____ ua or fa _____ leg _____

prev primary avf _____ date _____ arm _____ ua or fa _____ leg _____ side _____

_____

3. DECLOT/REVISION

site _____ patch angioplasty _____ interposition graft _____

extension graft _____

prev declot and/or revision _____ date ____

prev patch angioplasty _____ date _____

loc ____

prev interposition graft _____ date ____

prev extension graft _____ date _____

loc ____

_____

5. TENCKHOFF

tenckhoff catheter _____ moncrief _____

prev pd _____ prev date in _____ prev date out _____

prev moncrief _____ prev date in _____ prev date out _____

_____

6. DLC

DL catheter _____ perc site _____ cuffed, tunneled _____ type _____

site _____

catheter out _____ current catheter_____

type of current catheter _____

prev DL catheter _____ prev percutaneous _____ date in _____ date out _____

loc ____

prev DL catheter _____ prev percutaneous _____

date in _____ date out _____ loc _____

prev cuffed, tunneled _____ type _____

date in _____ date out _____ loc _____

prev cuffed, tunneled _____ type _____ date in _____

date out _____ loc _____

_____

7. FOLLOW UP

follow up date _____

patient alive @ f/u _____

functioning access @ f/u _____

access type at f/u _____

comments

_____

_____

_____

Guidelines for Restoring Patency of Occluded Central Lines with t-PA

Steps

1. Supplies (Needed prior to procedure)

__ (2) 5cc syringes

__ (2) 10cc vials of sterile, bacteriostatic normal saline

__ Sterile gloves

__ (2) Betadine wipes

__ (4) Alcohol wipes

__ (2) labels to be marked "t-PA Dwell"

2. Catheter Type and Lumen Volumes

Catheter Type	Volume (cc) Arterial Lumen	Volume (cc) Venous Lumen
Tesio (cuffed, separate lines)	2.3	2.3
Quinton (Mahurkar noncuffed)	1.2	1.3
Quinton Permcath (cuffed)	1.4	1.5
Ash Split (cuffed)	2.0	2.1

3. Aspiration of Heparin

Attempt to aspirate the heparin (0.5cc) from each port.

Heparin aspirated from:	Arterial Port	Yes / No (circle one)
	Venous Port	Yes / No (circle one)

4. Dwell Technique

1. Inject the port volume + 0.1cc
2. Reinject 0.1cc every 30 min - 1 hr for 3-5 hrs
3. Check patency by aspirating

    Dwell started at _____ (time)

    Reinjection _____ (time)

    Reinjection _____ (time)

    Reinjection _____ (time)

    Reinjection _____ (time)

    Arterial Port Occluded / Patent (circle one)

    Venous Port Occluded / Patent (circle one)

Infusion Technique: (See Standardized Order Sheet)

A. *Contraindications:*
- Coumadin use
- Systemic heparin use
- Bleeding disorder
- Plt < 90,000

B. *Preparation of t-PA*
- Pharmacy sends t-PA (5mg in 50cc saline) ready to infuse.

C. *Infusion*
- t-PA is infused at 12.5 ml/hr (1.25 mg/hr) over 4 hrs

D. *Monitoring*
- Patients monitored by direct observation and vital sign assessment for symptoms of bleeding at least every 15 minutes.

E. *End Infusion*
1. Attach syringe, unclamp catheter, attempt to aspirate blood.
2. If aspiration successful, remove 3-5 ml blood, clamp catheter, remove syringe, discard blood.
3. Replace syringe with 10 ml saline, unclamp catheter, flush with saline, reclamp.
4. Instill heparin (5000 units/ml) into each catheter lumen according to lumen volume. Lock-clamp catheter by closing clamp as last of heparin enters port.
- If patency has not been restored, notify interventional radiology immediately for catheter clearance.

Streptokinase/Urokinase/t-PA Protocol for Line Occlusions

t-PA—Fibrin specific. Hepatic cleared, same as urokinase and streptokinase.

Conversion from urokinase:	2mg t-PA ~ 5000 U urokinase
	Urokinase and streptokinase ~ 1:1
Usual urokinase infusion:	High dose - 250,000 U per hour
	Low dose - 125,000 U per hour
Usual t-PA infusion:	2mg bolus followed by 50mg infusion over short time.
	All t-PA should be mixed in saline.
Usual streptokinase infusion:	Screen patient for history of streptococcal infection.
	5000 U in 50 ml saline as bolus followed by:
	High dose - 250,000 U per hour
	Low dose - 125,000 U per hour

All patients should be monitored for bleeding per unit protocol for the duration of the infusion.

AIV

Ten-Year Experience With First Time Vascular Access

Ingemar J. A. Davidson, Jacek Dmochowski, Carolyn E. Munschauer

This abstract summarizes the editors 10 years of vascular access surgery. The most striking feature is the aging ESRD population. Second, there is an urgent need to improve the outcome after declotting procedures. Finally, measures to optimize appropriate native fistulae and dialysis catheter utilization are highly desirable.

This summary reports a decade's first time, first forearm vascular access placement by one surgical team (Table A.1), constituting 29% of the total vascular access operating room cases, based on current practice patterns at Medical City Dallas, Texas. Three time eras and three institutions are involved. First arm access represents 79% (233/295) of all PTFE or AVF placement. Notably, of all access surgery, catheters excluded, half of the cases involve access repair (N=297), most of which are PTFE graft thrombectomy procedures (Table A.2).

Table A.1. Ten years experience with first time forearm AV access

Era 1: 1/91-8/95	Parkland Memorial Hospital, Dallas, TX N=**415** PTFE, **85** AVF=	**500** access
Era 2: 1/96-2/97	SUNY, Buffalo, NY N=**53** PTFE, **25** AVF=	**78** access
Era 3: 9/98-12/00	Medical City Hospital, Dallas, TX N=**124** PTFE, **109** AVF=	**233** access
		Total: 811 access

Table A.2. Dialysis access OR activity

	All	First Time Forearm
PTFE	169	124
AVF	126	109
Totals	**295**	**233 (79%)**
Revisions	297	
Catheters		
Cuffed, Tunneled	170	
Peritoneal	39	
Total	**801**	**233 (29%)**

Medical City Dallas (Era 3) 1/1/98-12/31/00

Access for Dialysis: Surgical and Radiologic Procedures, 2nd ed., edited by Ingemar J.A. Davidson. ©2002 Landes Bioscience.

Methods

In addition to survival analysis for AV grafts and native fistulas, the current (era three) experience with hand ischemia from PTFE, as well as conversion of nondeveloping AVF to PTFE grafts were analyzed (Table A.3). Specific study features include one surgical team, outpatient surgery and the luxury of an access coordinator.

The survival curves were created according to the Kaplan-Meier technique. Tick mark represents the elapsed time to specific outcome events.

Table A.3. Access for dialysis outcome variables

PTFE	AVF
Patient survival	Patient survival
Graft survival	Fistula survival
Clot free survival	Fistula function

Special Outcome Report	
Arterial steal	Conversion to PTFE

Table A.4. Access for dialysis

Study Features
One surgical team
Outpatient surgery
Forearm first time access
Vascular access coordinator

Statistical Methods
Kaplan-Meier technique
Survival estimates

Results

Patient, graft and clot free survival rates have all declined in the last ten years, specifically, graft function at one year dropped from 92% to 76% currently, and clot free survival from 85% to 54% at present (Table A.5, Fig. A.1A-B).

The decline in PTFE outcome likely reflects two significant changes; first, both PTFE and AVF recipients are almost 15 years older in era three compared to era one; however, AVF patients continue to be ten years younger on average than PTFE patients (Fig. A.2A). Second, AVF utilization has increased from 17% at Parkland to 47% currently at Medical City Hospital (Fig. A.2B). This leaves the patients with less optimal anatomy (elderly, diabetics, obese) for PTFE placement.

With AV fistulae, one year function was unchanged, or 75% and 72% at one year; graft survival in case of AVF means usable for dialysis. The clot free data for AVF means fistula flowing or having a palpable thrill, but may not be usable. Therefore, so called clot free survival was higher than fistula function (survival) (Fig. A.3A-B, Table A.5).

Figures A.3A and A.3B depict the PTFE and AVF graft survival curves for Parkland (early 1990s) and Medical City Dallas Hospital (currently). Even though the selection or incidence of AVF was markedly different, 17% vs 47% (Fig. A.2B), the graft function with AVF is similar or less compared to PTFE (Fig. A.3A-B).

Table A.5.

Time Eras:			
Era 1: 1/91/-8/95			
	PTFE 415	**AVF 85**	**Total 500**
	PS	**GS**	**CFS**
PTFE (N=592)	94	92	85
AVF (N=219)	95	75	79
Era 2: 1/96-2/97			
	PTFE 53	**AVF 25**	**Total 78**
	PS	**GS**	**CFS**
PTFE (N=592)	90	78	62
AVF (N=219)	96	90	84
Era 3: 9/98-12/00			
	PTFE 124	**AVF 109**	**Total 233**
	PS	**GS**	**CFS**
PTFE (N=592)	89	76	54
AVF (N=219)	95	69	87

Total: 811

PS=Patient survival; GS=Graft survival; CFS=Clot free survival

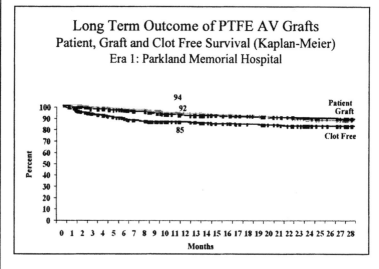

Fig. A.1A.

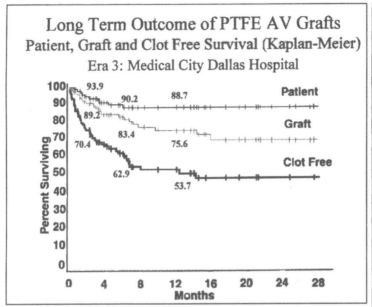

Fig. A.1B.

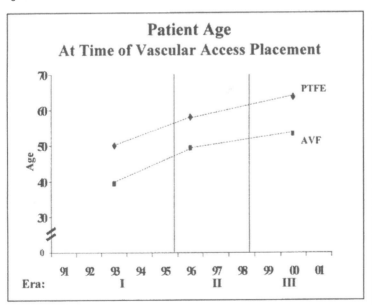

Fig. A.2A.

AV

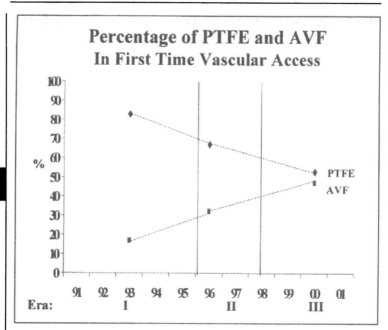

Fig. A.2B.

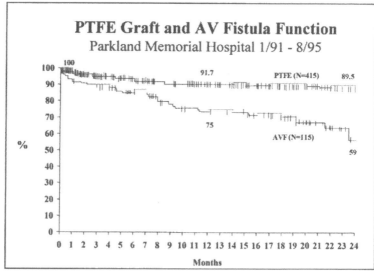

Fig. A.3A.

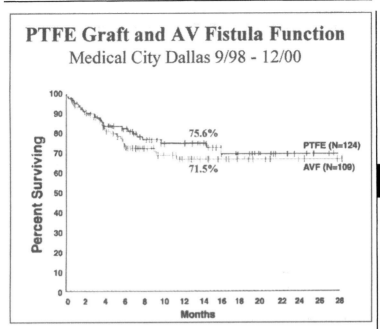

Fig. A.3B.

Secondary graft survival after surgical declotting has not changed or improved significantly. The three and six month secondary survival are about 60% and 50% respectively, which meets or exceeds that of the DOQI guidelines (Fig. A.4A-B).

Of the 27 failed AVF, 15 of 16 (94%) converted to PTFE graft in the same arm (utilizing the AV fistula vein) functioned at one year (Fig. A.5A).

Eleven PTFE patients (9%)developed hand ischemia and underwent banding at the apex of the loop, with 9 (82%) functioning at up to one year (Fig. A.5B). Banding therefore, rather than abandoning a stealing graft is a worthwhile procedure.

Summary Remarks

- The last ten years have brought marked changes in the ESRD population. Patients are 15 years older, half are diabetic, and the majority (65%) are obese.
- The PTFE and AVF patient groups are very different, due to the selection process. Most notably, the AVF patients are 10 years younger.
- The long term outcome or function with the stretch PTFE graft is the same or better than with the native AV fistula at the wrist, however, this occurs at a higher cost and morbidity severity.
- Outcome improvement efforts must involve all team players (not just surgery). Much of the increase in the AV fistula to PTFE ratio reported here is due to earlier referral from nephrology colleagues.
- Each and every ESRD patient must be evaluated and treated based on all current circumstances presented. This requires skills and knowledge, and most importantly a great deal of common sense and a positive attitude.

AV

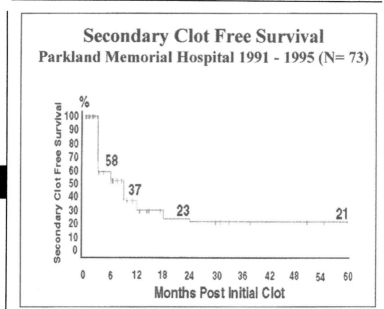

Fig. A.4A.

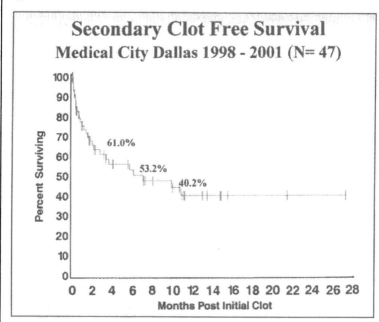

Fig. A.4B.

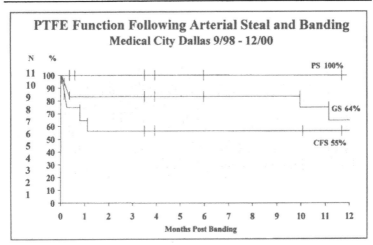

Fig. A.5A.

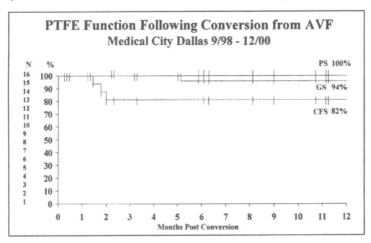

Fig. A.5B.

Selected References

1. Dawidson I, Ar'Rajab A, Melone D et al. Early Use of The Gore-Tex® Stretch Graft. In: Henry M, Ferguson R, eds. Vascular Access for Hemodialysis-III. W. L. Gore & Associate, Precept Press, Inc., 1995.

2. NKF-DOQI Clinical Practice Guidelines for Vascular Access. New York: National Kidney Foundation, 1997.

3. Davidson IJA, Ar'Rajab A, Balfe P et al. Long Term Outcome of PTFE AV Grafts. In: In: Henry M, Ferguson R, eds. Vascular Access for Hemodialysis VI. W.L. Gore & Associates, Precept Press, Inc., 1998.

4. Davidson IJA, Dmochowski J, Munschauer CE. Ten Years' Placement of 811 Consecutive First Time PTFE AV Grafts: A Single Surgeon's Experience. 2nd International Congress of the Vascular Access Society, London, UK, 2001

Index

Printed and bound by CPI Group (UK) Ltd, Croydon, CR0 4YY

28/10/2024

01780202-0001